Handbook on Tunnels and Underground Works

Handbook on Tunnels and Underground Works

Volume I: Concept – Basic Principles of Design

Edited by

Emilio Bilotta
Università di Napoli Federico II, Naples, Italy

Renato Casale
EnginSoft SpA, Trento, Italy

Claudio Giulio di Prisco
Politecnico di Milano, Milan, Italy

Salvatore Miliziano
Università di Roma Sapienza, Rome, Italy

Daniele Peila
Politecnico di Torino, Turin, Italy

Andrea Pigorini
Italferr SpA, Rome, Italy

Enrico Maria Pizzarotti
Pro Iter Srl, Milan, Italy

CRC Press
Taylor & Francis Group
Boca Raton London New York

CRC Press is an imprint of the
Taylor & Francis Group, an **informa** business

Cover image: Società Italiana Gallerie (SIG)

First published 2022
by CRC Press/Balkema
Schipholweg 107C, 2316 XC Leiden, The Netherlands
e-mail: enquiries@taylorandfrancis.com
www.routledge.com – www.taylorandfrancis.com

CRC Press/Balkema is an imprint of the Taylor & Francis Group, an
informa business

Library of Congress Cataloging-in-Publication Data
Names: Bilotta, Emilio, editor.
Title: Handbook on tunnels and underground works / edited by Emilio
Bilotta, Department of Civil, Architectural and Environmental
Engineering, University of Napoli Federico II, Naples, Italy, Renato
Casale, EnginSoft SpA, Italy, Claudio Giulio di Prisco, Tunnel
Engineering, Politecnico di Milano, Milan, Italy, Salvatore Miliziano,
University of Rome, Rome, Italy, Daniele Peila, Politecnico di Torino,
Turin, Italy, Andrea Pigorini, Società Italiana Gallerie, Milan, Italy,
Enrico Maria Pizzarotti, Pro Iter S.r.l., Milan, Italy.
Description: Leiden, The Netherlands ; Boca Raton, FL :
CRC Press/Balkema, 2022. | Includes bibliographical references. |
Contents: volume 1. Concept : basic principles of design—volume 2.
Construction : methods, equipment, tools and materials—volume 3.
Case histories and best practices.
Identifiers: LCCN 2021042162 (print) | LCCN 2021042163 (ebook)
Subjects: LCSH: Tunnels. | Underground construction.
Classification: LCC TA805 .H324 2022 (print) | LCC TA805 (ebook) |
DDC 624.1/93—dc23/eng/20211117
LC record available at https://lccn.loc.gov/2021042162
LC ebook record available at https://lccn.loc.gov/2021042163

ISBN: 978-1-032-18772-3 (hbk)
ISBN: 978-1-032-18774-7 (pbk)
ISBN: 978-1-003-25617-5 (ebk)

DOI: 10.1201/9781003256175

Typeset in Times New Roman
by codeMantra

Contents

Preface

The Italian Tunnelling Society is pleased and proud to introduce to the international tunnelling community this three-volume *Handbook on Tunnels and Underground Works*:

Volume 1: Concept – Basic Principles of Design
Volume 2: Construction – Methods, Equipment, Tools and Materials
Volume 3: Case Histories and Best Practices

The three sequential and integrated volumes have the purpose of offering a comprehensive and up-to-date scientific and technical content regarding the design and the construction of tunnels and underground works, useful for both universities and post-university studies and in the professional field.

As a handbook, it also aims to be a tool, in everyday tunnelling works, to rely on when facing specific dilemmas, such as:

- the assessment of a specific geotechnical risk or a computational analysis problem (Volume 1),
- the estimation of tunnel excavation rates or of the increased ground parameters achievable by a consolidation work (Volume 2),
- a search for suggestions or help in consulting data from previous important design and construction experiences (Volume 3).

The first volume 'Concept – Basic Principles of Design' describes the different stages of a project, related to the management of the uncertainties (risk management) in the design of underground works, starting from the analysis of initial data and models: functional data, geological and geotechnical model, and environmental framework.

The necessary theoretical elements are then provided for the development of computational calculations, with reference to the stability of the face and of the cavity, the analysis of the interaction with pre-existing structures and with the overall context and the design of temporary and permanent supports, both in static and dynamic conditions.

Specific detailed analysis is provided for monitoring activity during construction and for the design of refurbishment and maintenance of existing tunnels, always with reference to the definition of the theoretical basis and the main design elements.

Finally, we wouldn't neglect the stakeholder engagement process, which is essential for the acceptance of wide and complex infrastructures, such as those underground, to which an entire chapter is dedicated.

The second volume, 'Construction – Methods, Equipment, Tools and Materials', describes the main construction methods, the equipment, the machinery and the materials used for excavation, soil and rock enhancement, and construction of provisional and final supports, providing results according to geotechnical contexts and boundary conditions.

A specific detailed analysis is provided on monitoring tools and site plants and logistics.

The third volume, 'Case Histories and Best Practices', describes a number of tunnelling experiences, focusing mainly on how the risks identified in Volume 1 have been addressed in the design and construction phases, and refers to specific cases, in Italy and abroad, in order also to present the Italian underground works' know-how.

To sum up: design theory (Volume 1), construction and technology (Volume 2), and case histories and best practices (Volume 3).

We have entrusted the writing of these books to a scientific committee, whose members are:

• Professor Daniele Peila (Turin Polytechnic),
• Professor Claudio di Prisco (Milan Polytechnic),
• Professor Salvatore Miliziano (Sapienza Rome University),
• Professor Emilio Bilotta (Federico II Naples University),

either directors of postgraduate master's courses or holders of postgraduate courses in tunnelling (Turin and Milan Polytechnics) and in geotechnics applied to infrastructures (Rome and Naples Universities), held in Italian universities.

They have coordinated a group of about 50 experts from universities, engineering firms, construction companies and tunnelling industry and have been supported by a scientific secretariat. All of them invested knowledge, energies, passion and valuable time in this work: for that we are very grateful.

Furthermore, we would like to give a special thanks to Tatiana Todaro and Matteo Zerbi, two young engineers who hold the scientific secretariat.

The use of underground space is progressively growing, due to global urbanization and public demand for efficient transportation, together with energy saving, production and distribution. The increasing need for surface space, along with its ever-rising value and the challenges of meeting the goal of sustainable development, demands a greater and better underground space exploitation.

This will ensure a great support for sustainable and resilient transport infrastructure and more liveable cities and meeting the necessity to increase tunnelling knowledge, to rely on proper and valuable projects and on well-defined construction budgeting and scheduling.

We believe that the handbook will be a valuable reference text for university and master's degree students, tunnelling specialists, engineers, geologists and architects involved in underground planning, design and construction worldwide, to enhance the tunnelling culture.

The Editorial Board

Foreword I

The field of tunnelling is one of the most fascinating ones for an engineer, for we must work with the crust of the Earth itself. This material is only apparently docile, but instead ever changing and unexpected. Indeed, it is the true construction material with which underground works are made.

In the past decades, tunnelling has made giant steps as regards both conventional and mechanical digging, and thanks to new design approaches and new technologies, tunnelling has reached that goal which the specialists have always pursued: meeting the deadlines and costs of construction, in other words the industrialization of underground work sites.

However, the construction of tunnels remains an activity subject to uncertainties of various kinds that make it a particularly complex and difficult subject of civil engineering to address. Unlike the works built on the surface, the tunnels, in fact, have completely unique characteristics:

- They are built by subtracting, rather than adding material.
- The properties of the construction material (ground) aren't as well defined or known.
- The thrusts on their structures aren't previously known, nor is the response of the work in terms of resistance or deformability.

These are uncertainties that, if not addressed with adequate design and construction approaches, easily translate into higher costs and intolerable delays in construction times than those foreseen in the design stage.

In order to give a solution to the problem and to learn to approach the construction of large, long and deep tunnels with industrial criteria as required by the new means of transport and the growing demand for mobility, important progress has been made since the end of the 19th century, which in just over a hundred years has revolutionized the way of designing and building underground works. Considerable are the results achieved.

The first major revolution was undoubtedly the introduction by Sommeiller of the first drilling machine driven by compressed air for the excavation of the Frejus railway tunnel, which was followed between 1920 and 1960 by a second important revolution thanks to the studies of Terzaghi (the 'rock load' due to the weight of broken ground resulting from the excavation of the tunnel), Kastner and Fenner (the development of a 'plastic zone' in the rock mass surrounding the tunnel), Rabcewicz (NATM) and Lombardi (characteristic lines). The latter was the first to highlight how the static problem of the tunnels was a three-dimensional problem and not simply a two-dimensional one as previously considered. In 1950 Robbins, on the other hand, successfully experimented with the first modern full-face TBM in a rock mass, giving way to the birth of different types of TBMs, suitable for excavating not only stone materials, but also soft materials, above or below water table (hydroshield and EPB TBMs).

It was in this context that in the 1980s the second great revolution in the design and construction of tunnels took place, when with the introduction in Italy of the approach according to the ADECO-RS, having exploited the three-dimensionality of the stress field present deep in the rock mass, the importance of the 'core-face' was clarified, suggesting that the analysis and the control of the behaviour of the ground 'core' upstream of the excavation face are the secret to successfully build tunnels, economically and safely, in all stress–strain conditions, even the most difficult ones, guaranteeing certainty of construction times and costs. Consequently, the false dichotomy between conventional excavation and mechanized excavation has also been overcome, highlighting how the common key to success lies in the pre-confinement action of the cavity which, always advancing full face, is permanently ensured by acting on the core-face, an action that allows to constantly maintain the triaxial coaction $\sigma3$, originally present in the rock mass, on different and greater than 0 values, up to the implementation of the radial confinement of the cavity, generally operated through the timely installation of a load-bearing lining closed by the invert put in place close to the face.

The book that I have the honour to present here is the first volume of a very complete treatise. All the issues that preside over the realization of these difficult engineering works are examined, starting from the criteria for managing the uncertainties that are always present in these works, deriving from the initial data and models: functional data, geological model, geotechnical model and environmental aspects. The necessary theoretical elements are provided for the development of three-dimensional computational calculations (with reference to the stability of the core-face and of the cavity, to the analysis of interaction with pre-existing structures and to the environmental context) and for the design of temporary and final linings in static and dynamic conditions. Specific insights are provided for monitoring during construction and for the design of adaptation and maintenance interventions of existing tunnels, always with reference to the definition of the theoretical bases and the main design elements. Finally, even the stakeholder engagement process is dutifully considered, essential for the acceptance and sharing of complex works such as the infrastructural works in tunnel, to which an entire chapter is dedicated.

In short, an indispensable volume for all of us who dedicate ourselves to the noble and exciting task of designing and building tunnels!

The Italian Tunnelling Society (SIG) is to be congratulated for this excellent three-volume *Handbook on Tunnels and Underground Works*, which contains a wealth of information for researchers, students and professionals in the tunnel construction business. It comes at a time when the underground construction industry is experiencing a global boom, with growth rates significantly and consistently higher than the rates of the general construction industry. SIG is a highly qualified group to produce this book in view of the knowledge accumulated in Italy, a country actively involved in the construction of many heritage tunnels and in remarkable achievements in tunnelling during the 19th century, most importantly because Italy is the home of many developments of the modern tunnelling industry. One example is the design approach based on the analysis of controlled deformation of the ground mass. The book covers this innovation in sections dealing with both design and construction. The development of new materials and equipment because of that approach is also included.

Volume 1 establishes rational bases for tunnelling to be classified as a predictable engineering activity, rather than the almost heroic endeavour plagued with myriad geological uncertainties, which was the reality facing the industry in the past. Due to the recent progress in this topic, de-risking of projects has allowed contractors and private investors to feel more comfortable with the level of uncertainty to be faced in tunnelling operations. In addition, the topic of risk assessment is also contemplated in chapters devoted to modern computational methods for the analysis of the hydro-thermo-chemo-mechanical coupled processes which take place during construction and operation of tunnels. Quantitative and consistent probabilistic methods of analyses and the design of risk mitigating measures are also indicated.

The book is organized in such a way that the need for consultation of external references is minimized. For example, the chapters on input data related to geotechnics, geology and hydrogeology appropriately cover concepts of soil and rock mechanics.

In addition to the well-consolidated concepts of design that focus on structurally and geotechnically sound construction, the book covers the holistic approach to design, with chapters related to the minimization of both permanent and temporary negative environmental impacts as well as considerations regarding health and safety. Not only do these chapters provide guidelines for projects to be sustainable and useful for the society in a broader sense, but they also contribute to their acceptance by the society, a topic covered in Chapter 13. The chapter on occupational health and safety shows how a tailored design may promote those concepts. Information is given about how the number of casualties has decreased in underground construction as the industry aims at safer and predictable practices.

Volume 2 gives a broad overview of the latest construction techniques and the most modern systems and installations able to support the construction phase. The most modern technologies are described, which, at the present time, make use of highly automated apparatuses and processes and which allow companies to operate often without the direct involvement of workers in high-risk activities and with processes more and more similar to those of the manufacturing industry.

Finally, Volume 3 contains a rich series of case histories, which represent the application of the design and construction methodologies described and the most effective synthesis of the level reached by Italian engineering in the field of underground constructions. The described examples have been identified in such a way as to constitute a broad reference for each design and construction approach in the widest possible number of morphological, geomechanical, anthropic and environmental contexts.

This three-volume book will be a valuable tool in the search for better, safer and sustainable practices of underground construction.

Pietro Lunardi
SIG Honorary President

Tarcisio B. Celestino
Professor, University of Sao Paulo

Foreword 2

This three-volume series edited by SIG, which is the result of the World Tunnel Congress, held in Naples in 2019, not only collects, organizes and reports on the testimonies, the suggestions, the technical culture and the experiences of that fascinating 'other world' of underground work – that is the universe of tunnels, but also records, in the period after this international scientific event, the radical and unexpected historic change that has taken place in the international context. Therefore, we can say that the Congress itself took place in 'another world'.

The world before the pandemic – that saw a global outbreak which changed the behaviours and social life of millions of people, subverting their consolidated values and perception of the future, as well as the economic policy priorities (among others) of countries – and the world before the European Green Deal – with the new mottos of ecological transition, as an unavoidable feature of the new development, and of widespread digitalization, as a new mass technological literacy – constitute, in their inseparable interpenetration, the innovative binary paradigm for an accelerated transformation of the economy, manufacturing, services, education, the life of people and, above all, the new generations. It seems that, as in a sort of karst phenomenon, a lot of long gestating underground processes have suddenly emerged into the sunlight with an unpredictable potential for innovation that we are only now discovering in their disruptive scope, as they are coming out from our beloved WTC tunnels.

Faced with changes of this extent, the unavoidable topic is the new perspective of sustainable development based on a new meaning of growth, which calls for a closer dialogue between technical and humanistic disciplines. In this ecological transition context, a special role – for goods and passengers – is given to modal balance between road and rail and to sustainable mobility. In this context, train – of all types – represents the cornerstone in urban areas with underground railways, and also, on a large scale, with the TEN-T network, which will be an extraordinary continental metro, with cities as the stops and corridors as the interconnected lines.

This fundamental transport choice requires radical technological and infrastructural adjustments for thousands of kilometres of network, to be carried out in a tight time frame (within the 2030s) and in a coordinated manner to allow the 'system effect'. This effect cannot fully emerge without overcoming the main geomorphological constraints such as mountain chains, which risk becoming geopolitical constraints as well. Trains are efficient and profitable when they travel at plain surface. When there is a mountain, the only way to overcome it is to drill the mountain at the level of the plain and to build 'base tunnels' such as those – the longest in the world – of Mont Cenis and Brenner, currently under the responsibility of the authors.

A great innovation, compared to the past, is that today requirement is not only to work to ensure a strategic role in the long term, but also to be the creators of a solid immediate anti-recession action. The aim is to enhance all the economic and employment-related effects of the public expenditure envisaged for the system of European companies and the different countries, as well as, first of all, for the territories in which the work will take place.

The intervention area is potentially wide and covers not only construction activities themselves, but also the widespread added value that can be created in terms of services for people and companies such as accommodation, food services and, at a more sophisticated level, collaborating with schools and, in particular, with universities for training. The aim is to launch leading-edge disciplinary specializations as well as the interdisciplinary culture, to reach at the end the widespread innovation to be implemented using the extraordinary opportunity offered by our construction sites as laboratories for *in corpore vili* experimentation to solve today's problems while training a generation of technicians for tomorrow's increasingly global labour market.

In this context, the completion of the major Alpine tunnels currently under construction under our Alps represents, according to the concrete data available, a very important symbolic step.

It confirms the awareness of reaching environmental sustainability both in the construction phase, with the adoption of construction sites based on the criteria of circular economy, compatibility and integration with the territory, of environmental impact minimization and continuous implementation of safety, and in the operating phase, with the aim of complete transition towards diversification of transport modes and progressive decarbonization of this sector. These represent strategic projects for the entire European Green New Deal, which envisages a 90% reduction in CO_2 emissions from transport by 2050.

In addition, we would like to emphasize that these major projects represent the tangible realization of a technical-scientific and socio-economic dialogue between generations and peoples and constitute a laboratory of knowledge, experience and technological innovation for young engineers, young underground workers and young European citizens.

The work carried out in our construction sites, collected in this book from different perspectives, tells a story of miners, sometimes of their sons and grandsons, who dig the tunnels of our country, with courage and disdain for fatigue, inside the mountains, one volley after the other. And they do this colonizing a new universe of the underground each time and passing on a profession that, without excessive rhetoric, has the characteristics of an ancient art, even when the most modern technologies are used.

This work also tells the story of young technicians who, driven by their resolution, had the opportunity to complete their studies in our offices and construction sites, enriching the designs of these works with their enthusiasm, curiosity and knowledge, borrowing extraordinary experiences. These are young people cultivating ideal, in this case pursued with the language of knowledge combined with the language of work, new stories, probably among the most significant for our country.

Finally, it may be seen as a sign of the times that the legacy of the WTC of 2019 is being illustrated in books in 2021, the 'European Year of Rail'. The year has been declared by the EU Commission on the occasion of the 150th anniversary of the inauguration of the historic Mont Cenis tunnel, strongly advocated by Cavour, with the support of the great innovative engineers of that time (Sommeiller, Grattoni and Grandis).

Mario Virano
Tunnel Euralpin Lyon Turin – TELT
Board of Directors – CEO

Gilberto Cardola
Brenner Base Tunnel – BBT SE
Board of Directors – Italian CEO

About the editors

Emilio Bilotta graduated cum laude as Civil Engineer at the University of Napoli Federico II, and he got his PhD degree in Geotechnical Engineering from the University of Roma La Sapienza. He is currently Associate Professor at the Department of Civil, Architectural and Environmental Engineering of the University of Napoli Federico II, where he holds courses of fundamentals of geotechnical engineering for undergraduates and of tunnels and underground structures and underground constructions for postgraduates. His main research interests concern tunnelling in urban areas, tunnels under seismic actions and ground improvement for seismic mitigation.

He is Italian delegate of the Technical Committees TC 104 (Physical Modelling) and TC 219 (System Performance of Geotechnical Structures) of the International Society of Soil Mechanics and Geotechnical Engineering (ISSMGE); a member of the Italian Geotechnical Association (AGI) and of the Italian Tunnelling Society (SIG); a member of the editorial board of the journal *Rivista Italiana di Geotecnica* and of the scientific committee of the journal *Gallerie e grandi opere sotterranee*; and an associate editor of *Frontiers in Built Environment*. He is also co-founder of the academic spin-off and start-up Smart-G – innovative solutions in ground improvement and geotechnical engineering.

Renato Casale is currently CEO at EnginSoft SpA, based in Trento (Italy), and President of the Editorial Board at SIG (Italian Tunnelling Association) of *Handbook on Design and Construction of Tunnels*. Prior to this, he was Senior Advisor at SWS Group, based in Trento (Italy), President of the Organizing Committee of WTC 2019 at SIG (Italian Tunnelling Association) and Senior Engineer at Rocksoil (worldwide-known design company in the underground infrastructures). Prior to this, he was Vice-President and Chief of Rail Research Department at Ducati Energia S.p.A., based in Bologna (Italy), from June 2014 to June 2017 and, from May 2016 to May 2017, CEO at Telefin S.p.A., a railway telecommunication company, owned by Ducati Energia Group. Prior to this, Mr. Casale worked as an Executive with Ferrovie dello Stato Italiane Group from 1979 until April 2014. Mr. Casale was Director of special projects at Ferrovie Dello Stato Italiane SpA from October 2013 until April 2014. Prior to this, Mr. Casale was CEO of Italferr SpA, company of Ferrovie Dello Stato Italiane Group, from May 2007 until September 2013. Italferr, the engineering Company of Italian State Railways Group (Gruppo Ferrovie dello Stato Italiane), is a market leader in the field of multidisciplinary services

and comprehensive designs in the railways. Prior to this, Mr. Casale was Chief Investment Officer of Rete Ferroviaria Italiana (RFI) from April 2000 until April
2007, which is the company of the Ferrovie dello Stato Group with the public role
of Infrastructure Manager. Previously, he was Designer, Site Engineer and Project
Manager at Ferrovie Dello Stato Italiane, from October 1989 until March 2000.

Claudio Giulio di Prisco graduated at Politecnico di Milano. He developed his doctoral
thesis under the supervision of Prof. Roberto Nova at Politecnico di Torino. He
collaborated with Prof. Ioannis Vardoulakis and prof. Elias Aifantis. He is Full Professor of Geotechnics and Slope Stability at Politecnico di Milano since 2005, and
from 2015, at the same university, Director of the Master on Tunnel Engineering. He
was Director of the International Association Alert Geomaterials. He is the author
of more than 70 scientific papers in international journals and of many book chapters, mainly concerning the constitutive modelling of the mechanical behaviour of
granular materials, instabilities in geomaterials (localization and liquefaction) and
the design of shallow foundations, reinforcement of tunnel faces, piled embankments, geo-reinforced sand columns, etc.

Salvatore Miliziano, PhD, is Professor in Advanced Soil Mechanics, Director of Second Level Master in Geotechnical Design and Member of Teaching College of
PhD Course in Structural and Geotechnical Engineering of Sapienza, University
of Rome. His research activities span from theoretical studies to practical applications. He is the author of more than 100 scientific papers published in national and
international journals and conference proceedings. Salvatore is scientific head of
several research projects at Sapienza University involving, among others, several
tunnelling projects worldwide. He is founder and shareholder of the engineering
company Geotechnical Design Group. He is also founder and president of Geotechnical and Environmental Engineering Group (GEEG), an innovative start-up
of Sapienza. GEEG was set up with the aim of sharing the results of years of applied
research in mechanized tunnelling field with all stakeholders. Salvatore developed
advanced multidisciplinary knowledge in his 30-year career as civil and geotechnical engineer in the design and construction of engineering project in Italy and
worldwide. Referring to tunnelling, he is an expert in the design of tunnels (both
mechanized and conventional). He is a member of the Scientific Board of the Italian
Tunnelling Society and Editor of *Tunnels and Large Underground Works Journal.*

Daniele Peila graduated cum laude as Mining Engineer in 1987, at Politecnico di Torino
where, after a period as a field engineer in construction firms, he become research
assistant first and then professor. He presently is Full Professor at Politecnico di
Torino teaching 'tunnelling' and 'reinforcements of soil and rock'. He is Director of
the Postgraduate Master Course on 'Tunnelling and Tunnel Boring Machines' and
chair of the laboratory 'Tunnelling and Underground Space'.

He has been vice-president of International Tunnelling and Underground Space
Association and, presently, acts as vice-president of Italian Tunnelling Association
and a member of many editorial of international journals.

His fields of specialization are tunnelling (conventional and mechanized), mining and rockfall protection. In these fields, he acted as person in charge of many
research projects and is the author of more than 250 papers published on both

international journals and congress proceedings. Furthermore, he is often called upon to give worldwide lectures and seminars for events organized by international universities as well as by cultural associations and to act as keynote speaker.

Andrea Pigorini is an experienced mining engineer who has been working for more than 30 years in design and construction of important infrastructural projects in Italy and abroad, with a specific deep experience in tunnelling and underground works.

He formerly worked in construction companies, and he completed his professional background, working for a leading geotechnical engineering company, where he was in charge as designer and project manager of several road, railway and metro tunnelling projects.

He has been working at Italferr since 2002, and he has been responsible for the Tunelling Design Department from 2002 to 2015, and since 2015, he has been responsable for the Civil Infrastructural Engineering Department. From 2011 to 2017 he was also involved in the construction of the Brenner Base Tunnel, as project manager in charge of the supervision of the works of two construction lots.

Since 2013, he has been the President of the Italian Tunnelling Society, a scientific association sharing technical and scientific knowledge in engineering, design and construction of tunnels and underground works.

He has carried out several training sessions on tunnelling, and he has been the author of several papers on tunnel design and construction aspects, published in technical journals and international/national congress proceedings.

Enrico Maria Pizzarotti graduated at Politecnico di Milano in 1983. He has over 35 years of experience in planning and consultancy of large underground and geotechnical works, including both conventional and mechanized tunnels, underground caverns and stations, also in urban area, among which is the Brenner Base Tunnel (Italian side). He is Chairman of the BoD, Legal Representative and Technical Director of the engineering company Pro Iter S.r.l. based in Milan and a member of the ExCom and SG of SIG, of which he is the animator of the Working Group 2 – Research. He has been speaker in numerous seminars and conferences and is the author of numerous articles and technical publications and a co-author of the book *Engineering Geology for Underground Works* (Springer 2014).

Contributors

N. Antonias Italferr SpA Rome, Italy

M. Barbero Department of Structural, Geotechnical and Building Engineering, Politecnico di Torino, Turin, Italy

E. Bilotta Department of Civil, Architectural and Environmental Engineering, Università di Napoli Federico II, Naples, Italy

D. Boldini Department of Chemical Engineering Materials Environment, Sapienza Università di Roma, Rome, Italy

C. Callari Department of Biosciences and Territory, Università degli Studi del Molise, Campobasso, Italy

A. Carigi Department of Environmental, Land and Infrastructure Engineering, Politecnico di Torino, Turin, Italy

G. Dati Tunnel Euralpin Lyon Turin (TELT), Turin, Italy

A. Desideri Department of Structural and Geotechnical Engineering, Sapienza Università di Roma, Rome, Italy

L. Flessati Department of Civil and Environmental Engineering, Politecnico di Milano, Milan, Italy

P. Gattinoni Department of Civil and Environmental Engineering, Politecnico di Milano, Milan, Italy

A. Graziani Department of Engineering, Università degli Studi Roma Tre, Rome, Italy

A. de Lillis Department of Structural and Geotechnical Engineering, Sapienza Università di Roma, Rome, Italy

A. Luciani SWS Engineering, Trento, Italy

F. Martelli Italferr SpA, Rome, Italy

D. Martinelli Department of Environmental, Land and Infrastructure Engineering, Politecnico di Torino, Turin, Italy

A. Meda Department of Civil Engineering and Computer Science Engineering, Università degli Studi di Roma Tor Vergata, Rome, Italy

S. Miliziano Department of Structural and Geotechnical Engineering, Sapienza Università di Roma, Rome, Italy

R. Nebbia Department of Management and Production Engineering, Politecnico di Torino, Turin, Italy

S. Padulosi Italferr SpA, Rome, Italy

M. Patrucco Department of Sciences of Public Health and Pediatrics, Universita' degli Studi di Torino, Turin, Italy

D. Peila Department of Environmental, Land and Infrastructure Engineering, Politecnico di Torino, Turin, Italy

S. Pentimalli Department of Sciences of Public Health and Pediatrics, Universita' degli Studi di Torino, Turin, Italy

M. Pescara Tunnel Consult, Barcelona, Spain

A. Pigorini Italferr SpA, Rome, Italy

E.M. Pizzarotti Pro Iter Srl, Milan, Italy

G. Plizzari Department of Civil, Architectural, Environmental Engineering and Mathematics, Università degli Studi di Brescia, Brescia, Italy

C. di Prisco Department of Civil and Environmental Engineering, Politecnico di Milano, Milan, Italy

G. Russo Geodata Engineering SpA, Turin, Italy

S. Lo Russo Department of Environmental, Land and Infrastructure Engineering, Politecnico di Torino, Turin, Italy

G. Russo Department of Civil, Architectural and Environmental Engineering, Università di Napoli Federico II, Naples, Italy

S. Santarelli Consultant, Rome, Italy

L. Scesi Department of Civil and Environmental Engineering, Politecnico di Milano, Milan, Italy

A. Sciotti Italferr SpA, Rome, Italy

D. Sebastiani Geotechnical and Environmental Engineering Group, Rome, Italy

L. Soldo Pro Iter Srl, Milan, Italy

A. Sorlini Tunnel Euralpin Lyon Turin (TELT), Turin, Italy

G. Tiberti Department of Civil, Architectural, Environmental Engineering and Mathematics, Università degli Studi di Brescia, Brescia, Italy

C. Todaro Department of Environmental, Land and Infrastructure Engineering, Politecnico di Torino, Turin, Italy

Chapter 1

Introduction

C. di Prisco
Politecnico di Milano

D. Peila
Politecnico di Torino

A. Pigorini
Italferr SpA

CONTENTS

1.1 HISTORICAL OVERVIEW

Tunnels were 'invented' by man in the Stone Age to mine orebodies and have then been improved over the centuries to meet a large variety of economic, military and social needs. For example, the hydraulic tunnel of Eupalinos, on the Greek island of Samos, excavated more than 2,600 years ago and 1,250 m long, was bored to provide water to the fortified city. As is well known, there are numerous Roman tunnels to install services, such as for aqueducts, to remove waste waters, to drain wet areas and to allow a way through mountains, as shown in Figure 1.1. The genius Leonardo da Vinci clearly guessed the potential of tunnels for solving transportation problems in urban areas and improving urban livability. By designing his ideal city after the plague that caused an enormous number of fatalities in Milan, he promoted the use of the underground space for the transport of goods and for the supply of clean water and removal of waste water:

> Per le strade alte non de' andare carri [...] anzi, sia solamente per li gentili uomini; per le basse devono andare carri ed altre some ad uso e comodità del popolo. [...] Per le vie sotterranee si de' votare destri, stalle e simili cose fetide (along upper roads trucks should not travel [...], these should be reserved for noble men, in the lower roads trucks and other things needed by the people should go [...]. Underground, toilets, stables and similar dirty things should be realized).
> (Manuscript B – Institut de France, Paris, written between 1487 and 1490)

DOI: 10.1201/9781003256175-1

Figure 1.1 Cloaca Maxima (Rome) (above), a 600-metre-long tunnel constructed in Rome between 400 B.C. and 578 B.C. with the purpose of draining the wet areas north of the Palatine Hill, and the Furlo Tunnel, still in use, opened by Emperor Vespasiano in A.D. 76 (below).

Today, the use of underground space and tunnels, in both urban and extra-urban areas, is systematic: to cross natural obstacles, to improve the transport networks (rails, railways, metro and pedestrian) and for an incredibly wide number of applications. The modern success of the use of tunnels is also due to both their resiliency and durability, since, compared to other surface infrastructures, they are intrinsically protected against natural hazards such as floods, rockfalls, earthquakes and atmospheric action.

Generally speaking, tunnelling is the engineering process, very often implemented for transportation purposes, that, by means of the removal of large volumes of soil/rock from their original underground position, allows the creation of an underground void characterized by one prevalent dimension and a predefined cross section. The final product of the process consists in obtaining a tunnel that is capable of fulfilling, over a prescribed period of time, the provision of the necessary maintenance works, and the specific use for which it was planned and designed. This goal is achieved by considering a set of predefined 'constraints', strictly related to the specific site, and by minimizing both the time and costs of construction.

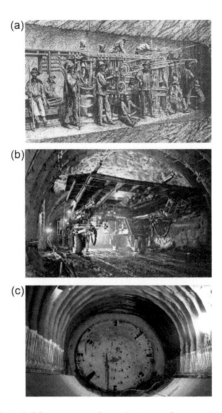

Figure 1.2 One hundred and fifty years of evolution of excavation techniques. The pictures show three Alpine tunnels close to each other: (a) Sommeiller's drilling device used in the excavation of the first Frejus Railway Tunnel (1857–1871), (b) the drilling device used for the excavation of the first Frejus Road Tunnel (1975–1979) and (c) the full-face TBM for the excavation of the second Frejus Road Tunnel (2011–2014).

With time, engineers have introduced solutions: to make the excavation safer for workers, the number of which, owing to more and more efficient mechanization, has also been progressively reduced; to safely excavate in more and more variable and 'poor' or potentially dangerous geological environments; to reduce costs and construction time; to improve the durability of the works; and, finally, to improve construction sustainability. Figure 1.2 makes clear the progressive improvements in mechanization in the excavation process that have taken place in the past decades: the three pictures refer to tunnels excavated nearby in three different epochs.

Generally speaking, tunnel engineering requires:

• an optimal merging of the design phase processes,
• the management of the construction/technological phases,
• a good knowledge of how the various techniques have been implemented in similar case histories.

In this editorial project starting with this volume, all these three different points of view are tackled. In fact, the second volume is mainly focused on the technological construction aspects and the third one provides a comprehensive overview of construction case histories.

For a long time, tunnelling has been seen as an art: both designers and builders had to be capable of accepting and dealing with the unpredictability of both geological and geotechnical conditions. They had to tailor their tunnel design to the current needs, these being discovered during construction and often evolving with each advancing step. Any inadequacy of their knowledge implied that the tunnel design would develop only during the tunnel construction. Today many things have changed, however, and no client is willing to accept the amount of time and the costs that would be needed to build tunnels without the help of a modern design approach, as well as modern excavation and supporting techniques. Moreover, no one can now accept the safety level that was associated in the past with tunnel construction. At the same time, although today's tunnels are larger, longer and deeper (or, in urban areas, shallower under surface structures), designers and construction engineers must manage the complexity of tunnelling within defined costs and time.

Numerous challenges are faced by designers. These are mainly related to:

•   the stability of both face and cavity,
•   settlements induced by the excavation,
•   safety during construction and operations,
•   the quality of the final product,
•   the feasibility of the construction process so markedly influenced by the hydro-thermo-chemo-mechanical behaviour of the natural soils and rock mass formations to be excavated.

Nowadays, a modern tunnel design protocol has to be based on suitably detailed geological and geotechnical studies (Chapters 3 and 4). This is necessary, as today numerical codes and sophisticated constitutive relationships allow, in most geological and geotechnical environments, reliable three-dimensional stress–strain analyses of the excavated soil/rock mass to be performed (Chapter 9): only a reliable preliminary investigation phase can provide the necessary and correct information for a rational and effective design. For this reason, the risk is still greater in all geological environments where investigations and previsions are more difficult, as in the case of long and deep tunnels (Chapter 7). As observed by Lunardi (2008),

> there are no tunnels easy or difficult because of the overburden or the ground to be tunneled, but only stress-strain situations in the ground in which it is, or is not possible to control the stability of the excavation, which will depend on our knowledge of the pre-existing natural equilibrium, on the approach to the design and on the availability of adequate means for excavation and stabilization.

Moreover, the design of an underground work should be developed according to the 'philosophy of doubt' (Pelizza, 1998; Barla & Pelizza, 2000), that is to say to find, on the one hand, the structural solution that minimizes the risk and, on the other, the construction technologies capable of dealing with them under safe conditions: it is necessary to assess and critically verify all the possible hazard scenarios. Therefore, tunnel design is basically a multi-stage decision (Chapter 2), where each decision deals with the uncertainties associated with a certain degree of residual risk.

1.2   DESIGN APPROACHES

During the past decades, different approaches have been used for design and construction, strictly linked to the upgrading and development of design tools, and excavations and means of support in conventional excavation. Starting from the traditional excavation by drilling and blasting in rock masses, the evolution of design and excavation techniques in more complex and poor ground has been great.

From the late 1800s, the Sequential Excavation Method was used: this multiple-drift tunnel construction technique was based on the design concept that a smaller excavation volume was easier to support by employing the poor supporting means available at that time (mainly wood). In the 1950s, the New Austrian Tunnelling Method was introduced (still used nowadays in several circumstances): this uses partialized face excavation and suggests multiple methods of excavation and means of support, to be tailored to the rock mass classification. This took good advantage of the use of shotcrete and bolts as readily available supports for excavation. In Italy, since the 1980s, the systematic use of pre-confinement and pre-support interventions (also named pre-reinforcement interventions) ahead of the tunnel face has become standard and has evolved in a design approach (named Analysis of Controlled Deformation in Rocks and Soils, or ADECO-RS). According to this approach, the unsupported, always full-face, tunnel core-face mechanical response (forecast of extrusions, pre-convergences and convergences) has to be three-dimensionally studied, the excavation operations and supports (Chapters 8–10 and Volume 2) have to be designed by performing three-dimensional analyses considering and using the core-face, improved if necessary by means of an adequate pre-confinement and/or reinforcement actions/interventions, as a stabilization tool, the monitoring, to be carried out during and after construction, has to be designed, and during construction, the design has to be updated (according to the project), by considering monitoring data in terms of deformation phenomena of core-face, cavity and ground level. This approach has widely been used for the construction in Italy and all around the world of more than a thousand kilometres of underground infrastructures. This also ensures reliable design predictions in terms of construction times and costs and improves the safety of workers. The wide use of this approach has pushed forward the development of new technologies and equipment, to install reinforcing elements in the core-face also for large lengths, and boosted the excavation of tunnels with a full section.

The greater the need for tunnel construction to deal with difficult stress–strain situations, the more important it is to operate with nearly circular full-face excavation rather than in a partialized face: this allows the invert lining to be built close to the excavation face and therefore a full round structural section to be achieved close to the tunnel face, better controlling the rock mass deformation. The extrusive phenomenon of the core-face is directly responsible for the evolution of the subsequent cavity convergence phenomena, and therefore of the tunnel's stability. The convergence is thus the last stage of the deformation response due to excavation, which begins upstream of the excavation face, due to the extrusive behaviour of the core, and then evolves into pre-convergence, which can increase and amplify the convergence downstream of the tunnel face itself (Lunardi, 2008, 2015).

At the same time, since the 1970s, there has been great development of the use of full-face TBMs (Tunnel Boring Machine). These started from gripper TBMs, able to excavate stable rock masses, and evolved to shielded TBMs, which are also able to apply a pressure at the face to counterbalance the water pressure and the soil pressure by means of different types of fluids. The excavation by means of shielded TBMs allows control of the deformation around the tunnel both at the face (with the applied pressure in the bulk chamber) and at the cavity, with the immediate support action of the shield and of the continuous installation of the full round segment lining, put in place inside the shield tail, close to the tunnel face.

1.3 GEOMETRY OF TUNNELS, LAYOUT AND ALIGNMENT

As previously mentioned, tunnels are built, as far as the alignment is concerned, with the final objective of being as short as possible and overcoming natural/anthropic obstacles. Tunnels have a prevalent longitudinal dimension and a transversal section, the size of which is strictly related to the use of the tunnel (inner section) and to the tunnel's structural design (excavation section) (Table 1.1). The longitudinal profile is strictly related to the access point elevation, but is also influenced by construction requirements. For instance, in the case of long conventionally excavated tunnels, to minimize the tunnel construction time, lateral adits are needed every 3–5 km. The longitudinal profile has also to take the tunnel use into account: railway tunnels are usually characterized by gentle slopes, varying from less than 1%–2%, while road tunnels may reach 5%–10%. Water management should also be considered so as to provide a natural water flux and avoid the need for pumping stations. In some cases, for long tunnels driven from both portals, humpback tunnels could be a solution to manage large amounts of water during construction. Higher slopes can be reached in rack tunnels (up to 20%–30%) and in hydraulic or service tunnels, these latter characterized in some cases by a vertical inclination (shaft). A transversal section may require a horseshoe excavation shape to reduce the tunnel excavated volume and guarantee available operational space in the tunnel crown (needed for electric traction and ventilation equipment for railway and road tunnels). On the other hand, a circular shape excavation section, despite the increment in the tunnel excavated volume, guarantees optimal management of high geotechnical and hydrostatic loads. Horseshoe shape tunnels may or may not have an invert depending on the ground properties: for example, recent Italian railway standards and codes indicate the invert as mandatory, to guarantee tunnel stability under any geotechnical condition and to avoid potential settlements in the railway embankment. Modern full-face machines are able also to excavate circular tunnels with very large diameters, and different layouts can be used to optimize the use of the section to host several utilities and transportation means in the same section (Table 1.2).

Table 1.1 Different tunnel functions, excavation methodologies and transversal sections (# refers to the number in Figure 1.3)

#	Project name	Tunnel function	Tunnel excavation methodology	Inner section data	Structural section data
1	Italian State Railway Standard Codes	Railway tunnel Single track $V \leq 200$ km/h	TBM	Radius 4.0 m	Excavation area 65 m^2 Excavation radius 4.55 m Segment thickness 40 cm
2	Italian State Railway Standard Codes	Railway tunnel Double tracks $V \leq 200$ km/h	Conventional	Radius (crown and sidewalls) 5.4 m	Excavation area 112 m^2 Preliminary support 20 cm Final lining (crown and sidewalls) 60 cm Final lining (invert) 70 cm
3	Virgolo Tunnel Bolzano – Italy	Railway tunnel Three tracks	Conventional	Radius 8.3 m (crown), 5.4 m (sidewalls)	Excavation area 185 m^2 Preliminary support 20 cm Final lining (crown and sidewalls) 80 cm Final lining (invert) 80 cm
4	S. Lucia Tunnel Florence – Italy	Highway tunnel	TBM	Radius 7.15 m	Excavation area 201 m^2 Excavation radius 8.0 m Segment thickness 55 cm
5	Santa Caterina Tunnel Sicily – Italy	Road tunnel	Conventional	Radius 6.45 m (crown and sidewalls)	Excavation area 154 m^2 Preliminary support 30 cm Final lining (crown and sidewalls) 90 cm Final lining (invert) 100 cm
6	Eurasia Tunnel Istanbul –Turkey	Double-deck road tunnel	TBM	Radius 6.0 m	Excavation area 147 m^2 Excavation radius 6.85 m Segment thickness 60 cm
7	Green Line Doha – Qatar	Metro tunnel	TBM	Radius 3.1 m	Excavation area 39 m^2 Excavation radius 3.5 m Segment thickness 30 cm
8	Gibe III Hydroelectric Project Ethiopia	Hydraulic power tunnel	Conventional	Radius 5.5 m	Excavation area 140 m^2 Preliminary support 10 cm Final lining (crown and sidewalls) 70 cm Final lining (invert) 80 cm
9	Stad Ship Tunnel Norway	Boat tunnel	Conventional	Height 49 m Width 36 m	Not available

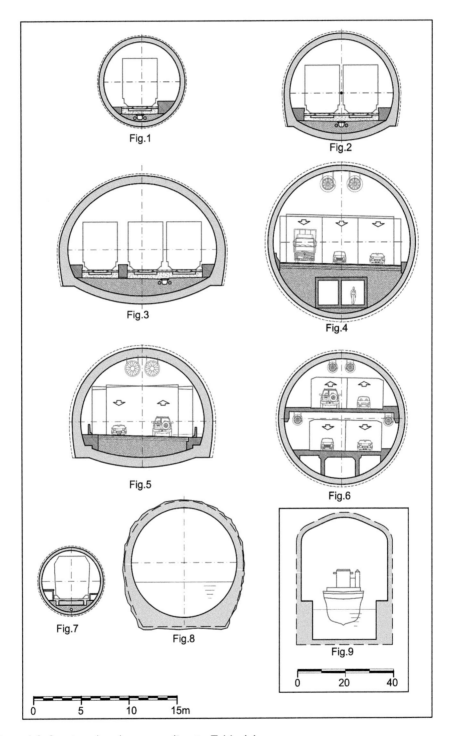

Figure 1.3 Section sketches according to Table 1.1.

Table 1.2 Different tunnel transversal sections – horseshoe without invert, polycentric with invert and circular shape (measurements expressed in m)

Project name	Tunnel function	Section peculiarity	Section sketch
Porrettana Railway Line (1864) Bologna–Florence Italy	Railway tunnel	Horseshoe shape tunnel without invert	0 5
High-Speed Railway Tunnel Milan–Genoa Italy	Railway tunnel	Polycentric shape tunnel with invert	0 5 10
Kat 3 Tunnel Istanbul Turkey	Mixed traffic Metro and road tunnel	Circular shape tunnel	0 5 10 15m

1.4   MANAGEMENT OF THE DESIGN PROCESS

The management of the design process of an underground work, passing through the different design phases and ending with the construction phase, plays a key role in the successful construction of an underground infrastructure. Costs and construction time are the pillars to be monitored and compared during the design process. From the very initial design stages, different alignments and different infrastructural solutions (requiring natural/artificial tunnels or open work stretches) have to be compared, by means of a multi-criteria analysis, so as to consequently perform the risk management on the chosen alignment, allowing the project managers to take full control of the decision-making process. The design process is usually divided into different design phases, starting from feasibility studies and preliminary design, passing through the detailed design phase, delivered for construction and ending with the tunnel construction (for-construction design). For each design phase (feasibility studies, preliminary, detailed and for-construction design), together with the project drawings and reports, the evaluation of costs and time is mandatory. Range and parametric construction costs and time are generally used for such an estimation in the early design phases (feasibility studies and preliminary design). On the contrary, a detailed bill of quantities and a detailed construction time estimation are used for detailed and for-construction design. For estimative computations, to take into account properly the cost level foreseen by the client for each specific work, cost lists should be provided by the client to the designer and the designer will provide his cost estimation by including production rates for materials, manpower, equipment and sustainability analysis.

The decision strategies, corresponding to the so-called conceptual design, may be implemented by employing probabilistic risk/statistical analyses, which offer the possibility of quantifying the project reliability by mathematically modelling the variability and uncertainty of the key parameters involved, and assessing the impact of parameter variations on time and costs (Chapter 7). Realistically, in tunnelling risks cannot be totally avoided and the detailed design should be developed on the basis of reducing risks. Nowadays, there is a need to manage the design process more efficiently, comparing the various alternatives from the very beginning, thus allowing the project managers to take full control of the decision-making process and construction process, rather than employing reactive crisis management, without disregarding the communication processes with the stakeholders and the populations involved in the construction (Chapter 13). It is clear that all the design choices require a set of basic information related to the tunnel's functional requirements, defined by the client according to the scope for which the tunnel is to be built; the geological and geotechnical properties of the ground encountered by the alignment (Chapters 3 and 4); the constraints due to the local environmental conditions (Chapter 5); and health and safety issues during construction (Chapter 6). The available design tools are used by the engineer to forecast the behaviour of the ground and its response to the excavation process, and to minimize the hazards by applying suitable countermeasures (Chapter 7).

In the early design phase, the excavation methods, the support tools and technologies, and the available ground reinforcements and improvement techniques have to be accounted for and critically compared, with the aim of checking their advantages and disadvantages with reference to the specific geotechnical context of the project

(Chapter 8 and Volume 2). Structural section types have to be defined, and preliminary calculations have to be performed by using simplified approaches. For the final detailed design, each structural section of the tunnel should be verified by using suitable computational methods (Chapter 10). In the past, tunnel design approaches were mainly focused on estimating the magnitude and distribution of loads applied to the tunnel supports and then individuating a lining capable of bearing these loads. Today, the design process starts from the identification of the main hazard scenarios and its development is governed by the aim of addressing and managing them. Therefore, nowadays the design is aimed not only at preventing catastrophic collapses and excessive loads on the structures, but also at ensuring structural durability throughout the tunnel's service life, by considering long-term hydro-chemo-mechanical phenomena and accidental loads (such as fire and earthquakes), without disregarding the influence of the construction method on the stability and final quality of the tunnel, and also by designing a proper maintenance and refurbishment policy (Chapter 12).

In conclusion, the designer has to assess what is the goal, in terms of risk reduction related to each mitigation measure (that is, the engineering design process), and then to discuss how to achieve, from the technological point of view, that goal (technological choices and design). Observation and monitoring during work (Chapter 11), which have also to be defined in advance for the specific project, allow the detection of any deviations of the tunnel's progress from the expected behaviour and therefore the adjustment of the project by applying the foreseen designed countermeasures. This procedure has to be considered as a planned design method, named the observational method (Chapter 10), and to be seen as a strategic approach for the management of the risks and for minimizing construction time and costs.

AUTHORSHIP CONTRIBUTION STATEMENT

The chapter was joined developed by the authors.

REFERENCES

Barla, G. and Pelizza, S. (2000). TBM tunneling in difficult ground conditions, Key Note Lecture, GeoEng2000, Melbourne.

Lunardi, P. (2008). *Design and Construction of Tunnels*. Berlin, Springer-Verlag Publishing.

Lunardi, P. (2015). Extrusion control of the ground core at the tunnel excavation face as a stabilisation instrument for the cavity. Muir Wood Lecture 2015. [Online] Available from: https://about.ita-aites.org/publications/muir-wood-lecture [Accessed 29th April 2021].

Pelizza, S. (1998). Selection of TBMs. Workshop on Selection of Tunnelling Methods. *ITA World Tunnel Congress '98*, Sao Paulo (Brazil).

Chapter 2

Risk management in tunnelling

C. di Prisco and L. Flessati
Politecnico di Milano

D. Peila
Politecnico di Torino

E.M. Pizzarotti
Pro Iter Srl

CONTENTS

2.1 INTRODUCTION

Tunnelling can be defined as the process of safe excavation and permanent stabilization of an underground cavity with specific cross sections, connecting two points along a predefined alignment, providing functionality and use/operational requirements of the system during its whole lifetime and minimizing construction time and costs (strictly related to each other).

DOI: 10.1201/9781003256175-2

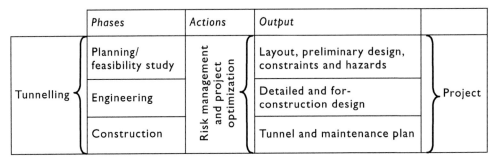

	Phases	Actions	Output	
Tunnelling	Planning/ feasibility study	Risk management and project optimization	Layout, preliminary design, constraints and hazards	Project
	Engineering		Detailed and for-construction design	
	Construction		Tunnel and maintenance plan	

Figure 2.1 Scheme of the tunnelling process

Conceptually, tunnelling can be subdivided into three phases (Figure 2.1):

- planning/feasibility study, aimed at individuating both project constraints and hazards, defining the tunnel layout and providing its preliminary design;
- engineering, dedicated to the development of the detailed and for-construction design;
- construction, implementing and eventually updating the for-construction design to build the tunnel and detailing the maintenance plan.

The inputs for these phases are characterized by numerous uncertainties and unknowns, never *a priori* completely avoidable (Guglielmetti et al., 2008; Gattinoni et al., 2014), which should be properly managed (Muir Wood, 2000). For this reason, during the whole tunnelling process a continuous update of input data is mandatory (AFTES, 2004) and, as is evident, by passing from phase 1 to 3, the number of input data increases and the output becomes progressively more detailed.

According to ITA WG 2 (2004), the actions necessary to pass from input data to output should be based on the systematic application of appropriate risk management strategies and on the project optimization. ISO/IEC Guide 73:2009 defines the risk as the convolution of the probability of occurrence of a specific hazard and its impact on exposed elements (vulnerability and exposure). A hazard is a situation or condition that has the potential for human injury, damage to property, damage to environment, economic loss or delay to project completion. A discussion on the impact of these aspects in the insurance strategies can be found in IMIA (2012).

The whole project can be considered as the final product of the subsequent output of the previously defined phases, including layout/preliminary design (phase 1), detailed and for-construction design (phase 2) and tunnel and maintenance plan (phase 3), as shown in Figure 2.1.

2.2 PLANNING/FEASIBILITY STUDY

The use and operational requirements are initially imposed and cannot be modified. These, along with the environment (nature and mechanical behaviour of materials to be excavated (Chapters 3 and 4), hydrogeology (Chapters 3 and 5), the presence of surface and underground infrastructures (Chapter 4), the environmental aspects (Chapter 5) and the technological aspects of the possible construction methods (Chapter 7), are the main input data for this phase (Figure 2.2) and govern all the subsequent choices.

	Input	Actions		Output
Planning/ feasibility study	Environment and use/operational requirements	Definition of tendering strategies		Layout, constraints (engineering, technological, logistical and sociopolitical) and preliminary design
		Constraint identification		
		Risk {	Assessment	
			Evaluation	
			Allocation	
		Definition of risk acceptability criteria		
		Choice of mitigation measure		

Figure 2.2 Scheme of the planning/feasibility study

This phase is intended to undertake five actions (Table 2): definition of tendering strategies (Section 2.2.1), constraints identification (Section 2.2.2), risk assessment/evaluation/allocation (Section 2.2.3), definition of risk acceptability criteria (Section 2.2.4) and choice of mitigation measures (Section 2.2.5).

2.2.1   Definition of tendering strategies

Tunnelling is a very complex and expensive process requiring clear and straight economical and financing strategies from all parties involved. In this perspective, tendering is fundamental for allowing the achievement of the final goal, defining the contract between the tunnel owner/client and contractors. Depending on the contractual models, these latter may assume the duty of designing/constructing the tunnel in a well-defined period of time, whereas the owner/client plays the role of carrying out the preliminary phases of the design process and of paying a fixed amount of money for the construction. Any uncertainty related to the tunnelling process may affect the contract and may represent a financial/legal hazard for both owner/client and contractor. Since in tunnelling uncertainties and related risks are unavoidable, the contract has to take them up front into consideration a procedure to find solutions, trying to avoid, if possible, applying for legal judgement. All these aspects, analysed in detail in the Emerald Book (FIDIC, 2019), are not considered in this volume, since even a non-comprehensive analysis of this item would require a too extensive discussion. More information on the various aspects of the Emerald Book contents can be found in Ericson (2019), Ertl (2019), Maclure (2019), Marulanda and Neuenschwander (2019), Neuenschwander and Marulanda (2019) and Nairac (2019).

2.2.2   Constraints identification

The second action of this phase is the identification of the constraints related to tunnelling, which can be subdivided into engineering, technological, logistical and sociopolitical.

The **engineering constraints** derive from use and operational requirements (usually input data given from the client) and very often are detailed in national/international standards (for instance geometry, maximum slope and curvature).

The **technological constraints** mainly derive from the tunnel geometry, geology and hydrogeology (Chapter 3), but also from specific aspects of the chosen excavation methods (Chapter 7). For instance, some excavation and support technologies cannot be adopted for excavations below the water table level, if the tunnel should be excavated in the proximity of buildings or underground infrastructures (e.g. induced displacements and vibrations) and in case of swelling soil/rocks.

The **logistical constraints** are related to the organization of the whole construction phase, affecting time and costs. For instance, the use of a certain excavation technique (Chapter 7) requires space availability, management of the excavated muck (Chapter 5) and supply and management of what necessary for implementing the chosen methodology/technology/machinery.

Sociopolitical constraints may affect tunnelling since tunnels are very often sensitive to the public opinion and stakeholders. This type of constraint must be identified and analysed before the construction, and a suitable strategy should be put in place to involve national and local communities.

2.2.3   Risk assessment/evaluation/allocation

After constraints identification, the planning/feasibility study passes to the risk assessment (identification), evaluation and allocation. For risk assessment, the designer must identify hazards, related exposure and potential damages. Numerous are the contributions concerning this item, but to be synthetic, in this chapter a detailed analysis of this topic is not included and some papers are suggested as reference: Duddeck (1987), Einstein (1996), Kalamaras et al. (2000), Lombardi (2001), British Tunnelling Society (2003), Chiriotti et al. (2003), Brown (2012), Guglielmetti et al. (2008) and Grasso and Soldo (2017); a comprehensive discussion on this topic can be found in Hudson and Feng (2015); a more detailed discussion is provided in Chapter 7.

Tunnelling hazards can essentially be subdivided into two categories: excavation and operational hazards. Excavation hazards (H_e) are those related to the construction phase, mainly affecting the spatial domain close to the tunnel ('near field') and far from the tunnel ('far field'). Operational hazards (H_o) are those potentially occurring during the tunnel operational life (long-term conditions). Both H_e and H_o can be subdivided into mechanical ($H_{elo, m}$), hydraulic ($H_{elo, h}$) and physical/chemical ($H_{elo, c}$).

For instance, in the category of mechanical hazards, we can identify tunnel face and cavity (unlined and lined) collapses, detachment of rock elements, activation of chimneys, initialization of dormant landslides, unacceptable tunnel cavity displacements ('convergence') and damages induced to existing structures and underground infrastructures either by excessive displacements or by vibrations related to the excavation processes. These hazards are significantly influenced by the mechanical properties of the materials, rock joint geometrical and mechanical characteristics, pore water pressure distribution (hydro-mechanical coupling) and chemical processes involving materials and groundwater (chemo-hydro-mechanical coupling).

In the category of hydraulic hazards, we consider separately those related to near field (for instance large water inrush, potentially putting at risk the safety of the workers) and those too far field (for instance variation in water table levels).

Chemical (or even radioactive) and physical hazards are those related to excavation in materials containing aggressive/dangerous agents, gases or pollutants (e.g. sulphates, chlorides, hydrocarbons, free hydrogen, methane, hydrogen sulphide, carbon dioxide, radioactive minerals or gases – radon, silica and asbestos) or at great temperatures (high overburdens). The chemical and radioactive elements may directly affect the safety of the workers and/or durability of structures (near field) and may also be released in the environment outside the tunnel (far field), in both atmosphere and groundwater (Chapters 5 and 6).

The hazards occurring during the tunnel operational life (H_o) are due to natural or anthropic events, for example:

- earthquakes;
- accumulation of displacements induced by movements of active landslides and faults;
- variation in pore pressure distribution induced by evolving hydrogeological conditions;
- fire and explosions induced by accidents.

Once risks are assessed, these are evaluated in terms of probability of occurrence and consequences in terms of effects on safety, time and costs.

Finally, the risk is allocated, assigning it to clients/owners, contractors or insurances, defining the contract clauses accordingly.

2.2.4 Definition of risk acceptability criteria

In the risk management, a crucial role is played by the definition of a set of criteria (in terms of probability and consequences of a given event) of risk acceptability (risk acceptability criteria). These are fundamental to decide whether mitigation measures are necessary and to make possible the risk allocation between the various parties involved in the whole design and construction process.

2.2.5 Choice of mitigation measures

The initial risk analysis and the definition of acceptability criteria allow us to individuate the cases in which risk is unacceptable and the implementation of suitable 'mitigation measures', acting on the probability of occurrence of an event and/or on its impact on the elements at risk, is necessary. The possible choices of mitigation measures to be implemented to reduce risks to an acceptable level must be detailed in the preliminary design.

2.3 ENGINEERING

The input for this phase (Figure 2.3) is both the input (environment and use/operational requirements) and the output (layout, constraints, hazards and preliminary design) of the previous one.

	Input	Actions	Output
Engineering	Environment and use/ operational requirements, layout, preliminary design, constraints and hazards	Definition of key performance indicators and relative threshold	Detailed and for-construction design
		Design of mitigation measures	
		Design of monitoring	
		Risk { Assessment / Evaluation	
		Forecasting of possible countermeasures	
		Preliminary maintenance plan and long-term monitoring design	

Figure 2.3 Scheme of the engineering phase

This phase is intended to undertake six actions (as shown in Table 3): (i) definition of key per-formance indicators and relative threshold (Section 2.3.1), (ii) design of mitigation measures (Section 2.3.2), (iii) design of survey and monitoring during excavation (Section 2.3.3), (iv) residual risk assess-ment/evaluation (Section 2.3.4), (v) forecasting of potential countermeasures (Section 2.3.5) and (vi) prelimi-nary maintenance and long-term monitoring design (Section 2.3.6).

2.3.1   Definition of key performance indicators and related threshold

The first step in the design of mitigation measures is the definition of key performance indicators (KPIs), a set of measurable quantities defining the system performance (for instance convergence, induced settlements and stresses in linings). The second step consists in individuating the KPI threshold values for attention and alarm, indicating whether and which mitigation measures are necessary.

2.3.2   Design of mitigation measures

The design of mitigation measures (Chapters 7 and 9; Figure 2.3) consists in achieving the goals set in terms of risk reduction (or limiting the KPI values under their own thresholds), defining how to implement the mitigation measure from a technological point of view. The mitigation measure to face a specific hazard is not unique, and different choices can be done, with reference to different residual risks and costs (Chapter 7 and Volume 2 of this handbook). The final choice is the result of an iterative procedure aimed at minimizing costs and maximizing benefits in terms of risk reduction.

2.3.3   Design during excavation survey and monitoring

As previously mentioned, tunnelling is affected by numerous unavoidable uncertainties and unknowns. For this reason, the designer must provide a detailed strategy for geological/hydrogeological/geotechnical survey during excavation and monitoring system (design during excavation survey and monitoring, Chapter 10 and Volume 2 of this handbook).

2.3.4   Residual risk assessment/evaluation

During the design phase, risk assessment and evaluation are to be performed continuously, to verify whether, after the implementation of the mitigation measure, the risk is acceptable (residual risk assessment/evaluation). In case of necessity, new mitigation strategies must be included in the design. It is also worth mentioning that, in some cases, the mitigation measures can potentially induce new hazards. For instance, when conventional tunnelling is adopted to excavate in coarse-grained materials, the improvement of the material by grouting techniques is very common to avoid the face collapse. These techniques, however, especially in case of shallow tunnels, can induce uplift of the ground surface, potentially damaging the existing buildings. In case the estimated induced heave values are not acceptable, additional (e.g. pressure relief pipes) or different mitigation measures must be adopted (e.g. mechanized tunnelling).

2.3.5   Forecasting of potential countermeasures

In many cases, during the construction phase, additional mitigation measures are necessary to face unexpected/unpredictable events (also called countermeasures). The output of the engineering phase should include the definition of emergency plans to implement these countermeasures.

2.3.6   Preliminary maintenance plan and long-term monitoring design

The design should finally include a plan of maintenance and long-term monitoring to provide and detect the long-term performance of the system, to be analysed and interpreted as a sort of guideline useful for the client/owner to ensure the durability of the system and maintaining its functionality throughout the service life.

2.4   CONSTRUCTION

The input for this phase (Figure 2.3) are both the environment and use/operational requirements and all the outputs of the previous phases (layout, constraints, hazards and detailed and for-construction design).

Input		Actions		Output
Construction	Environment and use/operational requirements, layout, detailed and for-construction design, constraints and hazards	Implementation of mitigation measures		Tunnel and maintenance plan
		During-construction survey and monitoring		
		Residual risk {	Assessment	
			Evaluation	
		Design/implementation of possible countermeasures and update of long-term monitoring		

Figure 2.4 Scheme of the construction process

This phase intends to undertake four actions (Figure 2.4): (i) implementation of mitigation measures (Section 2.4.1), (ii) during-construction survey and monitoring (Section 2.4.2), (iii) residual risk assessment and evaluation (Section 2.4.3) and (iv) design/implementation of possible countermeasures and update of long-term monitoring (Section 2.4.4).

2.4.1   Implementation of mitigation measures

This phase mainly consists in the tunnel excavation, according to what has been detailed in the for-construction design, and in the implementation of the mitigation measures, designed in the engineering phase.

2.4.2   During-construction survey and monitoring

According to the observational method (Chapter 9), due to the inevitable uncertainties and unknowns, the for-construction design must be verified and updated by using the data gathered during excavation. For instance, a continuous evaluation of KPI (e.g. convergence, surface subsidence and stresses in linings) allows us to verify the reliability of the assumptions made by the designers during the engineering phase and possibly to suggest the implementation of countermeasures.

2.4.3   Residual risk assessment/evaluation

The data gathered with the monitoring allow us, as already mentioned, to verify the assumptions made during the design and possibly to update the residual risk assessment/evaluation, according to what already indicated in the for-construction design.

2.4.4   Design/implementation of possible countermeasures and update of long-term monitoring

In case of necessity, to face unexpected/unpredictable events, the countermeasures indicated in the engineering phase have to be designed and implemented (design and implementation of possible countermeasures, Figure 2.4) and, accordingly, the long-term monitoring plan must be updated.

AUTHORSHIP CONTRIBUTION STATEMENT

The chapter was joined developed by the authors.

REFERENCES

AFTES (2004) Recommandation du GT 32: Risk geotechniques. *Tunnels et Ouvrages souterrains*, 185.

British Tunnelling Society (2003) The joint code of practice for risk management of tunnel works in the UK. London. BTS.

Brown, E.T. (2012) Risk assessment and management in underground rock engineering – an overview. *Journal of Rock Mechanics and Geotechnical Engineering*, 4 (3), 193–204.

Chiriotti, E., Grasso, P.G. & Xu, S. (2003) Analisi del rischio: stato dell'arte ed esempi. *Gallerie e Grandi Opere Sotterranee*, 69, SIG, Milano, Patron Editore, Bologna, 20–44.

Duddeck, H. (1987) Risk assessment and risk sharing in tunnelling. *Tunnelling and Underground Space Technology*, 2, 315–317.

Einstein, H.H. (1996) Risk and risk analysis in rock engineering. *Tunnelling and Underground Space Technology*, 11 (2), 141–155.

Ericson, G. (2019) The Geotechnical Baseline Report in the new FIDIC Emerald Book – suggested developments. In: Peila, D., Viggiani, G. and Tarcisio, C. (eds.) *World Tunnel Congress 2019: Tunnels and Underground Cities: Engineering and Innovation meet Archaeology, Architecture and Art, Proceedings of the ITA World Tunnel Congress*, 3–9 May 2020, Naples, Italy. London, Taylor & Francis.

Ertl, E. (2019) A new approach in the FIDIC forms of contract, and the Emerald Book's place in the Rainbow Suite. In: Peila, D., Viggiani, G. and Tarcisio, C. (eds.) *World Tunnel Congress 2019: Tunnels and Underground Cities: Engineering and Innovation meet Archaeology, Architecture and Art, Proceedings of the ITA World Tunnel Congress*, 3–9 May 2020, Naples, Italy. London, Taylor & Francis.

FIDIC – ITA/AITES (2019) Conditions of contract for underground works (Emerald book). FC-UW-A-AA-10, Geneva, FIDIC.

Gattinoni, P., Pizzarotti, E.M. & Scesi, L. (2014) *Engineering Geology for Underground Works*. Dordrecht, Springer.

Grasso, P.G. & Soldo, L. (2017) Risk analysis-driven design in tunneling: the state-of-the-art, learnt from past experiences, and horizon for future development, *Innovative Infrastructure Solutions*, 2 (1), 2–49.

Guglielmetti, V., Grasso, P.G., Mahtab, A. & Xu, S. (2008) *Mechanized Tunnelling in Urban Areas*. London, Taylor & Francis.

Hudson, J.A. & Feng, X. (2015) *Rock Engineering Risk*. London, CRC Press.

IMIA (2012), *A code of practice for risk management of tunnel works*, The International Tunnelling Insurance Group. [online]. Available from: https://www.imia.com/wp-content/uploads/2013/08/ITIG-TCOP-01_05_2012.pdf [Accessed 21st November 2020].

ISO (2009) *ISO/IEC Guide 73:2009 Risk management – Vocabulary*. Geneva, ISO.

ITA WG 2 (2004) Guidelines for tunnelling risk management. *Tunnel and Underground Space Technology*, 19, 217–237.

ITA WG 3 (2011) The ITA contractual framework checklist for subsurface construction contracts. [Online] Available from: https://about.ita-aites.org/publications/wg-publications/content/8-working-group-3-contractual-practices [Accessed 12th September 2020].

Kalamaras, G.S., Brino, L., Carrieri, G., Pline, D. & Grasso, P. (2000). Application of multicriteria analysis to select the best highway alignment. *Tunnelling and Underground Space Technology*, 15 (4), 415–420.

Lombardi, G. (2001) Geotechnical risks for project financing in non-urban areas. *Tribune*, 20, 19–23.

Maclure, J. (2019) The role of Engineer in the Emerald Book. In: Peila, D., Viggiani, G. and Tarcisio, C. (eds.) *World Tunnel Congress 2019: Tunnels and Underground Cities: Engineering and Innovation meet Archaeology, Architecture and Art, Proceedings of the ITA World Tunnel Congress*, 3–9 May 2020, Naples, Italy. London, Taylor & Francis.

Marulanda, A. & Neuenschwander, M. (2019) Contractual time for completion adjustment in the FIDIC Emerald Book. In: Peila, D., Viggiani, G. and Tarcisio, C. (eds.) *World Tunnel Congress 2019: Tunnels and Underground Cities: Engineering and Innovation meet Archaeology, Architecture and Art, Proceedings of the ITA World Tunnel Congress*, 3–9 May 2020, Naples, Italy. London, Taylor & Francis.

Muir Wood, A. (2000) *Tunnelling: Management by Design*. London, E&FN Spon.

Nairac, C. (2019) The claim, dispute avoidance and dispute resolution procedure in the new FIDIC Emerald Book. In: Peila, D., Viggiani, G. and Tarcisio, C. (eds.) *World Tunnel Congress 2019: Tunnels and Underground Cities: Engineering and Innovation meet Archaeology, Architecture and Art, Proceedings of the ITA World Tunnel Congress*, 3–9 May 2020, Naples, Italy. London, Taylor & Francis.

Neuenschwander, M. & Marulanda, A. (2019) Measuring the excavation and lining in the Emerald Book. In: Peila, D., Viggiani, G. and Tarcisio, C. (eds.) *World Tunnel Congress 2019: Tunnels and Underground Cities: Engineering and Innovation meet Archaeology, Architecture and Art, Proceedings of the ITA World Tunnel Congress*, 3–9 May 2020, Naples, Italy. London, Taylor & Francis.

Chapter 3

Input data: geology and hydrogeology

L. Scesi and P. Gattinoni
Politecnico di Milano

S. Lo Russo
Politecnico di Torino

CONTENTS

3.1 INTRODUCTION

The behaviour of rocks and soils being excavated depends on many factors, among which the most important are certainly the geological and hydrogeological ones. The knowledge of geological and hydrogeological conceptual models allows us to identify the hazards, estimate the risks and define their mitigation.

DOI: 10.1201/9781003256175-3

The reconstruction of the aforementioned conceptual models passes through:

1. The identification of the different lithological nature of rocks and soils, which determines the geotechnical and geomechanical behaviours of the materials. In particular, the risks related to the lithology are dissolution phenomena (in carbonate sedimentary rocks or in some evaporitic rocks); the lack of cement among grains (in clastic and/or pyroclastic sedimentary rocks); aggressive waters; the presence of gas; radioactivity; hazardous minerals; swelling and squeezing phenomena.
2. The characterization of the geomorphological setting: the underground works can interfere with the general environmental equilibrium on land surface. Particularly, shallow tunnels, entrance areas or tunnels close to the side of slopes are affected by meteoric events, weathering, superficial karst phenomena or landslides.
3. The characterization of the structural and tectonic setting. The tectonic phenomena, to which rocks were subjected during their geological history, can affect both the mechanical and hydrogeological behaviours of the rock mass.
4. The reconstruction of the hydrogeological setting. The water can represent damages both to the tunnel (i.e. tunnel inflow) and to the environment (i.e. water resource depletion) or to a valuable, exploitable water resource.

In this chapter, all geological and hydrogeological aspects, necessary for the reconstruction of the geological and hydrogeological conceptual models, will be presented and described. This chapter will also cover the seismic effects, the natural state of stress and all the geological investigations that can be performed to improve the knowledge of both the geological and hydrogeological conditions of the area of interest for tunnel construction.

3.2 ENGINEERING GEOLOGY FOR TUNNELLING AND UNDERGROUND WORKS

The construction of tunnels involves finding solutions to many complex technical problems depending on the geological and geoenvironmental conditions of the area in which the work has to be realized.

Only a careful analysis of all the geological and geoenvironmental issues and a correct reconstruction of the conceptual model can lead to optimal design solutions from all points of view and therefore ensure the safety of the workers, during construction and in the operation phase.

In order to obtain a cavity, during tunnelling, the natural ground (rock mass or soil) is removed. Before the excavation, the ground is in equilibrium under its original state of stress. The excavation modifies this state of stress by generating a stress deviation around the cavity, with stress concentration close to the boundary surfaces of the cavity. As a result, the ground undergoes deformations in order to reach a new equilibrium condition. These deformations and the related kinematics depend on the following:

- The depth of the excavation.
- The shape and the dimension of the cavity.

- The method, timing and technique of excavation.
- The geological model (different typologies of media, and geomorphological, geological, tectonic and hydrogeological features).

In particular, the knowledge of the geological and geotechnical models (Chapter 4) allows us to identify the hazards, to estimate the risks and to define their mitigation (Gattinoni et al., 2014).

The *identification of potential risks* (Chapter 7) involves the identification of hazard elements (e.g. a situation or a physical condition that potentially originates a damage or causes undesired consequences) and the evaluation of their causes and consequences. The most common geological hazard events in tunnelling are rockfall, sidewall instability (rock burst, spalling, slabbing, etc.), face collapse, groundwater interaction (tunnel inflow, water table drawdown and spring extinction), high temperature, surface settlements and sinkholes.

The *risk analysis* consists in the quantification of the likelihood of hazardous events and the magnitude of their consequences (in terms of costs, lead time, worksite safety, environmental impact, etc.). First, the possible hazards are identified and classified, and then they are analysed usually by filling in risk matrixes. The damaging events are listed according to their occurrence probability and impact, and they are taken into account in the following designing phases in order to define the measures apt to reduce and manage the risk.

The *risk evaluation* consists in comparing the results of the previous analysis using acceptability criteria, in order to determine which risks have to be reduced. In other words, it is possible to define the critical thresholds and the different levels of risk:

- Negligible/minor risk: no action required, but risk factors must be monitored.
- Significant, but acceptable risk: risk factors must be monitored, and the project may be supplemented by further investigations and/or a series of predefined measures.
- Major risk: it is not possible to start the construction until the risk has been reduced or removed.
- Unacceptable risk: if the risk cannot be controlled, the project may be abandoned or altered.

The *risk mitigation* implies the choice of mitigation measures to be applied and of their entity, followed by the verification of mitigation measures on the basis of monitoring data gathered during the construction phases (Chapter 10). If the threshold values are exceeded (major risks), modifications have to be implemented to take the risk level back below an acceptable threshold.

In the present chapter, the geological features and hazards will be taken into account.

3.3 THE GEOLOGICAL MODEL: FEATURES AND HAZARDS

The behaviour of the mass being excavated essentially depends on three main aspects: the lithological nature, determining the mechanical characteristics of the matrix; the structural features (stratification, schistosity, fracturing, etc.), governing the mechanical properties of the mass itself; the state of stress existing before the excavation. In particular, the variation of the above factors can induce a broad spectrum of unstable and deformation phenomena.

Moreover, as soils and rocks are multiphase media, groundwater can affect their behaviour during excavation.

Finally, the depth of the excavation with respect to the topographic surface, the presence of natural gas, aggressive water, weathering and swelling minerals, geothermal gradient, seismicity, radioactivity and hazardous minerals can be very relevant during tunnelling.

3.3.1   The identification of the different typologies of media

The geotechnical and geomechanical behaviours of soils (sediments) and rock masses depend primarily on their lithological features, e.g. their mineralogical and petrographic composition, and on the type of process which generated the lithology itself. Among the others, it is possible to distinguish:

Sediments: they are natural materials broken by the processes of weathering, chemical weathering and erosion, and subsequently transported by the action of wind, water or ice, and/or by the force of gravity. The sediments are composed of cobble, gravel, sand, silt and clay, and depending on the deposition environment, it is possible to distinguish glacial, alluvial, lacustrine sediments and talus slope.

In this kind of sediments, it is impossible to realize a tunnel without confining/strengthening interventions.

Rocks: they are natural aggregates of minerals and grains strongly and permanently bonded. The rocks, depending on their origin, can be magmatic (igneous), sedimentary and metamorphic.

The igneous rocks (with the exception of pumice and obsidian) and the metamorphic non-schistose rocks are generally the lithological types with the best strength characteristics; considering similar fracturing and weathering conditions, massive sedimentary rocks rank second, followed by the metamorphic schistose ones and highly stratified sedimentary rocks. The risks related to the lithology are as follows:

* Dissolution phenomena (in carbonate sedimentary rocks or in some evaporitic rocks).
* The lack of cement among grains leading to disintegration (in clastic and/or pyroclastic sedimentary rocks).
* Aggressive waters (these waters are rich in calcium sulphate, sulphuric acid, free carbon dioxide or chloride and magnesium sulphate due to the presence of anhydritic or gypseous lithotypes, chalk masses, evaporitic series or rocks linked with magmatic phenomena).
* The presence of gas (in carbonaceous rocks, scaly clays, organic clays, volcanic soils or igneous rocks).

- Radioactivity (in igneous rocks, such as lavas, tuff and pozzolanas, or other rocks containing uraniferous minerals or radium).
- Hazardous minerals (in ultrabasic rocks, e.g. peridotite, or in metamorphic rocks, e.g. serpentines and amphiboles).
- Swelling and squeezing phenomena (especially in ductile soft rocks).
- Rock burst (especially in brittle hard rocks).

3.3.2   The geomorphological issues

As far as the morphological conditions are concerned, it is important to distinguish between shallow and deep tunnels, as well as to consider specific problems related to tunnels close to the side of the slope and portals.

For a shallow tunnel (indicatively, overburden less than four times the excavation diameter), the disturbed area around it interferes with the ground surface, possibly affecting its general equilibrium. Specifically, shallow underground works, entrance areas or tunnels close to the side of the slope are affected by:

- Meteoric events leading to significant water inflows.
- Weathering zones.
- Superficial karst phenomena.
- Landslides (Figure 3.1).

For deeper tunnels, geomorphological conditions lose their relevance progressively; the only exceptions are when tunnelling in very steep slopes or along glacial valleys, as well as in areas characterized by deep-seated landslides or large karst phenomena.

3.3.3   Tectonic and structural issues

The lithosphere is subdivided into plates that may converge, diverge or scroll side by side, involving many geological effects (i.e. volcanism, earthquakes, continental drift, expansion of the oceans and orogenesis).

As a consequence, tunnelling in a tectonically active area (usually the margin of the plates) has to face a stress state depending on the plate kinematics.

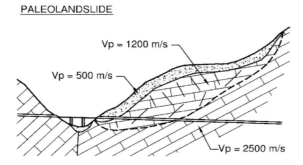

Figure 3.1 Paleo-landslide at the tunnel portal (Gattinoni et al., 2014).

Brittle tectonic structures as faults will be generated by divergent or transform tectonic movements. On the contrary, folds will frequently develop when movements are convergent.

In particular, either brittle or ductile deformations depend on the type of stress state (compression, tensile and shear); the physical and mechanical behaviour of the rocks (brittle or ductile); and temperature and pressure conditions.

Fractures or joints and faults are the result of brittle deformations of the earth's crust. Fractures are caused by stresses exceeding the rock strength; they are characterized by a specific orientation with respect to the main stress. In particular, a fault is a fracture characterized by a significant displacement (slip) of the two sides of the fault plane.

The presence of faults along the layout of an underground structure can cause significant problems.

Actually, if the shear stress along the discontinuity is very high, the rock mass can become completely disintegrated. Such deformations can interest wide strips of rock mass (named 'fault rocks', Figures 3.2 and 3.3), causing serious problems in tunnelling, because such a material often has limited, if any, self-supporting features.

The main challenges to be faced when an underground excavation crosses fault rocks are: very short stand-up time, face instabilities, high radial convergences, and potential squeezing, especially if stresses and/or water inflows are high.

Fracture zones can also involve preferential flow paths for groundwater, bringing water inflows along tunnels.

Figure 3.2 Examples of fault rocks.

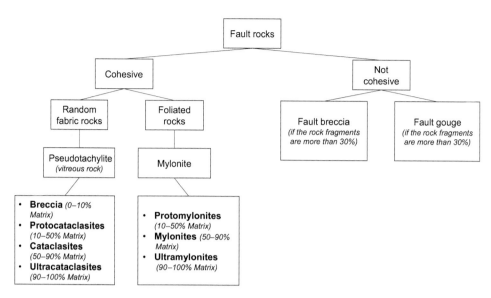

Figure 3.3 Classification of fault rocks. (Modified from Higgins 1971, Sibson 1977 and Wise et al. 1984.)

Folds are the result of plastic deformation of the earth's crust due to compression stresses. In tunnelling, they can involve residual stresses (compression in the core and tensile in the hinge), especially at high depths, as well as dissymmetrical loads and lithological heterogeneity.

Moreover, they are also relevant to the type of fold; actually, crossing a syncline core along its axial plane can involve strong lateral stresses and high water inflows, whereas crossing an anticline in its hinge can lead to releases and collapses at the ceiling.

The tectonic phenomena to which the rocks were subjected during their geological history, along with the different types of rock, can affect the technical and hydrogeological behaviours of the rock mass.

For these reasons, it is essential to collect all data related to the following structural characteristics: geometry (inventory of all brittle or ductile structures), kinematics (examination of the displacements and movements that lead to change of position, orientation, size and/or shape of rock bodies) and dynamics (reconstruction of the nature and orientation of the stresses producing the deformation). Based on these data, the behaviour of the material during excavation (Figures 3.4 and 3.5) can be predicted.

In the presence of bedded and/or fractured rock masses, the following parameters should be carefully evaluated, too:

- The layer thickness and/or the fracturing degree, i.e. the number of fractures per linear metre, or the distance between the discontinuities (strata or fractures).
- The joint characteristics (persistence, roughness, aperture, filling and alteration).
- The joint orientation relative to the walls of the underground cavity.

Figure 3.4 Overbreak controlled by the orientation of the discontinuities. (Courtesy of E.M. Pizzarotti.)

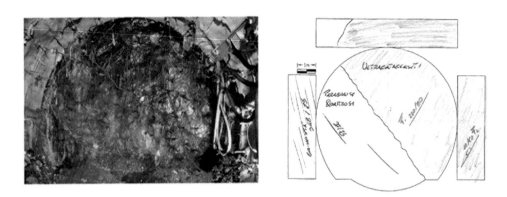

Figure 3.5 Example of geological survey on the tunnel face. (Courtesy of E.M. Pizzarotti.)

3.3.4   Geological issues

The identification of different lithologies (with their age, spatial distribution and orientation) of tectonic structures and of geomorphological data allows to reconstruct the geological setting concerning the area of influence of the underground work.

The main document summarizing all these issues is the geological map (Figure 3.6), whose interpretation is represented by the geological sections (Figure 3.7).

These documents allow us to identify critical areas where detailed geognostic surveys are requested.

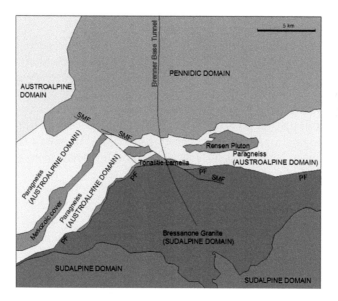

Figure 3.6 Example of geological map and tectonic setting (Gattinoni et al., 2016a).

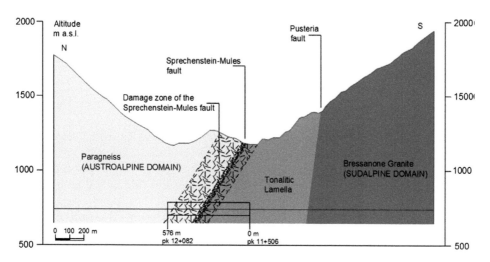

Figure 3.7 Example of a schematic cross section (Gattinoni et al., 2016a).

3.3.5   Hydrogeological issues

In case of underground excavations, hydrogeological issues are relevant (Table 3.1), as they can bring damages to both the tunnel (i.e. tunnel inflow) and the environment (i.e. water resource depletion).

From another point of view, the groundwater intercepted during the excavation of a tunnel can be considered as a valuable water resource that is efficiently exploitable and economically interesting.

For these reasons, groundwater can represent both a threat or a problem and an important resource. Therefore, in the modern design of underground works, the hydrogeological elements assume a centrality that must be considered not only during design and construction, but also in the operating phase.

In this subsection, a synthesis of the contribution of hydrogeology to tunnel design is presented. In particular, the following issues are addressed: (i) why you need a hydrogeological model in tunnel design and (ii) how the hydrogeological model has to be reconstructed and used for the tunnel design. In the following chapters, other issues related to the hydrogeological hazards in tunnelling will be addressed (i.e. how tunnel inflow can be assessed and how its impact on water resources can be quantified).

As far as the environment is concerned, tunnelling can have a relevant draining effect on both groundwater and surface water, bringing negative consequences, such as (Figure 3.8):

- Qualitative and quantitative changes in water resources and their flow path.
- Changes in vegetation.
- Changes in the hydrogeological balance at the basin scale, with possible drying up or depletion of springs and wells.
- Changes in slope stability conditions.

On its turn, a proper forecast of the interactions between tunnel excavation and groundwater is of paramount importance because water inflow could involve problems relevant to both the stability of the underground work and the safety of workers.

Several factors, related to both the tunnel features and the hydrogeological setting, affect the hydrogeological hazards in tunnelling, in particular:

- Shape and size of the tunnel (Farhadian et al., 2016).
- The excavation method (Gattinoni et al., 2014).
- The depth of the tunnel, especially with reference to the groundwater table (Cesano et al., 2000).
- The hydraulic conductivity of the ground (Farhadian and Katibeh, 2015).
- The feeding condition of the aquifer (Scesi and Gattinoni, 2009).
- The nearness of highly pervious geological (e.g. paleo-channel), geomorphological (e.g. karst phenomena) and tectonic (e.g. faults) features (Moon and Jeong, 2011).

This information is synthesized in the hydrogeological conceptual model of the system tunnel-aquifer, which, on its turn, depends on the design features, as well as on the geological and geotechnical model. Therefore, specific hydrogeological surveys are

Table 3.1 Examples of groundwater inflow during the tunnelling

Location	Year	Length (km)	Max inflow rate (m^3/s)	Main geomaterial	Reference
Sempione (ITA-CH)	1906	19.8	1.7	Limestone	Civita (2005)
M. Bianco (ITA-FRA)	1959–62	11.6	0.8	Granite	Civita (2005)
Gran Sasso (ITA)	1969–82	10.2	3	Limestone	Civita (2005)
S. Leopoldo (ITA-AUS)	1985	5.7	3.6	Limestone	Civita (2005)
Pont Ventoux (FR)	1999–2002	8.1	0.32	Gneiss	Perello et al. (2007)
Fiorenzuola (ITA)	1999–2003	15	1	Sandstone	Vincenzi et al. (2009)
Hsueh-Shan (Taiwan)	1991–2006	12.9	1.37	Sandstone	Chiu and Chia (2012)
Gotthard Base Tunnel (CH)	2000–2012	5	0.035	Gneiss	Masset and Loew (2013)
Qiyueshan Tunnel (China)		0.55	0.33	Limestone	Li et al. (2010)
Aica-Muse pilot tunnel (ITA-AUS)	2007–2010	10.4	0.23	Granite	Perello et al. (2014)

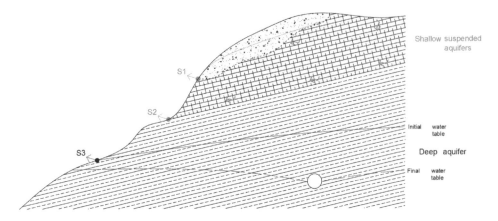

Figure 3.8 Water table drawdown induced by the opening of a tunnel.

needed in order to reconstruct a proper conceptual model, based on which the potential hydrogeological hazards can be identified.

Geognostic surveys in tunnelling have to be considered not as a cost, but as an investment, minimizing risks and uncertainties in ground conditions. The hydrogeologist must forecast and assess the risks related to groundwater, by evaluating the water flow and its interactions with the underground works, as well as the possible mitigation measures. For this reason, before the design phase of the tunnel, specific hydrogeological surveys (Table 3.2) have to be carried out in order to:

- Identify the type of aquifer (fractured, karsts, alluvial, etc.).
- Characterize the aquifer geometry in depth.
- Reconstruct the groundwater level and flow path at the basin scale.
- Determine the hydraulic conductivity of the geomaterials.
- Define the groundwater balance at different scales and the interactions with the surface water.

Generally speaking, if the tunnel is located above the water table, problems due to water inflow are connected to water reaching the excavation by infiltration or percolation.

Only in karstified rock masses a large, although temporary, inflow rate is possible even above the piezometric surface. On the contrary, if the excavation develops below the water table, water inflow can become continuous and make the excavation difficult. In such cases, a piezometric map showing both the maximum and minimum water table elevation is needed.

The map has to be updated (yearly/seasonally/monthly) for monitoring the water table trend, in order to avoid any potential hazard. Moreover, the piezometric map allows us to find out useful information about the hydrogeological conceptual model: not only the flow path and water table changes with time, but also the recharge and discharge areas of the aquifer, and therefore the groundwater boundaries (e.g. divides and aquicludes).

Table 3.2 Hydrogeological surveys in the different stages of the tunnel design

Project phase	Surveys	Objective
Preliminary design	Census of springs, water table levels, flow rate measures, planning of hydrodynamic tests, analysis of meteoclimatic data and chemical data on groundwater	Hydrogeological map, with reconstruction of the physical model and the groundwater flow path, assessment of the water balance and identification of the potential hydrogeological hazards
Detailed design	Geostructural and geophysical surveys; drillings and in situ permeability tests, depending on the type of aquifer: pumping tests mainly in soils, Lugeon packer tests in rock masses (with length of the packer interval equal to 2–3 tunnel diameter, as suggested by Moon (2011)), Kazemi method for dual-permeability geomaterials, Hantush method for anisotropic aquifers, etc.; laboratory tests (e.g. permeameters) and tracer tests (with both natural and artificial tracers); and monitoring of springs, water table and rainfall	Quantitative characterization of the aquifer system (permeability, transmissivity, etc.), tunnel inflow assessment, delimitation of the tunnel influence zone, design of the draining system and lining, and planning of the control system
Construction	Monitoring of water flow rate and pressure in tunnel, monitoring of springs and water table (not only within the tunnel influence zone, but also in the external zone) and chemical analysis of groundwater	Verification of the hydrogeological model, chemical control and control of the hydrogeological impact on the environment (by comparing monitoring data collected inside and outside the influence zone of the tunnel)
Post-construction	Monitoring of water (both in the tunnel and in the environment)	Control of the efficacy of draining/waterproofing system, and planning of control and alarm systems

Based on these boundaries, the groundwater basin, which is the area contributing to the aquifer supply and then the area which can be involved in the changes in the hydrogeological balance caused by the tunnel drainage, can be defined.

Moreover, a monitoring network of rainfall, spring discharge and groundwater table must be implemented before the excavation of the tunnel, since it is essential to tunnel impact evaluation and management. Actually, in order to properly evaluate the tunnel impact, the undisturbed conditions must be determined, before the beginning of the excavation works, and the natural fluctuations must be separated, by means of control points located outside the influence zone of the tunnel, from the induced variations.

In order to identify the potential hydrogeological hazards for tunnelling, it is important to envisage the conditions potentially leading to significant water inflow:

- High-permeability materials (such as granular soils and rocks that are permeable for porosity or fracturing degree), below the groundwater table.
- Sudden changes in permeability, such as the presence of buried river beds.
- Tectonic structures, such as faults or overthrusts and syncline folds, characterized by a significant water supply.
- Possible interconnection between shallow and deep aquifers (e.g. along shear zones).
- Karst phenomena, potentially causing significant tunnel inflow, even if the tunnel is located above the piezometric surface.

Tunnelling in granular soils below the water table leads to diffuse groundwater inflows and a generalized drawdown of the piezometric level. In spite of that, granular soils are not the most difficult hydrogeological environment to deal with when tunnelling. Actually, this kind of ground is easier to be modelled from a hydraulic point of view, as quite continuum and homogenous. On the contrary, the permeability in rock masses can assume a large range of values, spanning several orders of magnitude (from 1 to 10^{-10} m/s). In a fractured/karst system, water flow depends on the presence of discontinuities (faults, fractures, bedding and foliation planes with slow flow, acting as reservoirs) and conduits (characterized by quite rapid flows, and then acting as drains), and on the possible presence of a saturated zone. Considering that the water flow is ruled by the orientation and features of joints (Scesi and Gattinoni, 2009), the hydraulic characteristics of these geomaterials are neither homogeneous nor isotropic. As a consequence, in the conceptual model a representative equivalent permeability has to be defined. The first assumption is that the permeability of the rock matrix is very low and the effective permeability is ruled by fracture features and, in particular, by fracture orientation (dip and dip direction), aperture, spacing and persistence. More in detail, joint orientation rules the main permeability direction, whereas joint aperture mainly rules its magnitude. The latter can decrease with depth, because of the effect of stresses; moreover, the hydraulic aperture is smaller than the mechanical one, because of the joint roughness.

In hard rocks, groundwater inflows generally occur in few short stretches of the tunnel and most of the tunnel remains dry. Actually, at large scale, the water flow in hard rocks depends on the tectonic structure, which is the result of brittle deformations. A simple conceptual model for a fault structure involves shear lenses localized in the fault core, surrounded by a zone of distributed fractures in the damaged zone. The fault core (single or multiple) generally consists of gouge, mylonites, cataclasites or breccia (or a combination of these), and the damage zone generally consists of fractures over a wide range of length scale (from microfractures to macrofractures). Generally, the permeability in the damage zones is quite high, because of the high connectivity and the wide aperture of joints. On the contrary, the fault core is often characterized by lower permeability. The contrast in permeability governs the fluid flow: in the damage zone, the channelling of the water is favoured by the distribution of joints, converging towards the fault core, where often the hydraulic conductivity decreases, with a barrier effect (Gattinoni et al., 2016b). For example, in the Aica-Mules pilot tunnel for the Brenner Base Tunnel, the permeability is characterized by relevant variations along the fault zone, spanning over three orders of magnitude (Perello et al., 2014): the lowest value refers to the fault core of the Periadriatic Line, where the rock is reduced to a fine-grained fault gouge, whereas higher values represent the fault damage zone, where the fracture density varies from high to very high.

Finally, even if low-permeability geomaterials generally don't cause relevant tunnel inflows, they should not be neglected in the assessment of the environmental hydrogeological hazards arising from tunnelling. Actually, some Italian experiences demonstrated that even in the presence of aquitards, the tunnel drainage can lead to significant water resource depletion for the local water balance, and then for the environment (Vincenzi et al., 2009).

The hydrogeological conceptual model in tunnelling is aimed at ():

- reconstructing the groundwater flow under different conditions, assessing the tunnel inflow and the related radii of influence, in order to bound the area potentially affected by the hydrogeological hazard;
- assessing the probability that the tunnel inflow or the piezometric drawdown due to the excavation exceeds acceptable values, in order to quantify the hydrogeological risks both to the tunnel and to the environment.

3.3.5.1   The groundwater as a resource

The groundwater intercepted during the tunnel excavation must be managed both during the execution phase of construction and during the operating phases. A correct management of the infiltrations and of the concentrated and diffused inflow is essential to guarantee the safety of the construction, but above all to allow a normal and functional life cycle of the work during the operating phases (Lo Russo et al., 2019).

If the groundwater intercepted during the excavation of the tunnel is of good chemical and physical quality (case not so infrequent), it can actually be collected and conveyed outside to be reused for various purposes (human consumption and agricultural or industrial uses) (Furno et al., 2016; Dematteis et al., 2016).

From this point of view, the tunnel can therefore be considered not only as a method of crossing the rock mass, but also as a sort of linear draining system of the groundwater circulating in the rock mass itself.

For these reasons, at both project and construction stages, particular attention should be paid to the methods of collecting the individual water inflow, to their quantitative and qualitative monitoring and to their external conveyance. This collection mechanism should be developed with the use of building materials that do not change the characteristics of the intercepted water in order to be able to use them.

In addition, in many cases (such as alpine tunnels with high rock covers) in which the groundwater can reach temperatures that are considerable and close to those of the rock mass crossed, it can also be considered as a low-enthalpy geothermal resource exploitable in different ways (agriculture, fish farming, air conditioning, etc.). An important example of exploitation and the use of the warm water produced by the drainage of a tunnel is represented by the Lötschberg Base Tunnel in Switzerland (Hufschmied and Brunner, 2010).

Numerous are also the cases in which the groundwater resource, downstream of the use of heat transported outside and its cooling, can be used for human purposes and consumption (Link et al., 2015).

In conclusion, we can say that the groundwater intercepted during the excavation of a tunnel is both a problem to be overcome and a potential resource that can be exploited in the long term. Modern tunnel design must consider these elements in a comprehensive way and take into account positive externalities related to the presence and the drainage of the groundwater in the crossed rock masses.

3.4   GEOLOGICAL AND HYDROGEOLOGICAL INVESTIGATIONS

The main geological problems that are faced when carrying out an underground work derive from the definition of both the geological and geotechnical models:

1. The geological model identifies different typologies of geomaterials, including their spatial distribution, tectonic features and geomorphological/hydrogeological characteristics. It synthesizes the results of *geological studies and surveys* carried out on a large scale, based on the existing geological cartography. It provides a first layout of the geological (lithological, stratigraphic and tectonic) and hydrogeological features of the area, identifying the potential geological and hydrogeological hazards. At this aim, detailed geological and structural (Figure 3.9 and Table 3.3) and *geomechanical surveys*, as well as *geomorphological studies*, have to be carried out on rock outcroppings, identifying the possible instability phenomena (i.e. in portal areas and stretches with low overburden), karst phenomena and paleo-channels. Based on these surveys, geological, geomorphological and hydrogeological maps are produced and preliminary profiles of the tunnel (Figure 3.10) are constructed to identify the critical areas in which detailed geognostic surveys are required.

Figure 3.9 Example of geostructural survey.
Lithology: 'Arenarie di San Salvatore' (sandstones)
URV (unitary rock volume) (m^3): 0.22.

Table 3.3 Example of geostructural survey.

	Stratification	Discontinuities		
	S_0	K_1	K_2	K_3
Orientation	34°/76°	118°/82°	295°/82°	293°/31°
Spacing(cm)	38,17	78,53	78,53	106,58
Aperture (mm)	43,83	12,96	12,96	Closed
Persistence (%)	78	73,33	73,33	34
Filling	Clay	Absent	Absent	
Filling thickness (mm)	43,83			
Shape	Undulating	Undulating	Undulating	Undulating
Roughness (JCR)	4	3	3	4
Water	Damp	Wet	Wet	Dry

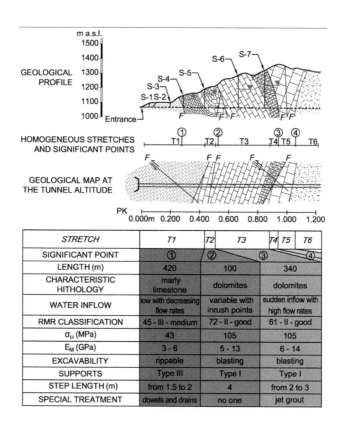

Figure 3.10 Example of geological section. (Modified from Gattinoni et al. 2014.)

2. The geotechnical model includes the lithotechnical, hydrogeological, geotechnical and/or geomechanical characterization of geomaterials, as well as the risk factors that may influence the excavation; this characterization is carried out by means of detailed geognostic surveys and geotechnical and/or geomechanical tests (see Chapter 4).

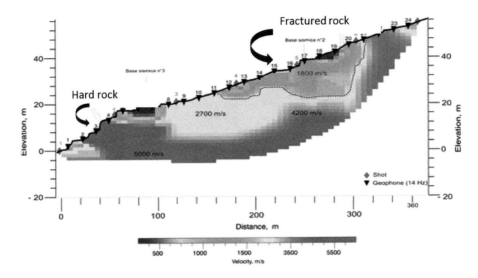

Figure 3.11 An example of seismic tomography for the reconstruction of the fracturing degree of the rock mass.

According to the overburden or the stretches considered (entrance or central part of the tunnel), the surveys will be different. For the characterization of *shallow overburden stretches*, the following are required:

- Geological, geomechanical and geomorphological surface surveys.
- Mechanical continuous core drilling boreholes.
- Geophysical surveys (seismic refraction, required to assess the depth of surface deposits or of weathered and fractured rocks as well as the elastic characteristics of the materials) (Figure 3.11).

For the characterization of *medium-to-high overburden stretches*, a survey carried out on the surface cannot be easily extrapolated at the excavation level. Drillings, whose costs and time increase with the increase in depth, provide punctual data (even so they are essential to obtain the samples for laboratory tests). Moreover, the data obtained by seismic refraction cannot always be correctly interpreted. Therefore, geognostic surveys referred to these stretches make use of other indirect methodologies. In this case, seismic reflection and other techniques such as seismic tomography can be very useful.

3.5   NATURAL STATE OF STRESS AND ITS ASSESSMENT

At depth, rocks are affected by significant stresses due to the weight of the overlying strata (lithostatic load), as well as to residual tectonic stresses.

The horizontal stresses acting within the ground are much more difficult to estimate than the vertical ones. Some measurements showed that the ratio between horizontal and vertical stresses tends to be high at shallow depth and that it decreases with depth (Hoek and Brown, 1980; Herget, 1988). Sheorey (1994) provided a simplified equation for estimating the horizontal-to-vertical stress ratio k:

$$k = 0.25 + 7E_h\left(0.001 + 1/z\right)$$

(3.1)

where $z(m)$ is the depth below the surface and E_h (GPa) is the average horizontal deformation modulus of the upper part of the earth's crust.

Unfortunately, Sheorey's theory does not explain the occurrence of vertical stresses higher than the overburden pressure, as well as the presence of very high horizontal stresses or why the two horizontal stresses are seldom equal. These differences probably arise from local topographic and geological features, strictly connected to the tectonic setting. The World Stress Map can give a first indication of the possible complexity of the regional stress field, showing the possible directions of the maximum horizontal compressive stress (Figure 3.12). Afterwards, in situ stress measurements can be used to refine the analysis; for instance, the in situ stresses near regional tectonic features (such as major faults) may be rotated with respect to the regional field and the magnitude of stresses may be significantly different from the values estimated from the general trends.

Relevant changes in the lithostatic stress state can also occur at great depth, because of the position of the tunnel with respect to the slope or the morphodynamic evolution of the site (e.g. the presence of former glaciers).

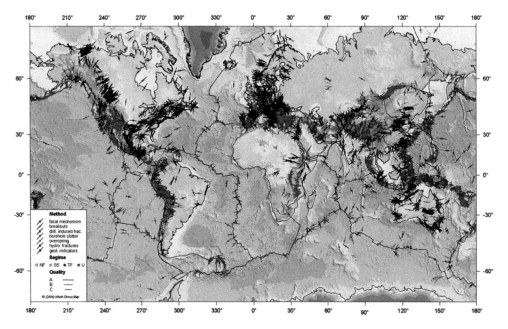

Figure 3.12 World Stress Map giving orientations of the maximum horizontal compressive stress (from www.world-stress-map.org).

When an opening is excavated in this rock, the stress field is locally disrupted and a new set of stresses is induced in the rock surrounding the opening. The knowledge of the magnitudes and directions of the in situ and induced stresses is an essential information for the underground excavation design, since, in many cases, the strength of the rock is exceeded and the resulting instability can cause serious consequences.

3.6 GEOTHERMAL GRADIENT

Temperature and pressure in the earth's crust increase with depth. The average pressure increase is about 27 MPa each 1,000 m, whereas the geothermal gradient is highly variable and ranges from 1.5°C/100 m to 5.0°C/100 m. This variability can be caused by cold water infiltration due to melting glaciers, by permafrost or by cooling magmatic masses at shallow depth, as well as by the local geodynamic evolution. The values of temperature and geothermal gradient measured in some tunnels are listed in Table 3.4.

The main problem arising from the increase in temperature is related to the working conditions; actually, workers operate under optimal conditions at temperatures below 25°C, whereas temperatures above 30°C involve the installation of effective air cooling systems.

3.7 SEISMICITY

Underground structures show a much lower seismic vulnerability than surface infrastructures, as they are usually flexible enough to withstand the strains imposed by the surrounding soil without reaching their breaking point. Obviously, a proper design can grant these structures the required seismic behaviour. Moreover, tunnel mass is generally small compared to the mass of the surrounding ground, and the confinement of the tunnel by the ground allows to significantly damp out the seismic perturbation. Anyhow, in spite of their usual good seismic behaviour, violent seismic events may become hazardous also for underground works, especially if the surrounding ground is affected by liquefaction phenomena or displacements along faults. Generally speaking, the vulnerability of underground structures to earthquakes depends on:

- geological conditions (i.e. the stiffness of rocks and soils),
- tectonic setting,
- tunnel depth (damage extent usually decreases with depth),
- location with respect to the valley side,
- tunnel size (the larger the section, the greater the seismic vulnerability),
- section type (with or without invert).

Table 3.4 Example of geothermal gradient observed during the excavation of some famous tunnels (Gattinoni et al., 2014)

Tunnel	Geothermal gradient	Temperature at different depths
Lötschberg (Switzerland)	~2°/100 m	34° at 1,500 m of depth
Sempione (Italy–Switzerland)	~2.5°/100 m	56° at 2,000 m of depth
Gottardo (Switzerland)	~2°/100 m	30° at 1,500 m of depth
Monte Bianco (Italy–France)	~1.5°/100 m	30° at 2,000 m of depth

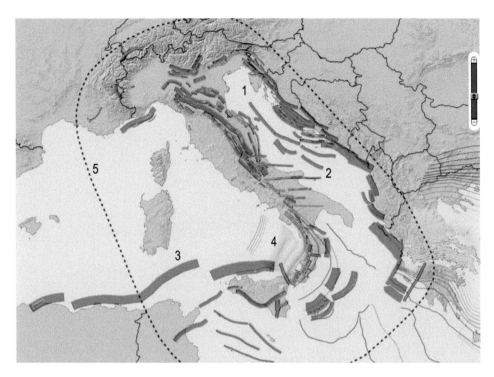

Figure 3.13 An example of a seismogenic map. A seismogenic map of Italy: 1 = active faults; 2 = active fold; 3 = composite seismogenic source; 4 = subduction slab depth; 5 = area of relevance. 'Istituto Nazionale di Geofisica e Vulcanologia' (modified from http://diss.rm.ingv.it/diss/).

For the above reasons, it is necessary to evaluate the seismicity of the area of interest for an underground excavation and to define the relationship between the seismicity and the geological and tectonic setting of the area. In this way, it is possible to:

• forecast the behaviour of rocks and soils during earthquakes to locate the areas susceptible to liquefaction phenomena or where there is a possibility of new faults being generated, and
• identify which faults are responsible for the seismic activity. In this case, it is possible to refer to seismogenic databases and/or to maps realized by the geological and/or geophysical services of different countries of the world. An example of a seismogenic map regarding the Italian territory is reported in Figure 3.13.

3.8 SWELLING, SQUEEZING AND ROCK BURST

3.8.1 Swelling

Some geomaterials tend to expand or contract depending on the changes in environmental conditions (wet and dry conditions and temperature).

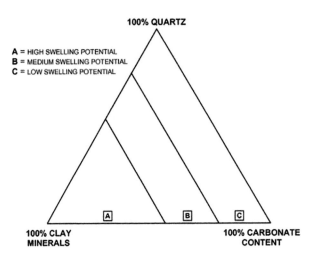

Figure 3.14 Swelling potential chart (Gattinoni et al., 2014).

Adsorption or absorption of water (due to differences in concentration), unsaturated or partially saturated bonds and differences in potential are frequently associated with time-dependent increases in volume, leading to swelling phenomena.

In general, these phenomena are related to the lithological features of rocks and soils and to their mineralogical composition: the most common swelling materials are those containing clayey minerals as the group of the smectites, of the illites or of the kaolinites (phyllosilicates). Moreover, swelling phenomena can occur even when anhydrite ($CaSO_4$) turns into gypsum ($CaSO_4*2H_2O$) in consequence of water imbibition.

These geomaterials, in consequence of the cavity opening and the subsequent deprivation of their natural confinement, tend to a significant increase in volume.

Generally, if a material contains a percentage of clay minerals ranging from 50% to 100% and a percentage of quartz and carbonate ranging from 0% to 50%, the swelling probability is very high; if a material contains a percentage of clay minerals ranging from 25% to 50% and a percentage of quartz and carbonate ranging from 50% to 75%, the swelling probability is medium; if a material contains a percentage of clay minerals ranging from 0% to 25% and a percentage of quartz and carbonate ranging from 75% to 100%, the swelling probability is low (Bonini et al., 2009) (Figure 3.14).

3.8.2 Squeezing

Squeezing is a time-dependent deformation associated with the plastic behaviour of geomaterials; it is governed by both material properties and overstress conditions around the tunnel.

Some effects of squeezing are evident immediately after excavation (i.e. convergences occurring even when the excavation doesn't advance), but normally, the long-term effect is prevalent, including continued ground movements and/or a gradual build-up of load on the tunnel support system (Figure 3.15).

Figure 3.15 Example of squeezing behaviour in the S. Martin La Porte exploratory adit of the Lyon–Turin Railway Base Tunnel.

The prediction of squeezing conditions is very important to define a stable support system of the tunnel. Generally, it is possible to say that (Barla, 2002):

1. Squeezing behaviour is associated with poor rock mass strength parameters in relation to initial stress.
2. Squeezing implies the occurrence of yielding around the tunnel. The onset of a yielded zone surrounding the tunnel causes a significant increase in tunnel convergence and face displacements (extrusion). These are generally large, increasing with time, and represent the more significant effects of the squeezing behaviour.
3. Orientation of discontinuities (i.e. bedding planes and schistosity) plays a very important role in the onset and development of large deformations around tunnels, and therefore in the squeezing behaviour, too. In general, if the strike of main discontinuities is parallel to the tunnel axis, the deformation will be enhanced significantly, as observed in terms of convergence during face advance.

   1. The pore pressure distribution and the water table head can also influence the rock mass stress–strain behaviour. Drainage systems causing a reduction in the groundwater table often help in reducing ground deformations.
   2. Construction techniques (i.e. the excavation sequences and the number of excavation stages, including the stabilization methods) may influence the overall stability conditions of the excavation.

3. Large deformations associated with squeezing may also occur in rocks susceptible to swelling. Although the factors causing either of these behaviours are different, it is often difficult to distinguish between squeezing and swelling, as the two phenomena may occur at the same time and induce similar effects. For example, in overconsolidated clays, the rapid stress relief due to the tunnel excavation causes an increase in deviatoric stresses, with simultaneous onset of negative pore pressure. However, due to the negative pore pressure, swelling may occur with a more sudden onset of deformations under constant loading. Therefore, if swelling is restrained by means of early invert installation, a stress increase may take place with probable onset of squeezing.

Among different methods that can be used for a quick qualitative estimation of the risk of squeezing, it is possible to remember the methods of Singh et al. (1992) (Figure 3.16), Goel et al. (1995) (Figure 3.17), Hoek and Marinos (2000), Jethwa et al. (1984), Bhasin (1994) and Panet (1995).

3.8.3   Rock burst

Squeezing (previously described) and rock burst are the two main modes of underground instability caused by the overstressing of the ground, the first occurring in ductile soft rocks and the second in brittle hard rocks.
     According to Cai and Kaiser (2018):

- A **rock burst** is defined as the damage to an excavation that occurs in a sudden and violent manner and is associated with a mining-induced seismic event. 'Rock burst' is a generic term and is independent of the cause of damage and failure process.

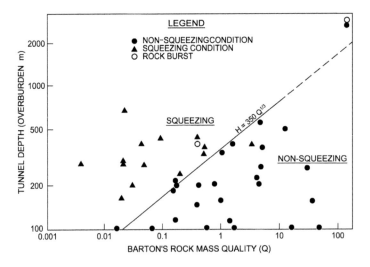

Figure 3.16 Empirical approach to predict squeezing. The squeezing occurs above the line having the following equation: $H = 350 \cdot Q^{1/3}$, where Q represents the rock mass quality of Barton (Gattinoni et al., 2014).

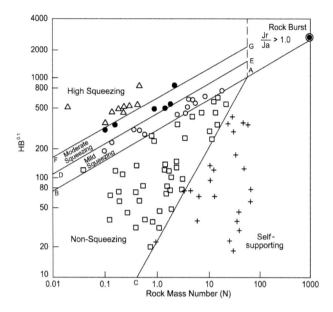

Figure 3.17 Empirical approach to predict squeezing conditions. This method is based on the rock mass number N, defined as stress-free Q ($N = (Q)_{SRF=1}$), the tunnel depth H and the tunnel span or diameter B (Gattinoni et al., 2014).

Strain, pillar and fault-slip bursts are all rock bursts if they cause damage to an excavation or its support. The seismic event may be remote from or co-located with the damage location. A seismic event that does not cause any damage is not a rock burst.

- A **strainburst** is a sudden and violent failure of rock near an excavation boundary caused by excessive straining of a volume of stiff and strong rock (burst volume). The primary or a secondary seismic source is co-located at the damage location. Strainburst could be triggered or dynamically loaded by a remote seismic event.

Moreover, as remarked by Diederichs (2005):

- **Spalling** is a mode of damage and overbreak in tunnels at depth mainly in hard and low-porosity rocks. It is defined as the development of visible extension fractures under compressive loading near the boundary of an excavation and can be violent (evolving in rock burst or strainburst) or not and time dependent. While even relatively weak rocks can spall, the ability to store energy, typical of strong rocks, is required for strainbursting.

Generally, these phenomena take place at great depths in hard low fractured rocks (brittle behaviour), but they can also be induced at shallower depth where high horizontal stresses are acting. Selmer-Olsen (1964, 1988) has experienced that, in the hard rocks in Scandinavia, such anisotropic stresses might cause spalling or rock burst in tunnels located inside valley sides steeper than 20° and with the top of the valley reaching height higher than 400 m above the level of the tunnel.

Rock burst can consist of sudden failures associated with high energy release, potentially causing the projection of rock volumes from the tunnel wall, whose dimensions range from small rock fragments to slabs of several cubic metres. However, they cause significant problems and reduced safety for the tunnel crew during excavation (Chapter 7).

3.9   AGGRESSIVE WATERS, GAS, RADON, AND ASBESTOS

3.9.1   Aggressive waters

Tunnelling can intercept water chemically attacking the concrete, leading to a complete breakdown of the final lining, with significant economic losses. Therefore, the identification of aggressive waters (water with pH lower than 6.5) during the design phase is of paramount importance.

This potential hazard is directly related to the lithological features of the ground, since aggressive substances are released into groundwater by the geological materials in which the water flows. Obviously, the forecasting study of aggressive waters should cover a stratigraphic succession larger than the one directly affected by the work, as they may originate from grounds other than those directly intercepted by tunnelling.

The most frequent types of aggressive waters are listed below, together with the geomaterials generally responsible for their presence:

- Selenitic water: rich in calcium sulphate, it is released by anhydritic or gypseous lithotypes.
- Water rich in sulphuric acid: it is typical of peaty soils or clayey soils containing pyrite, as well as in chalk masses.
- Water rich in free carbon dioxide: it is typical of surface deposits covered with forests and ground hosting mineralized waters related to magmatic phenomena.
- Water rich in chloride and in magnesium sulphate (always associated with sodium chloride): it can be met in grounds belonging to evaporitic series.

3.9.2   Gas

During tunnelling, gas retrieval can cause risky situations for the worker safety, especially if the gas is under pressure. The potential hazard related to gas can be identified on the basis of the lithological nature of the ground and/or the presence of open fractures, which may constitute preferential paths conveying such gases into reservoir rocks different from source rocks. Therefore, the forecasting study of gas hazard should cover a stratigraphic succession larger than the one directly affected by the work, as it may originate from grounds other than those directly intercepted by tunnelling.

Listed below are the gaseous substances that can be found during tunnelling, together with the geomaterials commonly responsible for their presence:

- Methane (CH_4): this gas is generally contained in carbonaceous rocks, marshy deposits, as well as flysch formations rich in clay or belonging to the 'scaly clay' rock type. It is very hazardous because it is odourless and colourless and can explode when mixed with air (in proportions ranging from 5% to 14%, called 'firedamp' or 'grisou').

- Carbon dioxide (CO_2): it is frequently associated with methane, usually in carbonaceous, organic clayey or volcanic soils. It is poisonous to humans and very aggressive towards concrete.
- Carbon monoxide (CO): it is mostly present in carbonaceous rocks and is very poisonous.
- Nitrogen oxides: these gasses are common in carbonaceous rocks or in rocks containing decaying organic substances, or even within volcanic soils. It is not toxic, but can accumulate in the ceiling of the tunnel causing death by asphyxiation.
- Hydrogen sulphide (H_2S): this gas is typically related to volcanic exhalations, but can also be produced by bacterial reduction from the decomposition of sulphates or sulphur, or released from water containing putrefying organic substances. It is toxic, combustible and explosive if mixed with air.
- Sulphur dioxide (SO_2): it can be found in igneous rocks. It is highly toxic and aggressive towards concrete.

When the presence of gas is envisaged, tunnelling must proceed with caution, carrying out a continuous and thorough monitoring of the air quality, using shielded flameproof machines and implementing an effective ventilation system (Chapter 6).

3.9.3   Radon

Some zones are characterized by a high natural radioactivity, resulting from high concentrations of radioactive minerals such as uranium, thorium and all elements originating from their radioactive decay.

Among these elements, radon is quite common, in particular its Rn-222 isotope (decay time 3.82 days), belonging to the decay chain of uranium U-238.

Radon is produced by some igneous rocks (e.g. granites, lavas, tuff and pozzolanas) and by some marbles, marls and flysches that contain uraniferous minerals, as well as radium. The amount of radon depends on permeability, density and grain size, as well as on soil conditions (dry, wet, frozen and snow covered) and weather conditions (ground and air temperature, atmospheric pressure, wind speed and direction). The unit of measure for radon is Bq/m^3, indicating the number of nuclear disintegrations taking place in a second in a cubic metre of air.

Under standard temperature and pressure conditions, radon is colourless, odourless and water-soluble. Even if its concentration in atmosphere is typically extremely low, as it disperses quickly (diffusion and convection only allow radon migration on a centimetre–metre scale), when dissolved in a fluid this gas can be conveyed very far from its release point.

Radon migration in water is affected by the ground permeability, and therefore by a number of geological features (e.g. karstification, fracturing degree and lithology).

High concentrations of radon in a tunnel can be managed by enhancing ventilation processes, whereas if radon is in the water, specific water treatments have to be implemented.

3.9.4 Asbestos

Asbestos is a set of natural silicate minerals (e.g. chrysotile, anthophyllite and trem-olite) belonging to the group of serpentine and amphibole. It can be found in veins or small dispersed fibres in ultrabasic rocks (e.g. peridotite) or in metamorphic rocks (serpentines and amphiboles).

Dusts containing asbestos fibres are very dangerous to human health. Therefore, when the geological conceptual model indicates the possible presence of asbestos-containing rocks, specific procedures have to be followed:

* Systematic watering (eventually using water with surfactant additives) in order to retain dust at the excavation face and during the loading and transport phases. Afterwards, the water must be treated in specific plants, and the excavation face must be immediately covered with shotcrete, in order to isolate the tunnel from the origin of a potential asbestos release (Chapter 5).
* Means of transport with air-conditioned cabs, dust filters and closed load area.
* Air quality monitoring next to the excavation face and the grinder, with extractor fans equipped with dust filters.
* Adequate storage sites for excavated materials.

3.9.5 Karst

The karst phenomena derive from the dissolution of soluble rocks such as limestone, dolomite and gypsum. These phenomena can occur on the topographic surface, char-acterized by specific 'shapes' such as limestone pavement or karren, sinkholes and vertical shafts, or under the topographic surface, essentially characterized by caves, more or less large.

As regards the underground works, the karst phenomena can produce many prob-lems, such as water inrush, tunnel collapse and/or surface subsidence (Figures 3.18 and 3.19 and Table 3.1).

Sometimes, this is also due to the unexpected location, irregular geometry and unpredictable dimensions of karst structures.

To avoid both structural and hydrogeological risks during the construction of a tunnel and to ensure the safety of the workers, a good geomorphologic survey to identify all the evidences that can suggest the presence of a deep karst, immediately followed by a geological, hydrogeological, geostructural and geomechanical survey, is mandatory. As regards the identification of the caves and their underground develop-ment, the geophysical survey is more useful (seismic refraction, seismic reflection or other techniques such as seismic tomography).

3.10 SUMMARY

The aim of this chapter was to present a synthesis of both geological and hydroge-ological issues related to the underground construction. As is well known, a good design necessitates first the definition of the geological/hydrogeological conceptual models. These models allow us to identify the lithology, the spatial distribution of materials, the tectonic structures and the geomorphological and hydrogeological

Figure 3.18 An example of water inflow in the Simplon Tunnel (historical photo) (Gattinoni et al., 2014).

Figure 3.19 Example of karst phenomena intercepted during the excavation. (Courtesy of E.M. Pizzarotti.)

data concerning the area of interest for the underground work. These models also constitute the basis for the geotechnical model, giving information about the litho-technical and hydrogeological characterization of materials, together with their geomechanical characterization. This model also allows us also to individuate the risk factors that may influence the mechanical behaviour of materials and the safety of the underground work.

In particular, in this chapter, besides the identification of the lithological nature of rocks and soils, besides the discussion of the role played by geological, geomorphological, hydrogeological, tectonic conditions, and geological investigations, the authors have taken many different risk phenomena into account, such as dissolution in carbonate sedimentary rocks or in some evaporitic rocks, debonding in cemented materials, weathering induced by aggressive waters, swelling, squeezing and rock burst, diffusion of gases and radioactive agents.

AUTHORSHIP CONTRIBUTION STATEMENT

The chapter was developed as follows. Scesi: chapter coordination; Scesi and Gattinoni: jointly developed the chapter; Lo Russo: § 3.3.5 The editing was managed by Scesi.

REFERENCES

Barla G. (2002) Tunneling under squeezing rock condition. In: *Tunneling Mechanics-Eurosummerschool, Innsbruck*. Verlag Berlin, Berlin, pp. 169–268. ISBN 3897228734.

Bhasin R. (1994) Forecasting stability problems in tunnels constructed through clay, soft rocks and hard rocks using an inexpensive quick approach. *Gallerie e Grandi Opere Sotterranee* 42: 14–17.

Bonini M., Debernardi D., Barla M., Barla G. (2009) The mechanical behaviour of clay shales and implications on the design of tunnels. *Rock Mechanics and Rock Engineering* 42: 361–388.

Cai M., Kaiser P.K. (2018) *Rockburst Support Reference Book*. Available to download from: http://www.imseismology.org/imsdownloadsapp/webresources/filedownload/Cai-Kaiser (2018) ockburstSupport-Vol.1.

Cesano D., Olofsson B., Bagtzoglou A.C. (2000) Parameters regulating groundwater inflows into hard rock tunnels – a statistical study of the Bolmen tunnel in southern Sweden. *Tunnelling and Underground Space Technology* 15(2): 153–165.

Chiu Y., Chia Y. (2012) The impact of groundwater discharge to the Hsueh-Shan tunnel on the water resources in northern Taiwan. *Hydrogeology Journal* 20: 1599–1611.

Civita M. (2005) *Idrogeologia applicata e ambientale*. Casa Editrice Ambrosiana, p. 794, Milan.

Dematteis A., Gilli P., Parisi M.E., Ferrero L., Furno F. (2016) Maddalena exploratory adit: feedback on hydrogeological and geothermal aspects. *Acque Sotterranee – Italian Journal of Groundwater*, AS14066: 37–43.

Diederichs, M.S. (2005) Design methodology for spalling failure and rockburst hazards. *Summary of Meetings with GEODATA Engineering Torino*, September 5–9.

Farhadian H., Katibeh H. (2015) Groundwater seepage estimation into Amirkabir tunnel using analytical methods and DEM and SGR method. *International Journal of Civil and Environmental Engineering* 9(3): 296–301.

Farhadian H., Katibeh H., Huggenberger P., Butscher C. (2016) Optimum model extent for numerical simulation of tunnel inflow in fractured rock. *Tunnelling and Underground Space Technology* 60: 21–29.

Furno F., Barla M., Dematteis A., Lo Russo S. (2016) Methodological approach for sustainable management of water inflow and geothermal energy in tunnels. *Acque Sotterranee – Italian Journal of Groundwater*: 37–43.

Gattinoni P., Pizzarotti E.M., Scesi L. (2014) *Engineering Geology for Underground Works.* Springer, Dordrecht, pp. 1–305.

Gattinoni P., Pizzarotti E.M., Scesi L. (2016a) *Geomechanical Characterisation of Fault Rocks in Tunnelling: The Brenner Base Tunnel (Northern Italy).* Tunnelling and Underground Space Technology n. 51. Elsevier, pp. 250–257.

Gattinoni P., Rampolla C., Francani V., Scesi L. (2016b) Hydrogeological structures of fault zones triggering slope instability in central Valtellina (northern Italy). *International Multidisciplinary Scientific GeoConference Surveying Geology and Mining Ecology Management SGEM*, vol. 1: 839–846.

Goel R.K., Jethwa J.L., Paithakan A.G. (1995) An empirical approach for predicting ground conditions for tunneling and its practical benefits. *Proceedings on 35th US Symposium on Rock Mechanics*, pp. 431–435.

Herget G. (1988) *Stresses in Rock*. Balkema, Rotterdam, p. 175.

Higgins, M.W. (1971) Cataclastic rocks. Professional Paper, United States Geological Survey, no. 687, p. 97.

Hoek E., Brown E.T. (1980) *Rock mass properties*. In Hoek's corner. Rocscienceinc (http://www.rocscience.com/education/hoeks_corner).

Hoek E., Kaiser P.K., Bawden W.F. (1995) *Support of Underground Excavations in Hard Rock*. Balkema, Rotterdam, p. 215.

Hoek E., Marinos P. (2000) Predicting tunnel squeezing problems in weak heterogeneous rock masses. Tunnels and Tunneling International.

Hufschmied P., Andreas Brunner A. (2010) The exploitation of warm tunnel water through the example of the Lötschberg Base Tunnel in Switzerland. *Conference: 59th Geomechanics Colloquium*.

Jethwa J.L., Singh B., Singh B. (1984) Estimation of ultimate rock pressure for tunnel linings under squeezing rock conditions—a new approach. In: Brown E.T., Hudson J.A. (eds) *Design and Performance of Underground Excavations. ISRM Symposium*, Cambridge, pp. 231–238.

Li S., Li S., Zhang Q., Xue Y., Liu B., Su M., Wang Z., Wang S. (2010) Predicting geological hazards during tunnel construction. *Journal of Rock Mechanics and Geotechnical Engineering* 2(3): 232–242.

Link K., Rybach L., Imhasly S., Wyss R. (2015) Geothermal energy in Switzerland – Country update. *Proceedings World Geothermal Congress 2015*, Melbourne, Australia, 19–25 April 2015.

Lo Russo S., Taddia G., Cerino Abdin E., Parisi M.E. (2019) "La Maddalena" exploratory adit - base tunnel of the Turin-Lyon high speed rail project: hydrogeological monitoring data analysis. *Rendiconti online della Società Geologica Italiana* 47: 58–63.

Masset O., Loew S. (2013) Quantitative hydraulic analysis of pre-drillings and inflows to the Gotthard Base Tunnel (Sedrun Lot, Switzerland). *Engineering Geology* 164: 50–66.

Moon J., Jeong S. (2011) Effect of highly pervious geological features on ground-water flow into a tunnel. *Engineering Geology* 117: 207–216.

Olsen R. (1988) General engineering design procedures. Norwegian *Tunnelling Today*, Tapir 1988, pp. 53–58.

Panet M. (1995) Calcul des tunnels par la method convergence-confinement. Ponts et chausses. Paris 1995.

Perello P., Baietto A., Burger U., Skuk S. (2014) Excavation of the Aica-Mules pilot tunnel for the Brenner base tunnel: information gained on water inflows in tunnels in granitic massifs. *Rock Mechanics and Rock Engineering* 47: 1049–1071. https://doi.org/10.1007/s00603-013-0480-x.

Perello P., Venturini G., Delle Piane L., Dematteis A. (2007) Groundwater inflows in tunnels excavated in faulted rock mass. *Rock Mechanics* 25(4): 28–34.

Scesi L., Gattinoni P. (2009) *Water Circulation in Rocks.* Springer, Netherlands, pp. 1–165.

Selmer-Olsen R. (1964) *Geology and Engineering Geology (in Norwegian).* Tapir, Trondheim, Norway, 409 pp.

Sheorey P.R. (1994) A theory for in situ stresses in isotropic and transversely isotropic rock. *International Journal of Rock Mechanics and Mining Sciences and Geomechanics.* Abstract 31(1): 23–34.

Sibson R.H. (1977) Fault rocks and fault mechanisms. *Journal of the Geological Society* 133: pp. 191–213. doi: 10.1144/gsjgs.133.3.0191.

Singh B., Jethwa J.L., Dube A.K., Singh B. (1992) Correlation between observed support pressure and rock mass quality. *Tunnelling and Underground Space Technology* 7: 59–74.

Vincenzi V., Gargini A., Goldscheider N. (2009) Using trace tests and hydrological observations to evaluate effects of tunnel drainage on groundwater and surface waters in the Northern Apennines (Italy). *Hydrogeology Journal* 17(1): 135–150.

Wise D.U., Dunn D.U., Engelder J.T., Geiser P.A., Hatcher R.D., Kish, S.A., Odom A.L., Schamel S. (1984) Fault-related rocks: suggestions for terminology. *Geology* 12: 391–394.

Chapter 4

Input data: geotechnics

E. Bilotta
Università di Napoli Federico II

D. Boldini
Sapienza Università di Roma

M. Barbero and D. Martinelli
Politecnico di Torino

A. Graziani
Università degli Studi Roma Tre

A. Sciotti
Italferr SpA

CONTENTS

DOI: 10.1201/9781003256175-4

4.1   INTRODUCTION

Reducing the risks associated with tunnel construction is the main purpose of geotechnical investigations and characterization. It is therefore essential that geotechnical investigation be carried out since the very beginning of the project. This requires collecting all existing information and data. At the same time, the relevant design parameters should be assessed from the collected data to: choose a suitable alignment for the tunnel, analyse the stability of the face, define the associated ground support and improvement, design the tunnel lining and carry out a pre-construction risk assessment on buildings adjacent to tunnelling works. On such basis, the geotechnical characterization should permit us to assess the feasibility of alternative schemes, to select the most appropriate construction method, to anticipate possible risks during construction and to undertake any possible mitigation.

In this chapter, after recalling the main definitions of geotechnical properties, laboratory tests and site investigation for geotechnical characterization of tunnelling are presented and the geotechnical model is discussed.

4.2   FUNDAMENTAL CONCEPTS FOR THE HYDRO-MECHANICAL CHARACTERIZATION OF GEOTECHNICAL MATERIALS

4.2.1   Elasticity

Elasticity represents a theory allowing us to define a simple constitutive model for describing the material mechanical behaviour assuming the stress tensor $\left(\sigma_{ij}\right)$ to be a single-valued function of the strain tensor $\left(\varepsilon_{ij}\right)$.

$$\sigma_{ij} = f\left(\varepsilon_{ij}\right) \tag{4.1}$$

The strain history does not affect the stress state, and the material deformation is fully reversible along cyclic stress paths. If Equation 4.1 is linear, the material is *linear elastic*. If the material behaviour is independent of the loading direction, then it is *isotropic*.

4.2.2   Plasticity

The *plasticity* theory assumes that beyond a given stress threshold, defined by means of the yield function, plastic/irreversible strains, remaining after loading–unloading, can develop. In case yield surface coincides with the failure envelope, the term perfectly plastic is used. Elastic–plastic models assume the yield function to coincide with the boundary of the elastic locus and, when inside the yield function the mechanical response is assumed also to be rigid, the model is defined rigid perfectly plastic.

In elastic–plastic models, the strain rate tensor $\left(\dot{\varepsilon}_{ij}\right)$ can be decomposed into two additive components: *elastic* $\left(\dot{\varepsilon}_{ij}^{e}\right)$ and *plastic strain rates* $\left(\dot{\varepsilon}_{ij}^{p}\right)$. Within the yield surface, the material behaves as elastic. Once the yield surface is reached (*yielding*), a *plastic strain rate* develops. The direction of the plastic strain rate tensor is defined by means of the *flow rule* according to which the increment in the plastic strain vector is normal to an additional function: the *plastic potential*. If the yield function and the plastic potential coincide, the plastic strain increments are normal to the yield surface ('*normality*') and the flow rule is called *associated*.

In case the yield function does not evolve, the term perfect plasticity is used; otherwise, the evolution of the yield function is called *plastic hardening* and in particular *strain hardening* in case its evolution is governed by the accumulation in plastic strains. When a progressive shrinkage of the yield surface takes place, the term *strain softening* is employed.

4.2.3 Brittleness

The mechanical behaviour of soils and rocks is strongly dependent on the confining pressure. Under low confining pressures, the mechanical response (that is the stress–strain curve) is characterized by a *peak,* followed by a drop in stresses (*softening*). This behaviour, typical of dense sands, over-consolidated clays and hard rocks, is called *brittle* and is associated with *dilation* (increase in volume under shear) and with the formation of localized *shear bands.* By increasing the confining pressure, a less evident or even no peak is observed: the stress–strain curve is characterized by a monotonically increasing trend, producing plastic *hardening* and a *ductile* failure after an important volumetric *contraction.*

4.2.4 Viscosity

The above definition of elasticity and plasticity implies the material behaviour to be independent of the strain rate imposed. Although the ground behaviour can be in many cases idealized as *rate independent,* in some cases it depends on strain rate. *Creep* at constant stress or *relaxation* of stress at a fixed strain is related to strain rate dependence. Yielding may also be affected by the strain rate. As the property of fluids to resist against shearing at constant rate is called *viscosity,* similarly, rate-dependent effects in geomaterials can be modelled as *viscous effects.* Therefore, *viscoelasticity* and *viscoplasticity* can be used to model the rate dependence of both reversible (elastic) and irreversible (plastic) deformations, respectively.

4.2.5 Anisotropy

An isotropic material exhibits the same behaviour along any direction. However, soil deposition and layering, rock bedding and discontinuities may affect the mechanical properties of the ground along different directions, making the mechanical behaviour of the geomaterial *anisotropic.* In this case, the same stress change can produce very different strain increases if applied along different directions. Hence, the adoption of an elastic–plastic model with anisotropic elasticity (Section 4.3.2) and anisotropic definition of both yield function and plastic potential may be necessary.

4.2.6 Effective stresses

Soil is an inherently multiphase system: it is made of an assembly of solid particles and voids (*interparticle pores*). The pores are filled by one or more fluids (air, water and possibly others). The soil particles are organized in a layout that is called *solid skeleton.* Particles' size, shape, roundness and possible interparticle bonds affect the mutual interaction between particles. Pore fluids influence the stress transmission at the interparticle contacts. Since the soil is a multiphase system, any applied load is shared between the solid skeleton and the pore fluid. Under saturated conditions, only a single fluid (generally water) fills the pores. Its pressure, u, may change as a consequence of a variation of boundary conditions. This affects the way the external loads are shared between the solid skeleton and the pore fluid. The *effective stress* is defined as the difference between the *total stress* resulting from the total loads applied to the soil volume, $[\sigma]$, and the *pore pressure*, u:

$$\left[\sigma_{ij}{}'\right] = \left[\sigma_{ij}\right] - u \cdot \left[\delta_{ij}\right] \tag{4.2}$$

The effective stress principle (Terzaghi, 1923) states that only the effective stress $[\sigma']$ is responsible for any 'measurable' effect in the soil, that is deformability and strength. For this reason, when dealing with soils, all the above-mentioned constitutive equations should be written in terms of effective stresses. For rocks, the validity of the principle was experimentally proved for the strength of materials of also very low porosity (Handin, 1963; Sulem & Ouffroukh, 2006); in contrast, a modified version of Equation 4.2 should be considered when dealing with deformability, due to the non-negligible influence of solid matrix compressibility, C_s, in comparison with that of the solid skeleton, C (Biot & Willis, 1957; Geertsma, 1957; Skempton, 1961):

$$\left[\sigma_{ij}{}'\right] = \left[\sigma_{ij}\right] - \alpha \cdot u \cdot \left[\delta_{ij}\right] \tag{4.2bis}$$

with $\alpha = 1 - C_s/C$.

4.2.7   Hydraulic conductivity

The coefficient of *hydraulic conductivity* or *permeability, k*, is used in the Darcy's law to express the seepage velocity as a function of the hydraulic gradient. In the general formulation of seepage, hydraulic conductivity is described by a tensor, considering the possible anisotropy of permeability in geomaterials. Darcy's permeability is a function of the average *grain size* (namely of the finest part of the grain size distribution), soil *porosity*, *viscosity* and *density* of the permeating fluid and somehow of the *shape of voids* that influences the tortuosity of the flow path. Since the permeating fluid is in most cases water, and the variability of soil porosity is generally limited in a rather small range, the soil grain size distribution is the main factor affecting the coefficient of permeability. The range of variability is extremely high, reaching values as low as 10^{-12} m/s in high-plasticity clays and intact rocks and as high as 1 m/s in coarse gravels and weathered rocks.

It is worth noticing that the actual permeability of ground depends very much on ground conditions and 'structure' at the scale of the problem (e.g. *soil layering* or *rock discontinuities*); hence, it deserves in most cases to be carefully determined on site.

4.2.8   Drained and undrained conditions

The ability of pore water to move quickly in the intergranular voids when the solid skeleton undergoes a volumetric deformation is a key factor affecting the behaviour of saturated soils. In this respect, there is a significant difference between fine- and coarse-grained soils, mainly due to the difference in permeability in comparison with the loading rate (associated with tunnel excavation and construction rates generally not greater than 1 m/h). Where the coefficient of permeability is larger than 10^{-7} to 10^{-6} m/s (coarse-grained soils) excess pore water pressures generated during construction dissipate immediately: these conditions are called '*drained*'. For lower permeability (in fine-grained soils), the excess pore water pressures take time to dissipate and the so-called undrained conditions occur in the short term. Volumetric strain is not allowed under undrained conditions, since the pores are saturated of water that cannot drain elsewhere.

It is worth noting here that under undrained conditions excess pore pressure may increase or decrease, depending on the tendency of the soil to change its volume under shearing under drained conditions. Loose sand and normally consolidated clays that tend to contract under drained conditions will increase pore pressure under undrained conditions, while dense sands and over-consolidated clays that tend to dilate under drained conditions will decrease pore pressure under undrained conditions. Sandy materials are generally under drained conditions during construction, but they may be undrained under dynamic loading. For stiff, low-porosity argillaceous formations (i.e. marls, argillites, claystone and shales) a one-phase approach (i.e. dry medium) is typically adopted considering that the large absorption capacity of many clay minerals, combined with the slow development of hydration reactions, can completely use up the small amount of free water available in deep formations of low porosity (Lord et al., 1998).

4.3 MECHANICAL PROPERTIES OF SOIL AND INTACT ROCK ELEMENTS

4.3.1 Strength

Strength is the ability of a material to carry stress. In soils, the shear stress τ can reach a limiting value, τ_f, that cannot be exceeded (*failure*). This is called *shear strength* and, according to the Mohr–Coulomb failure criterion, is defined by the sum of two contributions: one independent of stresses (*cohesion*) and the other proportional to the normal stress σ_n acting on the failure plane (*friction*):

$$\tau_f = c' + \sigma'_n \cdot \tan\phi' \tag{4.3}$$

For *cemented soils* and *rocks* some tensile strength, commonly put in relation to shear strength parameters c (*cohesion*) and ϕ (*angle of shear strength* or *friction angle*), is also available. It is worth noting that, since many soils are unbonded, the effective cohesive term is in most cases just the consequence of interpreting as linear (over a wide range of confining stress) the actual non-linear strength criterion under long-term (drained) conditions.

Under strain-controlled conditions, soil specimens at large strains exhibit a *stationary* response, characterized by no change in stresses and volume. This state is called *critical* (Figure 4.1). In case of very dense (or interlocked) granular soils and over-consolidated fine-grained soils (that is when a soil has been subjected to a previous history of loading and unloading) this state is reached after a *peak* in stresses.

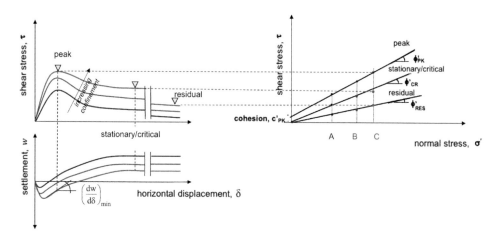

Figure 4.1 Shear strength (typical direct shear test results; see Section 4.9.4).

Hence, two set of parameters should be defined: one at peak, c_p and ϕ_p, and the other at the critical state, c_{cs} and ϕ_{cs}. The latter set is independent of the initial state of the soil (i.e. initial density or over-consolidation ratio). The cohesive term at the critical state, c_{cs}, can be assumed null. In clayey materials the friction angle further reduces with a very large shear strain. Such a state is called *residual*, and the corresponding lower value of friction angle is named residual friction angle, ϕ_r (Figure 4.1).

The excess pore water pressure generated by the perturbation in the short term, occurring by definition under undrained conditions, is often hard to be computed; therefore, it is a common practice formulating the strength criterion in terms of total stresses:

$$\tau_f = c_u \tag{4.4}$$

where c_u (or S_u) is the so-called *undrained cohesion* or, better, *undrained shear strength*. Since it is related to an equivalent single-phase formulation of the continuum, c_u depends on the state of soil and cannot be considered an intrinsic parameter of the material.

For specimens of rock matrix (the rock matrix is assumed macroscopically undisturbed, i.e. no fracture is visible to the naked eye and only microfractures can affect it), a non-linear criterion is mandatory to well interpret experimental evidences. The strength criterion for rock is usually plotted in the principal stresses plane. The generalized Hoek–Brown criterion allows a good interpolation of the laboratory test results on rock specimens, and it is representative of both compressive and tensile strengths:

$$\sigma'_1 = \sigma'_3 + \left(m_i \sigma_{ci} \sigma'_3 + s_i \sigma_{ci}^2 \right)^{\alpha} \tag{4.5}$$

where m_i and s_i are two parameters representing the lithotype and the fracturing degree of the *rock mass*, respectively, and σ_{ci} is the uniaxial compressive strength of the *intact rock* or *rock matrix*.

In the case of intact rock, $s_i = 1$ and $\alpha = 0.5$. When experimental data are available, m_i and σ_{ci} parameters are obtained by representing the pairs $\sigma_3-\sigma_1$ at failure in the plane X–Y, where $X = \sigma'_3$ and $Y = (\sigma_1-\sigma_3)^2$. In this reference plane, in fact, the Hoek–Brown equation is linear. So, the experimental data are simply interpolated by a straight line, which intersects the Y-axis at σ_{ci}^2 and has an angular coefficient equal to m_i:

$$(\sigma'_1 - \sigma'_3)^2 = m_i \sigma_{ci}\sigma'_3 + s_i\sigma_{ci}^2 \rightarrow Y = m_i \sigma_{ci}X + \sigma_{ci}^2 \tag{4.6}$$

If no experimental data are available, some estimated values of m_i are suggested in the literature.

If necessary, the Hoek–Brown criterion can be linearized in a given range of the minimum principal stress, resulting in the Mohr–Coulomb criterion:

$$\sigma'_1 = \frac{1+\sin\phi'}{1-\sin\phi'}\sigma'_3 + \sigma_c, \tag{4.7}$$

where σ_c is the uniaxial compressive strength.

Many rock materials (particularly, foliated metamorphic rocks such as schists, shales and slates) are characterized by closely spaced weakness planes, which make the strength to vary with the loading direction (*anisotropy*). Usually, these rocks can be considered *transversely isotropic* (with the isotropy plane coincident with the foliation orientation). The strength decreases with the increase in the deviation of the loading direction from the normal to the isotropy plane (Jager, 1960; Singh et al., 1989).

4.3.2   Stiffness

The behaviour of a continuum, isotropic linear elastic material is described by a stress–strain relationship that follows Hooke's law and requires only two material constants to be defined: the *Young's modulus, E*, and the *Poisson ratio, v*, are the more common choice. Hence, the stress–strain relation reads:

$$
\begin{pmatrix} \varepsilon_x \\ \varepsilon_y \\ \varepsilon_z \\ \gamma_{xy} \\ \gamma_{yz} \\ \gamma_{xz} \end{pmatrix} = \frac{1}{E} \begin{bmatrix} 1 & -v & -v & 0 & 0 & 0 \\ -v & 1 & -v & 0 & 0 & 0 \\ -v & -v & 1 & 0 & 0 & 0 \\ 0 & 0 & 0 & 2(1+v) & 0 & 0 \\ 0 & 0 & 0 & 0 & 2(1+v) & 0 \\ 0 & 0 & 0 & 0 & 0 & 2(1+v) \end{bmatrix} \begin{pmatrix} \sigma_x \\ \sigma_y \\ \sigma_z \\ \tau_{xy} \\ \tau_{yz} \\ \tau_{xz} \end{pmatrix} \tag{4.8}
$$

The *shear modulus, G*, is defined as a combination of the previous two parameters:

$$G = \frac{E}{2(1+v)} \tag{4.9}$$

The *stiffness* of a material affects its response to stress change in terms of strain. Formally, it is expressed by a *stiffness matrix* $[D]$ that relates the strain increments, $\delta\varepsilon$, to the stress increments, $\delta\sigma$:

$$\left\{\delta\sigma_{ij}\right\} = [D]\cdot\left\{\delta\varepsilon_{ij}\right\} \qquad (4.10)$$

The definition of a linear elastic material was given in the previous Section 4.2.1. However, the stress–strain behaviour of the soil is not linear; hence, stiffness changes with the strain level and accordingly the elements of the stiffness tensor change. When the stress–strain curve is non-linear, two stiffness values can be defined: the *secant* or the *tangent stiffness*. For instance, in terms of shear modulus:

$$G_s = \frac{\tau}{\gamma} \quad \left(\text{secant shear stiffness}\right) \qquad (4.11)$$

$$G_t = \frac{\delta\tau}{\delta\gamma} \quad \left(\text{tangent shear stiffness}\right) \qquad (4.12)$$

The dependency of stiffness (either secant or tangent) on the strain level is generally represented by employing a plot of the normalized stiffness modulus (with respect to its maximum value, exhibited at very small strain) as a function of strain. A typical curve in terms of normalized shear modulus, G/G_0, varying with shear strain, γ, is shown in Figure 4.2. The range in which different laboratory tests are able to characterize the soil stiffness at different strain levels is shown in the figure. At the same time, the typical range of strain mobilized in geotechnical construction processes is also identified. In most cases, the level of shear strain involved in tunnelling falls between 0.01% and 1%.

It is worth noting that the confining effective stress affects the soil stiffness; hence, in tunnel applications stiffness increases with the overburden stress. Stiffness is also

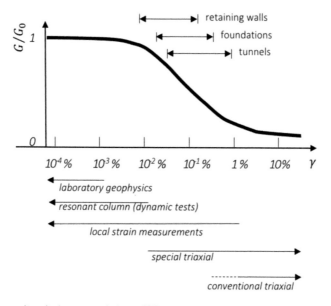

Figure 4.2 Normalized shear modulus, G/G_0, varying with the shear strain, γ. (Modified after Atkinson and Sällfors (1991).)

affected by the previous stress history, being larger in over-consolidated soil than in normally consolidated ones, at the current stress level. These aspects must be taken into account when planning and carrying out ground investigation and mechanical characterization.

Depending on whether 'drained' or 'undrained' conditions apply, the stiffness parameters should be defined in effective stresses (*drained conditions*) or in total stresses (*undrained conditions*).

The stress–strain curve of a rock matrix subject to unconfined compression usually shows a linear trend in a range of stress at the turn of half the compressive strength value. The characteristic elastic modulus E_i is the tangent one calculated in that range. Correspondingly, a Poisson ratio, v_i, is calculated. They are called E_{t50} and v_{t50}, respectively.

In case the ground is characterized by an anisotropic mechanical behaviour, this affects the stiffness matrix. The condition of horizontal ground layers, which generally implies symmetry around the vertical axis, can be modelled by using a cross-anisotropic matrix: the behaviour is isotropic in the horizontal planes, while a different stiffness value corresponds to the axis perpendicular to the stratification. In this case, five parameters are needed to fully describe the stiffness matrix: two normal stiffness moduli along vertical and horizontal directions, E_v and E_h, two Poisson ratios in the vertical and horizontal planes, v_{vh} and v_{hh}, and one independent shear modulus G_{vh}.

4.4   ROCK MASS AND ROCK DISCONTINUITIES

4.4.1   Definitions

The *rock mass* is made up of two elements: the *rock matrix* and the *discontinuities*. The term 'discontinuities' refers to all the weakness planes of different origin. Typically, *sets of discontinuities* and *main discontinuities* can be distinguished: the former are groups of planes with similar attitude, geological origin and characteristics, and the latter are single important fracture planes (e.g. faults), usually showing a complex morphology.

The rock mass can be considered as a fractured medium whose behaviour is highly influenced by the mechanical characteristics of these two elements. Strength, deformability and permeability, in fact, depend on both (i) a complex combination of the contributions of these two main elements and (ii) their interactions. The purely discontinuous feature of the rock mass and the consequent non-negligible scale effects make its mechanical characterization a complex process.

4.4.2   Description of discontinuity systems

The following characteristics are representative of a rock discontinuity (ISRM, 1978b):

*   Orientation: the orientation of a rock discontinuity in the geographical space is measured under the assumption that each discontinuity is represented by a mean plane whose position is uniquely determined by two angles: the *dip direction α*,

representing the facing direction, measured clockwise from the north (0°) to the line of maximum dip of the inclined plane (it is generally expressed by an angle between 0° and 360°) and the *dip ψ*, providing the degree of inclination with respect to the horizontal plane (thus ranging between 0° and 90°). Discontinuities characterized by similar orientations, provided a similar geological origin, are assumed to belong to the same set.

- Spacing: it is the distance between two adjoining discontinuities belonging to the same set; it is measured along the direction orthogonal to the discontinuities. The mean value of spacing is assumed to be characteristic of the set of discontinuities. The number of joint sets, the orientation and spacing of each set define the shape of the typical *rock block*, which characterizes a rock mass of *regular* structure.
- Persistence: this parameter gives a measure of the extent of a discontinuity interface inside the rock mass, i.e. it takes into account the presence of *rock bridges*. It should be evaluated as the ratio between the summation of the discontinuous areas and the total area of the discontinuity. It is usually represented in percentage, with 100% persistence representing a continuous fracture with no rock bridges. Because of the difficulty in measuring this parameter, it is usually assumed to be the length of the discontinuity trace observable on an outcrop. If the discontinuity terminates against another one or in the rock matrix, the persistence reduces.
- Roughness: three levels of roughness characterize the contact faces of a discontinuity, depending on the scale of observation: *microscopic* scale (crystallographic texture), *laboratory* scale (measurable in laboratory) and *site* scale (waviness, measured with respect to the mean plane of the discontinuity). To determine the roughness, photographic and photogrammetric methods as well as laser or mechanical profilometers (*Barton comb*) are used, depending on the scale of observation. All three of the roughness levels contribute to the shear strength of the discontinuities.
- Wall strength: it is the compressive strength of the discontinuity faces. It can be lower than the uniaxial compressive strength of the intact rock because of the exposure to atmospheric agents and alteration in the discontinuity walls. A *Schmidt hammer* is usually used to estimate this parameter. The wall strength influences the shear strength of the discontinuity.
- Aperture (or separation, or opening): the aperture of a discontinuity is the perpendicular distance between its faces. The gap between the two faces of the interface can be filled with air, water or other materials. This parameter influences the hydraulic characteristics of the rock mass and is highly affected by the stress level. In fact, the water flow primarily occurs into the discontinuities, as their permeability is greater than that of the intact rock.
- Infilling (or gouge): the filling material lays in between the two faces of the discontinuity. It can be sand, silt, clay, breccia, etc. Both its thickness and mechanical characteristics influence the shear strength of the discontinuity especially if it is high in comparison with the surface roughness.

The orientation, persistence and spacing of the discontinuities define the *shape* and *size of rock blocks* in which the rock matrix is subdivided. The orientation of the discontinuities with respect to that of the tunnel planes (face, roof, invert and walls) determines the probability that the rock block mobilizes.

4.4.3   Shear strength of discontinuities

The *shear strength of discontinuities* is severely influenced by their roughness and wall strength. At the laboratory scale, the micro- and laboratory levels of roughness are taken into account. The test used to estimate the shear strength is the *direct shear test*, where a rock sample containing the discontinuity is subjected to a constant normal stress and a continuously increasing tangential/horizontal displacement. Both *smooth* and *rough discontinuities* can be tested, and the test results can be represented on the Mohr (σ'–τ) reference plane. In case of smooth discontinuities, the experimental data are fairly aligned, so a linear interpolating strength criterion (Mohr–Coulomb criterion) can be used.

- Two friction angles can be defined, depending on the interface characteristics:
- artificial ('*saw-cut*') smooth discontinuity: ϕ'_b = *base* friction angle,
- natural smooth discontinuity: ϕ'_r = *residual* friction angle.

The *peak shear strength* data obtained by testing rough discontinuities are arranged in a highly non-linear trend when plotted in the σ'–τ plane. So, a non-linear strength criterion is required to correctly describe the shear strength envelope. The non-linear Barton criterion is commonly used:

$$\tau' = \sigma'_n \cdot \tan\left[\text{JRC} \cdot \log_{10}\left(\frac{\text{JCS}}{\sigma'_n} \right) + \phi'_r \right], \tag{4.13}$$

where JRC = joint roughness coefficient; JCS = joint wall compressive strength; and ϕ'_r = residual friction angle (alternatively, the base friction angle ϕ'_b can be introduced).

JRC is a coefficient related to the *roughness* profile of the discontinuity interface planes, ranging between 0 (*smooth surfaces*) and 20 (*highly rough*).

JCS is the *uniaxial compressive strength* of the faces of the discontinuity. It is estimated on the basis of the condition of the discontinuity: (i) if the walls of the discontinuity are not weathered, JCS is assumed equal to the uniaxial compressive strength of the intact rock σ_{ci}, obtained by *uniaxial compressive tests* or *point load tests* on rock matrix samples; (ii) if the walls are weathered, JCS is estimated by using the *Schmidt hammer tests* on the interface walls; and (iii) if the rock mass is at an advanced stage of alteration, the JCS value is provided by uniaxial compressive tests or point load tests on rock samples taken from the weathered part of rock mass.

When the shear strength criterion has to be used at the field scale, scale effects have to be taken into account. The non-linear criterion is modified as follows:

$$\tau' = \sigma'_n \cdot \tan\left[\text{JRC}_n \cdot \log_{10}\left(\frac{\text{JCS}_n}{\sigma'_n} \right) + \phi'_n + i_u \right] \tag{4.14}$$

where i_u is the waviness angle and JRC_n and JCS_n are calculated by using the Bandis relationships:

$$\text{JRC}_n = \text{JRC}_0 \left(\frac{L_n}{L_0} \right)^{-0.02 \cdot \text{JRC}_0} \tag{4.15a}$$

$$JCS_n = JCS_0 \left(\frac{L_n}{L_0} \right)^{-0.03 \cdot JRC_0} \tag{4.15b}$$

where JRC_0 and JCS_0 are measured on laboratory rock samples, L_n is the in situ characteristic length of the discontinuity, and L_0 is the laboratory discontinuity length (e.g. the sample diameter).

4.4.4 Rock mass classification systems

The rock mass classification systems permit us to assign a *quality class* to the rock mass on the basis of its global mechanical properties. The area of interest for the excavation can thus be subdivided into homogeneous zones, characterized by the same quality class. They can be used to estimate the *equivalent mechanical characteristics* of the rock mass at the site scale, when it can be assumed as equivalent continuum and homogeneous medium. In tunnelling, the rock mass classification systems can help in evaluating the type of support systems needed for the tunnel safety and other parameters useful for tunnel design (such as self-supporting distance or rock support pressure).

Different classification systems are provided in literature, which differ on the basis of the parameters chosen as representative for the global quality of the rock mass. In the following, the most used are described: Q (Barton et al., 1974), RMR (Bieniawski, 1989) and GSI (Hoek, 1994).

Q system
Q value is estimated from the following expression:

$$Q = \frac{RQD}{J_n} \cdot \frac{J_r}{J_a} \cdot \frac{J_w}{SRF} \tag{4.16}$$

where RQD (*rock quality designation*) is defined as follows:

$$RQD = \frac{100 \cdot \left(l_{total} - l_{10} \right)}{l_{total}} \tag{4.17}$$

with l_{total} = the total length of drilling and l_{10} = Σ(length of cores shorter than 10 cm).

J_n is the rating for the number of discontinuities sets, J_r is the rating for the roughness of the discontinuities sets most unfavourable to the tunnel stability, J_a is the rating for the degree of alteration or infilling of the most unfavourable set of discontinuities, J_w is the rating for the water inflow and water pressure effects, and SRF is the rating related to the presence of weak zones in the excavation area (faulting, squeezing or swelling, strength/stress ratios in hard massive rocks, etc.). The values to be assigned to each rating are suggested in the charts provided by Barton et al. (1974).

RMR (rock mass rating)
Five parameters are considered to characterize the rock mass quality (RMR_5): P_1, uniaxial compressive strength of intact rock (or strength index obtained by point load tests); P_2, RQD; P_3, spacing of the discontinuities; P_4, condition of the discontinuities (roughness, persistence, alteration, aperture and infilling); and P_5, the presence of water.

Table 4.1 Influence of orientation on RMR_6

Dip direction	Dip	Rating	P_6
Perpendicular to tunnel axis, drive with dip	45°–90°	Very favourable	0
	20°–45°	Favourable	−2
Perpendicular to tunnel axis, drive against dip	45°–90°	Mediocre	−5
	20°–45°	Unfavourable	−10
Parallel to tunnel axis	45°–90°	Very unfavourable	−12
	20°–45°	Mediocre	−5
All	0°–20°	Mediocre	−5

As suggested in the charts provided by Bieniawski in 1989, a rating is assigned to each of the five parameters. The sum of ratings is the value of RMR. On the basis of the RMR value, a class of quality is assigned to the rock mass, ranging between 0 (worst quality) and 100 (best quality) (Bieniawski, 1989).

In tunnelling, RMR_6 can be used, where *a further parameter* (P_6) is considered, which takes into account the influence of the orientation of the main discontinuities sets on the tunnel stability (Table 4.1).

GSI (geological strength index)
The RMR classification index includes a parameter for groundwater conditions, P_5. Consequently, it cannot be directly adopted for estimating the strength of rock masses in the context of static calculations without the risk of double-counting their effect, since hydraulic conditions are there explicitly considered, and the strength is typically expressed in terms of effective stresses. Similar considerations apply to the rating adjustment for joint orientations.

With the purpose of defining an index for the geotechnical characterization of rock masses, Hoek et al. (1995) suggested considering the 1976 version of the RMR classification index and assumed $P_5 = 10$ (dry conditions) and very favourable joint orientations, that is $P_6 = 0$ (see Table 4.1). The final rating, called RMR_{76}, can then be used to estimate the value of the new classification system *GSI (geological strength index)*:

$$GSI = RMR_{76} \tag{4.18}$$

Bieniawski's 1989 classification can also be used to estimate the value of GSI in a similar manner to that described for the 1976 version. In this case, $P_5 = 15$ and, again, $P_6 = 0$. The final rating, called RMR_{89}, can be used to estimate the value of GSI as follows:

$$GSI = RMR_{89} - 5 \tag{4.18 bis}$$

4.5   MECHANICAL PROPERTIES OF THE ROCK MASSES

4.5.1   General framework

Discontinuities split the rock mass into several rock blocks whose size can be small or big in comparison with the dimensions of the boundary value problem taken into consideration, and consequently, the interaction between the rock mass and the engineering work can be quite different. In order to choose an appropriate geotechnical model, the ratio between the representative block size and the size of the engineering structure to be analysed (rock sample diameter, tunnel diameter, landslide thickness, etc.) has to be considered.

Three different models can be adopted (Chapter 9):

- *Continuum model* is adopted if the size of the geotechnical work is very small with respect to the characteristic block size (drilling of a rock block, for example). In this case, the rock matrix mechanical characteristics are required.
- *Discontinuum model* is adopted if the block size is of the same order of that of the structure or when one of the discontinuity sets is significantly weaker than the others. In these cases, the stability of the structure should be analysed considering the failure mechanisms involving either sliding or rotation of blocks and wedges defined by intersecting discontinuities. The characterization of both matrix and discontinuities is required;
- *Equivalent continuum model* is adopted if the block size is small compared with that of the structure. Global mechanical characteristics have to be estimated taking indirectly into account the interaction between matrix and discontinuities. This means that the parameters obtained at the laboratory scale cannot be used at the field scale as they are; the transition from the laboratory to the field scale is an empirical procedure based on rock mass classifications.

If the rock mass can be modelled as an equivalent continuum, the transition from the rock volume unit to the rock mass-scale characterization requires the *rock mass classification*. Both the RMR (Bieniawski, 1989) and the GSI (Hoek, 1994) can be used to estimate the Hoek–Brown strength parameters and the deformability modulus of the rock mass.

4.5.2   Rock mass strength

The non-linear generalized Hoek–Brown criterion is used with the same expression employed for the rock matrix, where parameters m_b and s_b are corrected on the basis of the rock mass classification index and a parameter D, introduced by Hoek et al. (2002). This depends on the degree of disturbance – ranging between 0 (undisturbed rock mass) and 1 (very disturbed rock mass) – induced to the rock mass by external dynamic stresses and stress relaxation:

$$\sigma_1' = \sigma_3' + \left(m_b \sigma_{ci} \sigma_3' + s_b \sigma_{ci}^2 \right)^{\alpha} \tag{4.19}$$

where

$$m_b = m_i \cdot e^{\frac{GSI-100}{28-14 \cdot D}} \tag{4.20a}$$

$$s_b = e^{\frac{GSI-100}{9-3 \cdot D}} \tag{4.20b}$$

$$\alpha = \frac{1}{2} + \frac{1}{6}\left(e^{\frac{-\text{GSI}}{15}} - e^{\frac{-20}{3}} \right) \tag{4.20c}$$

4.5.3 Rock mass stiffness

The deformability of rock masses can be estimated by employing in situ test results (see Section 4.10) as well as by means of the classification index GSI. Different empirical relationships have been formulated over the years, initially based on RMR index and then on GSI (Hoek & Brown, 2019), such as:

$$E_d = \left(1 - \frac{D}{2}\right)\sqrt{\frac{\sigma_{ci}}{100}} \cdot 10^{\frac{\text{GSI}-10}{40}} \quad [\text{GPa}], \tag{4.21}$$

which is to be used when the compressive strength of the rock matrix is lower than 100 MPa and σ_{ci} is in MPa (Hoek et al., 2002), or

$$E_d = E_i\left(0.02 + \frac{1 - \frac{D}{2}}{1 + e^{\frac{60 + 15 \cdot D - \text{GSI}}{11}}} \right), \tag{4.22}$$

with E_i being the Young's modulus of the rock matrix (Hoek & Diederichs, 2006).

4.6 IN SITU STRESS CONDITIONS

The stress state in the ground before the excavation is in equilibrium with the gravity field. The corresponding stresses are called *initial stresses*. The process of tunnelling (excavation and subsequent construction of support) induces a stress redistribution ('arching'). Stresses will change during the various construction phases until a final stationary equilibrium is reached. The new equilibrium stresses in the ground, after tunnelling, are called *induced stresses*.

The estimation of initial stresses is needed to calculate the stresses induced on the tunnel support system. The full overburden may act on a very shallow tunnel, while ground arching that develops during tunnel construction shall be carefully taken into account at larger depths.

It is worth noting that initial stresses increase with depth. Around a shallow tunnel, such an increase is large compared to the average value at the tunnel axis, enough not to be neglected in calculations. On the contrary, the increase in stresses with depth around a deep tunnel may be neglected with good approximation.

Below the groundwater table, two stress components have to be considered: the effective stress and the pore water pressure. Although they both affect the load acting on the lining structure, the former only directly affects the mechanical response of the soil.

The ratio between the natural horizontal and vertical effective stresses in one-dimensionally compressed soil deposits (both principal stresses in this case) is called *coefficient of lateral earth pressure at rest*, k_0. It commonly ranges between 0.2 and 2.

For normally consolidated ('virgin') soils, the following assumption is common (Jaky, 1944):

$$k_{0,\text{NC}} = 1 - \sin\phi'$$

(4.23)

with φ' being the friction angle of the soil.

In case of over-consolidated soils (that is when the soil experienced larger stresses in the past), k_0 is likely higher than 1. It may be assessed by using the equation (Mayne & Kulhawy, 1982):

$$k_{0,\text{OC}} = k_{0,\text{NC}} \cdot \text{OCR}^{\sin'\phi}$$

where the over-consolidation ratio, OCR, is commonly assessed from either oedometer compression test or in situ test results (see Sections 4.9 and 4.10).

Although sediments are deposited under gravity in horizontal layers, sedimentary rock layers are often deformed. Tectonic stresses may have induced geologic structures such as folds, joints and faults. In such conditions, the value of the coefficient of lateral earth pressure k_0 can vary depending on the position of the tunnel relative to the geological formation. Principal stresses are likely to be rotated and not coincident with the vertical and horizontal stresses. In these cases, their ratio is not easily assessed and needs to be measured. Moreover, site morphology affects stress intensity distribution and principal stress orientation, such as at valley bottom and near the slope surface.

4.7 SPECIAL ISSUES IN GEOTECHNICAL CHARACTERIZATION FOR TUNNELLING

Obtaining the largest possible amount of information about both the ground conditions and the mechanical behaviour of soils and rock masses is the primary goal of ground investigation and geotechnical characterization. However, it is difficult to define a standard characterization valid for any tunnel project, since several and different site-specific geotechnical issues have to be encountered. In this section, some special issues, to be addressed by site investigation and characterization, are put in evidence.

4.7.1 Soil behaviour in shallow tunnelling

Shallow tunnels frequently involve the challenge of soft ground tunnelling. In this case, recent *sedimentary soils* and heterogenic *anthropic fills* are encountered in the project, which may have rather variable mechanical properties. A base rock may underlie the soft ground layers, at varying depths: the position of the *soil/rock contact* along the tunnel route needs to be determined to optimize the choice of the tunnel depth. Fortunately, in many urban environments, a large amount of information on material properties and their spatial distribution may be easily available from previous underground works; hence, in many cases site investigation is corroborated by existing geotechnical profiles. Moreover, compared to deep tunnels, investigation can be more easily carried out from ground surface, thus achieving a reliable description of the ground conditions before tunnelling.

On the other hand, the *non-linear constitutive behaviour* of soils and weathered soft rocks, found at shallow depth, may require a complex mechanical characterization in order to assess the *deformation field* induced by tunnelling. This issue is of the utmost importance for urban tunnels, since their construction at shallow depth interferes with the built asset. Here, predicting accurately the *ground and structural response to tunnelling* is crucial to assess the associated risks and to design any necessary *mitigation action*. Hence, factors such as the dependence of soil stiffness on both the current stress and strain levels and the previous stress history (see Section 4.3.2), as well as on the onset of plastic deformation upon yielding (see Section 4.2.2), require special care when planning and interpreting geotechnical tests (see Section 4.9). Furthermore, consolidation and associated *long-term ground deformation and settlements* need to be taken into account in fine-grained soils.

4.7.2   Swelling and squeezing conditions

Some minerals may increase their volume when absorbing water. This process is called *swelling* (Chapter 3). Upheaval of the invert as a consequence of swelling, due to groundwater around the tunnel or ingress from portals, may last several years.

It is worth distinguishing between *physicochemical swelling* (osmotic processes, intracrystalline adsorption of free water and hydration), which is a volume increase due to water attraction and absorption by minerals, and *mechanical swelling*, that is a consequence of a release of confining stresses. Both can be revealed with appropriate tests, as described later in this chapter (i.e. *swelling test* and *oedometer test*). If the increase in volume is prevented, the pressure on the confining structure progressively increases. Swelling in tunnels can be either limited (*rigid lining*) or allowed (*flexible* or *ductile lining*).

Swelling in tunnels can be often confused with *squeezing*, when rock converges into an underground opening primarily because of overburden pressure. Different from swelling, squeezing is not associated with a significant increase in volume. The processes occurring in squeezing rocks may be considered as *viscous*. The volume of ground around a tunnel interested by squeezing is generally larger than that affected by swelling. As a consequence, extremely *high convergence* values can be observed, up to metres if the in situ stress exceeds about three times the uniaxial compressive strength of a squeezing rock (Thomas, 2009).

4.7.3   Rock properties for spalling prediction

The design of tunnels in hard rocks must consider the potential for stress-induced spalling failures around the excavation boundary. In situ observations in both massive and fractured hard rocks have shown that spalling/slabbing initiates on the boundary of the excavation. The stress required to initiate spalling occurs at magnitudes that are considerably lower than the peak strength of intact samples obtained in the laboratory.

When evaluating the stress–strain response of laboratory samples under compression, the initiation of cracking occurs well below the peak strength (UCS, for uniaxial tests).

Nicksiar and Martin (2012) evaluated different strain-based methods for establishing the onset of cracking in laboratory compression tests on sedimentary, metamorphic and igneous rocks and demonstrated that crack initiation occurred at approximately 45% of the peak strength. This value is only slightly below the spalling strength observed in situ. Hence, when attempting to assess the potential for spalling on a new tunnelling project, determining the *crack initiation (CI)* stress magnitude, e.g. by means of accurate measurements of the volume deformation of the samples tested under uniaxial compression, can serve as a lower-bound estimate for 'in situ scale' spalling strength. If the circumferential stress acting around the boundary of the tunnel exceeds the laboratory CI stress, the next step will be the prediction of the extent of the V-shaped notch (Figure 4.3a) created by spalling.

It is clear from the examination of case histories that the depth of spalling is stress-driven and proportional to the tunnel size. The initial modelling attempts based on both classical elastic–perfectly plastic and elastic–brittle constitutive laws did not capture either the localization or the depth of the spalling. However, the in situ observations have suggested the idea that this failure mode can be better represented by a process of initial *cohesion weakening* (CW), followed by progressive *friction strengthening* (FS) as a function of plastic strain (used as a 'damage' parameter). This CWFS model, possibly refined by the use of the Hoek–Brown instead of the Mohr–Coulomb failure envelope, has given satisfactory predictions. For the well-documented case of the Mine-by Test Tunnel in granitic rock, Martin (2014) found reasonable results (Figure 4.3b) by applying Hoek–Brown 'spalling' parameters determined from CI stress and Brazilian tensile strength T_{BT}:

$$a = \frac{CI}{UCS} \tag{4.24a}$$

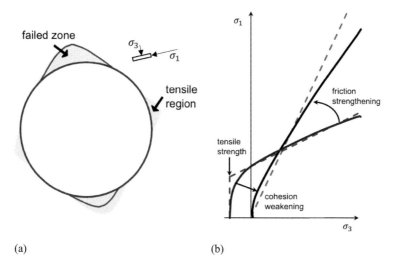

(a)   (b)

Figure 4.3 Spalling: (a) failed V-shaped notch; (b) examples of Hoek–Brown and Mohr–Coulomb strength criteria used for CWFS model. (Modified after Martin (2014).)

$$s = \left(\frac{\text{CI}}{\text{UCS}} \right)^{\frac{1}{a}} \tag{4.24b}$$

$$m = s \frac{\text{UCS}}{|T_{\text{BT}}|} \tag{4.24c}$$

4.7.4   Importance of the post-peak behaviour for the rock mass around deep tunnels

The evaluation of the strength of a rock mass has always been considered the most difficult problem in the design of underground excavations (Hoek, 1983) because it is often impossible to carry out large-scale in situ tests and, although widely used, the correlations between strength parameters and quality indexes (for instance GSI or Q index) are still affected by considerable uncertainties.

The adoption of strain-softening models for an equivalent continuum representing a jointed rock mass is widely used in the analysis and design of underground exca- vations, because this approach is the simplest way to represent the global behaviour (load-carrying capacity) of an annulus of failed rock around the excavation and its effect on the excavation stability.

The strain-softening behaviour beyond peak strength is schematized by means of a constant softening modulus until residual strength conditions are reached. The residual strength and the softening modulus of a rock mass are even more difficult to evaluate for the engineering design.

The analysis of the results obtained from tests on fractured laboratory samples (Ribacchi, 2000) can be a useful contribution to the modelling approach in which the strength parameters of a rock mass are evaluated, starting from the characteristics of the rock material and the fracturing conditions.

Jointing tends to reduce the strength more than the stiffness. The drop in strength parameters from peak to residual conditions seems to be independent of the degree of initial damage (initial fracturing conditions); moreover, the drop in the cohesion term σ_{cd} is larger than that in the friction term m_d:

$$\sigma_{\text{cd}}^r = 0.20 \cdot \sigma_{\text{cd}}^p \tag{4.25}$$

$$m_d^r = 0.65 \cdot m_d^p \tag{4.26}$$

$$\frac{s^r}{s^p} = 0.04 \tag{4.27}$$

The ratio between the post-peak and pre-peak slope of the stress–strain curves shows a tendency to decrease at increasing confining stresses and typically ranges between 0.5 and 0.1.

According to Vermeer and de Borst (1984), dilatancy angles of rock under near- peak conditions decrease at increasing confining stresses and are typically about $20°$ lower than friction angles.

Some authors have also suggested the use of the GSI system for the estimation of rock mass residual strength. For this purpose, the peak GSI value is reduced based on the reduction in the two major controlling factors in the GSI system, i.e. the residual block volume $V_{b, res}$ and the residual joint condition factor $J_{C, res}$, to obtain the residual GSI_{res} value. For the former, a value of 10 cm^3 is suggested; for the latter, half of the peak value can be assumed.

4.7.5 Problems in the characterization of fault rocks, tectonized rocks and fault filling materials

Difficult ground conditions frequently occur during the excavation of *'mountain'* tunnels (Figure 4.4). Typical situations include the crossing of faults or tectonized formations as well as the presence of complex mélanges, flysches or soil–rock mixtures (often named bimrocks or bimsoils, e.g. Medley (1994), Xu (2008)). Cataclastic rocks and fault gouge present highly variable properties, from those of soil-like materials to those of competent rock mass.

The application of the classical classification systems is almost impossible when the rock mass is not characterized by a regular structure. Alternative approaches to the evaluation of the GSI, based on the direct observation of rock mass faces without attempting to define quantitative parameters were thus proposed for tectonized and sheared rock masses (Hoek et al., 1998; Marinos & Hoek 2000) and heterogeneous rock masses (Marinos et al., 2005).

Similar difficulties are faced in the presence of mixed soil–rock formations. A very conservative approach in the geotechnical characterization is that of neglecting the presence of the rock component since this latter is associated with larger values of both strength and stiffness than the soil matrix. However, a more reliable estimate

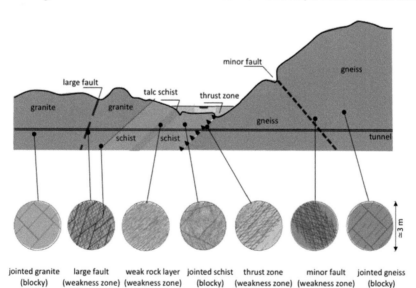

Figure 4.4 Some examples of rock mass conditions. (Modified after Stille and Palmström 2008.)

of the mechanical properties requires the execution of non-standard laboratory and in situ tests, often on artificially prepared samples. Specifically designed direct shear tests, named bimtests by Coli et al. (2011), allow us to consider the contribution of rock blocks to the shear strength, provided the interpretation is carried out accordingly by improved limit equilibrium methods (e.g. Zhang et al., 2020) or specifically designed DEM numerical models (e.g. Graziani et al., 2002).

4.7.6   Identification and hydraulic characterization of water-bearing structure

For deep tunnel projects, the detection and hydraulic characterization of water-bearing zones ahead of the advancing tunnel face is of paramount importance. Risky situations may occur, such as at the intersections with non-cohesive fault core zones under high hydraulic head, shear zones and water-filled karstic cavities in carbonate formations.

The initial assessment of rock mass permeability is usually based on the interpretation of borehole tests (see Section 4.10): Lugeon tests in hard rocks and Lefranc tests in soils or soft rocks.

Pesendorfer and Loew (2007) compared tunnel inflows during excavation with pre-excavation investigation borehole inflows. They observed that: (i) structures with low permeability are often not recognized in the pre-excavation exploratory programme; (ii) the detection efficiency of a water-bearing structure increases with its flow rate and can be significantly enhanced by drilling parallel boreholes through the face; and (iii) the exact prediction of inflow points depends upon the orientation of the conductive zones (e.g. steeply dipping structures across the tunnel axis are more easily detectable).

The analysis of transient inflow rates recorded at the tunnel face (in horizontal boreholes) can be very fruitful for 'updating' the major design parameters such as permeability and storage parameters of the aquifer to dig through. By back-analysing the same data, information can be obtained also on the characteristic dimensions of the water-bearing zone and the spatial distribution of water pressure.

4.8   PLANNING AND EXECUTION OF IN SITU SURVEYS AND INVESTIGATIONS

4.8.1   Site investigations

Due to the complexity of soil and rock behaviour, *site investigations* are a fundamental component of any civil engineering project, including that for a tunnel construction, to acquire geological, geotechnical and other relevant information which might affect its construction, performance and maintenance. The consistency of design calculations and the efficiency of construction are strictly related to the amount and quality of knowledge about the subsurface characteristics gathered before and, if necessary, during the excavation. Hereinafter, the focus is on the geotechnical characterization and the geotechnical model.

Planning and execution of the site investigation programme should consider the complexity of the site with the aim of preventing unidentified risks during tunnelling

(Chapter 7) and to define a reliable subsoil model, in which possible uncertainties are clearly highlighted. A conceptual relationship (Carter, 1992) between the knowledge of ground conditions at the design stage, cost of investigations and risk of unexpected events during construction is plotted in Figure 4.5. The consequences of limited site investigations have a direct impact on the non-budgeted costs, time and the number of bids raised by tunnelling constructors. A sufficient total length of borehole drillings has to be guaranteed, capable of being significantly cost-effective and thus implicitly limiting the related schedule delays. Obviously, this threshold is intended to be project specific, since the extent and types of site investigations depend on a number of factors including the geology, the project characteristics and use, the project stage, the construction method and the environmental issues.

A site investigation programme typically includes the following steps: review of existing information (*desk study*), *topographic survey*, *geological survey* and *mapping*, *geophysical investigations*, *drilling* and *sampling*, *in situ testing* and *monitoring*. This process is completed by *laboratory testing* on the samples collected during borehole drilling or, on some occasions, in the form of blocks.

Each design stage (i.e. *conceptual* or *feasibility*, *preliminary* and *detailed* or *final*) is typically associated with a specific site investigation programme that has the aim of refining the level of knowledge acquired in the previous steps, acquiring the information needed for the project and eventually planning the next investigation campaign (Figure 4.6).

At the level of the feasibility study, the main concern is to collect enough data to evaluate different options in terms of tunnel alignment, construction method and design

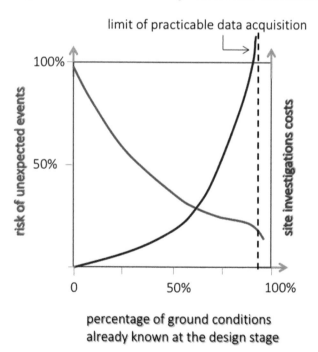

Figure 4.5 Link between the knowledge of ground conditions, cost of investigation and risk. (Adapted after Carter 1992.)

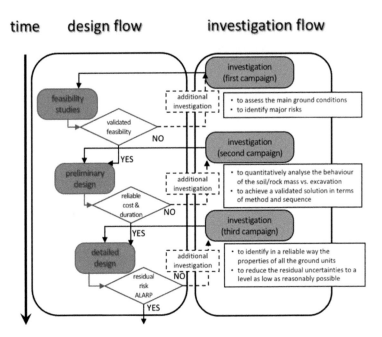

Figure 4.6 Scope of site investigations in relation to the design stages of a tunnel project. (Adapted after ITA 2015.)

approach, and to make a first estimate of costs and construction time. The preliminary design stage is the one during which most of the site and laboratory investigations should be planned, executed and interpreted in order to quantitatively assess the ground properties, to set up the geological and geotechnical models and to finalize the technical solutions for excavation and support. All these outcomes are then validated and integrated during the detailed design stage with the aim of (i) minimizing and quantifying the project risks and (ii) defining proper strategies for risk management during construction.

A tunnel excavation, due to its notable length compared to other geotechnical structures, possibly associated with high depth, very often requires additional site investigations to be implemented during the construction. This can be recommended to further reduce the risk, but is often essential, also in view of contractual claims between the client and the contractor, in order to update the design with respect to the encountered soils and rocks and to their response to tunnelling. Specific monitoring measurements are carried out in the context of the so-called observational method (Chapter 10), which provides some flexibility in the application of the different 'section types' as a function of the actual ground conditions (Chapter 11).

4.8.2 Survey

The *site* or *walkover survey* is typically conducted after the completion of the feasibility desk study and, a second time, once the first planning of site investigation is prepared, to confirm its consistency or to propose integration. Field inspection should provide

	Application														
	Ground investigation		Logging or in situ testing techniques				Depth range			Tunnel components					
	Soft ground	Hard ground	Geophysical borehole logging	Logging of drilling parameters	SPT / CPT	Packer testing	Shallow	Intermediate	Deep	Portal site	Site compound	Shaft site	Shallow tunnel	Intermediate tunnel	Deep tunnel
Percussive (physical – invasive)															
Dynamic probing	●				●		●	○		●	●	●	●		
Window sampling	●				●		●	○		●	●	●	●		
Static cone penetration test	●				●		●	○		●	●	●	●		
Percussion (shell and auger) boring	●	○	○		●		●	●		●	●	●	●	○	
Rotary drilling (physical – invasive)															
Rotary percussion	●		○	○			●			●		●	●	●	●
Rotary open hole	●		○	○		○	●	●		●		●	●	●	●
Vertical rotary coring	●		○	○	○	●	●	●	●	●	●	●	●	●	●
Oriented core drilling	●		○	○		○	●	●	●	●		●	●	●	●
Excavation (physical – invasive)															
Trial pit	●						●			●	●	●			
Trial shaft		●					●	○	○	●	○	●	●	○	○
Trial adit		●					●	○	○	○			●	○	○
Geophysical (non-invasive)															
Ground probing radar	○	○					●	○		○	○	○	○	○	
Electrical resistivity	○	○					●	○		●	●	●	○	●	
Seismic refraction	○	○					●	○	●				●	●	
Seismic reflection		○						○	●				○	○	●
Cross-borehole seismic		○					○	●						○	○

Key: ○ Applicable to some ground types or project circumstances
      ● Fully applicable

Figure 4.7 Ground investigation for tunnelling projects. (Modified after BTS/ICE 2004.)

information about geology, geomorphology and hydrogeology features as well as on site accessibility for drilling and in situ test equipment and on the damage of existing structures, which can help detect the possible ongoing deformative processes in the ground. It is crucial to highlight the existence of faults within the tunnel corridor, even with the help of a helicopter survey to reach the most inaccessible locations.

For tunnelling in rock masses, the so-called *geomechanical survey* is also fundamental to obtain quantitative information about specific characteristics of discontinuities such as spacing, orientation, persistence, roughness, wall strength, weathering, aperture, infilling and seepage. It is performed along *selected scanlines* within areas of limited size, depending on the availability of rock outcrops. The geomechanical survey is also regularly carried out *at the tunnel face* during the construction to evaluate the conditions of the rock mass at depth and along all the alignments (Figure 4.7). Safety issues can sometimes reduce the possibility of accessing the face, and *non-contact techniques* based on digital image photogrammetry and laser scanning can be used. These technologies allow the fracture system to be mapped at a distance from the face with semi-automatic sampling procedures. Such procedures can reduce the bias associated with the manual gathering of fracture data and increase the amount of information collected in a field survey.

4.8.3   Exploratory boreholes and sampling

Exploratory *boreholes* are executed to obtain information about ground profile and to collect samples for laboratory testing. In addition, after the drilling, boreholes can be employed for in situ tests and for the installation of monitoring instruments.

Boring can be carried out without intact core retrieval, by *percussion drilling*, *auger drilling* and *rotary drilling* depending on ground characteristics and borehole depth, or with intact core retrieval by *core rotary drilling*. Cores in soil strata are typically accommodated in *single-tube barrels*, while *double-barrels* are employed in rocks to obtain high-quality cores useful for an at-depth assessment of discontinuity characteristics, such as spacing, roughness, wall strength, weathering and infilling, and for obtaining samples for laboratory testing. A *borehole televiewer* is sometimes employed to get the missing data in terms of discontinuity orientation, aperture and seepage. *Special triple-barrels* can be used to recover the material in highly fractured rock masses. As a general indication, Figure 4.7 (adapted from BTS/ICE, 2004) summarizes the ground investigation techniques that the British Tunnelling Society considers fully appropriate for most tunnel projects, and those that are of limited use and can be used to supplement the former ones.

Ideally, exploratory boreholes should be drilled to below the tunnel invert, for about 10 m. Nowadays, high-capacity rigs are capable of drilling depths up to 1,000 m, but the lack of direct investigation is still an issue for deeper tunnels, as for example those excavated below the Alps. In these cases, the exploration of the effective ground conditions should be performed during the excavation by drilling *horizontal boreholes* ahead of the tunnel face, often in combination with geophysical tests. Frequently, a *pilot tunnel* of smaller dimensions, later employed for technological and safety facilities, is excavated in advance to collect data for the final design of the main tunnel.

The selection of the *number of boreholes* along the tunnel route follows the general rules discussed at the beginning of the paragraph. A slight offset of the drillings with respect to the tunnel cross section is advisable if monitoring instrumentation needs to be installed.

Samples of soil material for laboratory testing are collected with *samplers*, the most used ones being *thick-wall samplers* (e.g. Raymond), *thin-wall samplers* (e.g. Shelby and Osterberg) and *double-core samplers* (e.g. Denison). Good-quality undisturbed samples needed for the mechanical tests can be obtained only from fine-grained soils, unless special sampling techniques are used (e.g. freezing).

4.8.4   In situ testing and monitoring

In situ testing, together with laboratory testing, allows the quantitative determination of the physical, mechanical and hydraulic properties of both soils and rock masses affected by tunnel excavation. The most suitable and common tests are described in Section 4.9. Special tests, for example for TBM design or for the reuse of the muck as a construction material, are particularly relevant in tunnelling. A comprehensive site investigation should include both in situ and laboratory testing, according to the nature of the ground, given the inherent advantages and disadvantages of the two approaches. In fact, in situ tests involve a larger volume of ground and are faster than laboratory tests. However, the latter permit a control of stress–strain paths and drainage conditions and frequently provide the knowledge of the actual state of stress.

The interpretation of in situ tests (Section 4.10) as well as the definition of the geotechnical model (Section 4.11) requires the assessment of the in situ hydraulic conditions. Monitoring of the pore pressure regime is carried out by using different piezometers in relation to the ground permeability and expected variation rate, as extensively discussed later in this book. Less frequently, inclinometers are installed during the design stage to verify the *presence of active movements* in sloping areas.

4.9 LABORATORY TESTS

Laboratory testing is the main tool to characterize the mechanical behaviour of soils and rock matrices. In this section, a short overview of a selection of laboratory tests that can be carried out on specimens of soils and rocks (Table 4.2) is presented. This selection is limited to the main tests required for ground characterization in tunnelling projects, and it should not be considered as exhaustive. Further laboratory tests can be carried out for the special needs or requirements of the project. The reader is therefore invited to refer to additional readings on the general topic of geotechnical laboratory testing for further details (e.g. AGI, 1977; ISRM, 1978a; ISSMGE, 1998).

4.9.1 Uniaxial compression tests

Uniaxial compression tests are performed to measure the compressive unconfined strength and the elastic parameters related to the stress–strain behaviour of mostly rock samples. Cylindrical samples with a height-to-diameter ratio of about 2.5 are subjected to a continuously increasing axial load until failure occurs. Strain- or stress-controlled conditions are used for tests. Axial load, and axial and radial strains are measured during the test. The results are given in plot of the axial, ε_a, and radial, ε_r, strains versus the axial stress, σ_a. The uniaxial compressive strength is calculated by dividing the maximum axial load by the original cross-sectional area of the sample; average secant and tangent Young's modulus E and Poisson's ratio v are determined from the average slopes of the quite linear portion of the axial and radial strain–axial stress curves. In rocks, this linear portion is usually located at half the compressive strength. The ratio between elastic modulus and uniaxial compressive strength can be also used to determine the quality class of rock matrix. Details on the uniaxial compressive test on rock samples are given by ISRM Suggested Methods for Determining the Uniaxial Compressive Strength and Deformability of Rock Materials (ISRM, 1978a).

The uniaxial compressive strength of rock matrix can also be estimated by employing quick tests, performed in the laboratory as well as in the field. The point load test provides a strength index of a rock sample, correlated to uniaxial compressive strength. The test, performed by portable equipment, consists of applying a concentrated increasing compressive load, by two truncated conical platens, to a sample, characterized by any shape (cores, cut blocks and irregular lumps) and a size ranging

Table 4.2 Laboratory tests

Properties	Test
Physical properties	Identification and classification tests: physical indexes, sonic wave speed[*], grain size distribution[**] and Atterberg limits[**]
Mechanical properties	Uniaxial and triaxial compression tests, direct shear tests, tensile tests, one-dimensional compression and swelling tests[**], sonic wave propagation tests[*], point load tests[*], pocket penetrometer[**]
Permeability	Permeability test[**]

Note: [*]rock only; [**]soil only.

between 15 and 100 mm. The load at failure P and the distance between the two specimen–platen contact points D are measured, and the strength index I_S is calculated as the ratio between P and D^2. In anisotropic rocks, a strength anisotropy index I_a, calculated as the ratio between the I_S values measured along the directions giving the maximum and minimum values, can also be estimated. The mean value of strength indexes obtained by a sufficient number of tests (more than 10), multiplied by a number (usually ranging between 20 and 25), according to the lithotype, provides a good preliminary estimation of the uniaxial compressive strength. This test is often performed on weak and complex rocks, where obtaining regular samples for uniaxial tests is difficult (ISRM, 1978a).

Another very quick test, providing a preliminary estimation of the compressive strength of a rock discontinuity surface (Section 4.4.3), is the Schmidt hammer test, performed by means of a sclerometer specifically for rocks. A concentrated dynamic compressive impulse is applied to a rock surface, and a rebound index is read on the instrument. This index is related to the uniaxial compressive strength, for a given rock density. In order to get a representative mean strength value, many tests have to be performed at different locations on the rock surface.

4.9.2 Triaxial tests

The goal of triaxial tests is to assess the *strength* and *stiffness* of a soil (and rock) in terms of *effective stresses* (e.g. E', v', ϕ' and c'). In the standard compression triaxial test, the specimen is subjected to cylindrical compression stress path by applying a deviatoric stress at constant radial stress. However, more general stress paths can be imposed in suitably designed triaxial cells. The test may be either strain- or stress-controlled. Drainage conditions are controlled, thus allowing the soil to be characterized under either 'drained' or 'undrained' conditions. In the last case, 'undrained', equivalent mechanical parameters can be determined, to be used in total stress analyses (e.g. E_u, v_u and c_u). The results of triaxial tests are often produced in a graphical form, that is in plots where deviatoric stress q is put in relation to either the axial strain, ε_a, or the total and effective isotropic stresses p and p', and void ratio, e (or equivalently specific volume, $v = 1 + e$) with ε_a. Furthermore, Mohr's circles of total and effective stresses are also produced. Details on the execution of these tests are available elsewhere (e.g. ASTM D2850, D4767 and D2166).

Measurements of elastic waves velocity propagating in the specimen can be carried out during triaxial tests in a specially equipped apparatus, to achieve measurements of stiffness at very low strain and different confining stress levels (e.g. *bender elements*).

4.9.3 Tensile tests

The tensile strength of the rock matrix is usually measured by performing the *indirect tensile test* called 'splitting test' or 'Brazilian test'. It consists in applying a compressive load along the diameter of a rock sample with diameter not less than 54 mm and thickness approximately equal to the sample radius, until its failure. The failure occurs along a fracture connecting the two press–sample contact points, that is when the rock tensile strength is reached at the centre of the sample. The corresponding compressive load P is measured, and the tensile strength is calculated as the ratio $2P/\pi tD$, with t being the thickness and D the diameter of the sample. Details on the testing procedure can be found in the ISRM Suggested Methods for Determining Tensile Strength of Rock Materials (ISRM, 1978a).

4.9.4 Shear box tests

Direct shear tests are used to assess the *drained shear strength* of a soil or weak rock sample, one-dimensionally pre-compressed, at *peak, stationary state* and *residual conditions*. The specimen is contained in a split rigid box (*Casagrande box*): the lower half-box is pushed to slide horizontally, while the soil specimen within the box is offering resistance. Details on the testing procedure can be found in ASTM D 3080 or AASHTO T236. During shearing, the applied shear force, the relative horizontal displacement δ_h between the two halves of the box and the vertical displacement of the specimen top cap δ_v are measured. The results are typically shown in the $[\tau \cdot \delta_h]$ and in the $[\delta_v \cdot \delta_h]$ planes. The stress values at failure are plotted in a $[\tau \cdot \sigma_v]$ plane, to be fitted by means of a linear Mohr–Coulomb criterion; hence, ϕ' and c' can be determined.

Although allowing to test soils under a large number of conditions, the test has some limitations: the failure is constrained to develop along an imposed surface (hence, the shear strength can be overestimated); drainage conditions cannot be controlled; horizontal strains (localized along the failure surface) cannot be determined from the horizontal displacements; and horizontal stresses cannot be measured; hence, Mohr's circle is not known.

This test is also performed along *rock discontinuities* by using an appropriate shear box (for example *Hoek box*) in which the lower half-box is fixed and the upper half can move along the horizontal direction. The core specimen containing the natural or artificial discontinuity is moulded in a position such that the discontinuity is oriented with the shear direction. A constant normal load and a continuously increasing shear force are applied until the irreversible sliding occurs along the discontinuity. Normal and shear forces, and normal and shear displacements are measured, and the results are plotted in the shear displacement/shear stress and shear displacement/normal displacement planes. *Base, peak* and *residual shear strength* are obtained for *smooth, artificial* and *natural discontinuities*, respectively. Details are given by ISRM Suggested Methods for Determining Shear Strength (ISRM, 1978a).

4.9.5 One-dimensional consolidation and swelling tests

One-dimensional consolidation tests can be carried out in an *oedometer apparatus*. These tests are useful in fine-grained soils to obtain information on their non-linear and stress-dependent compressibility in the long term (*oedometric modulus, E_{oed}*) and on the time dependency of the material mechanical behaviour (*coefficient of vertical consolidation, c_v*). Furthermore, their *permeability* can be estimated. Oedometer tests are generally carried out on clayey specimens that can be obtained from undisturbed samples retrieved on site.

The oedometer apparatus can be also used to reveal clayey rock *swelling potential*, by allowing water access to specimen: if the normal stress on the specimen is kept constant, it expands with time; if the volume strain of the specimen is prevented, the stress increases with time.

The ISRM recommends three swelling tests, providing different parameters: the *swelling pressure*, the *swelling strain* and the *unconfined swelling strain* (ISRM, 1989). The aim of the *swelling pressure test* is to measure the pressure necessary to constrain an undisturbed rock specimen at constant volume when immersed in water. The *swelling strain test* measures the swelling strain under a given constant stress, when the specimen is immersed. In this case, the unloading path starts from the overburden in situ stress, preliminarily reached along a one-dimensional loading path. The free swelling of a rock specimen over a period of time can be measured in an *unconfined swelling strain test*, where the immersed specimen is not constrained, nor any load is applied on it. In this case, both axial and radial strains with time are measured.

4.9.6   Cyclic and dynamic mechanical tests

The decay of soil stiffness (see Figure 4.2) from very small strain level ($\gamma = 10^{-3}\%$) to large strains (1%) may be important to calibrate constitutive models for an accurate prediction of displacements induced by shallow tunnels at the ground surface (and possible effects on buildings) or for the dynamic analysis of tunnels (Chapters 9 and 10). For this purpose, laboratory tests can be carried out such as: *resonant column (RC) test*, *torsional shear (TS) test*, *cyclic simple shear (CSS) test* and *cyclic triaxial (CTX) tests*. Special arrangements can also be used in conventional triaxial cells (such as using bender elements and local deformation transducers). Table 4.3 summarizes the typical results of such tests and, for each of them, the corresponding range of shear deformation which the resulting shear moduli refer to.

4.9.7   Small-scale prototype tests – extrusion tests

Specifically developed (Broms & Bennemark, 1967; Attewell & Boden, 1971; Lunardi, 2008) to assess the pre-confinement pressure necessary to control pre-convergence at the face, *triaxial cell extrusion test* can be carried out in a modified triaxial apparatus. The soil sample is confined by the fluid pressure in the cell: this mimics the isotropic component of stress at the tunnel depth, before the excavation. A cylindrical cut is made within the specimen before testing, called *extrusion chamber*. This is coaxial with the sample axis, and it intends to simulate the face: this is stabilized by a fluid pressure, applied through a membrane. Such a pressure is gradually decreased, and the extrusion of the cut is measured. The results are generally plotted as a curve of normalized

Table 4.3 Typical field of application of cyclic and dynamic laboratory tests on soils

	Test	Results
Cyclic	Cyclic triaxial or simple shear	G_o or $G(\gamma)$ ($\gamma > 10^{-2}\%$)
	Torsional shear	G_o or $G(\gamma)$ ($\gamma > 10^{-4}\%$)
Dynamic	Resonant column	$G(\gamma)$ ($\gamma > 10^{-4}\%$)
	Bending elements	G_o ($\gamma < 10^{-3}\%$)

extrusion (i.e. the ratio between the extrusion and the cylindrical cut diameter) and the applied pressure (*extrusion curves* or *core-face characteristic lines*). For tunnel with medium to deep overburden, where the influence of the soil self-weight on the extrusion at the face can be neglected, such a curve can be used to assess the supporting pressure required to limit the pre-convergence to a prescribed value.

It must be noticed that the test is limited by the specimen size, that is very small. The specimen is therefore representative of the homogeneous matrix, rather than of the whole ground. Any influence of the soil structure at larger scale is overlooked. Moreover, the stress conditions are simplified as they neglect the effect of both gravity field (that is a strong limitation for shallow tunnels) and stress anisotropy. In some cases, *centrifuge extrusion tests* (Lunardi, 2008) may be used to overcome such a limitation, although much more expensive.

4.9.8   Index tests for cutting tools optimization

These tests are carried out to assess the abrasivity potential on the one hand and the hardness and brittleness of the rock mass, on the other. Some tests might be applied in soil as well, in case the grains have a mineralogy which can rise up the problem of wear of the metal parts. The results of these tests are helpful to reduce this risk and are essential for applying prediction models to the TBM design. An overview of the more common index tests is shown in Table 4.4.

Indicators of the performance of cutters, known as NTNU/SINTEF drillability indices (Dahl et al., 2010), are defined on the basis of the index properties described in Table 4.3. The *drilling rate index (DRI)* is assessed from the Brittleness Value S20 and Sievers' J-value SJ (NTNU, 1998; Dahl, 2003): for a Sievers' J-value of 10, which is common for granite, the DRI equals the S20 value. The *Bit Wear Index (BWI)* is assessed by considering the DRI as the abrasion value AV (NTNU, 1998; Dahl, 2003), and it is used to estimate the life of drill bits (measured in $\mu m/m$). The *cutter life index (CLI)* is assessed from Sievers' J-value and the Abrasion Value Cutter Steel, AVS (NTNU, 1998), and expresses life in boring hours for cutter disc rings of steel for tunnel boring machines.

4.9.9   Clogging potential in TBM tunnelling

In case of a TBM excavation involving clayey strata, an assessment of the stickiness and *clogging potential* has to be carried out. These soils, especially the so-called plastic clays which have the tendency to aggregate and be bound by cohesion forces, may behave differently in a TBM plenum, depending on their consistency and plasticity. An important role in this context is played by the water content w, as the consistency of the clays is often assessed by using the *consistency index I_c*:

$$I_c = \frac{w_L - w}{w_L - w_P} \tag{4.28}$$

where w_P is the *plastic limit* and w_L is the *liquid limit* of the considered clay. Both these limits are intrinsic parameters of a clay (also known as *Atterberg limits*), and they are

Table 4.4 Overview of index tests for TBM cutting tool optimization

Property	Material	Definition	Index	Test	Test description
Hardness	Rock	Resistance to indentation	Sievers' J-value (SJ)	Sievers' miniature drill test	The test is performed on a pre-cut rock sample. The Sievers' J-value is the drill hole depth after 200 revolutions of the miniature drill bit, measured in 1/10 mm. The SJ value is the mean value of 4–8 drill holes.
Brittleness		Ability to resist crushing by repeated impacts	Brittleness Value (S20)	Brittleness test	The sample volume corresponds to 500 g of rock chips of density 2.65 g/cm^3 from the fraction passing a 16 mm mesh sieve and retained by an 11.2 mm mesh. The Brittleness Value S20 equals the percentage of material that passes the 11.2 mm mesh after the aggregate has been crushed by 20 blows in the mortar. The Brittleness Value S20 is the mean value of 3–5 tests.
Abrasivity		Ability to cause surface wear by friction	Cerchar Abrasivity Index (CAI)	Cerchar abrasivity test	A standardized metal cone under a constant load of 7 kg scratches the rock over a length of 1 cm; this procedure is repeated six times in various directions running against the main structure of the rock, always using a fresh metal tip. The wear of the tip is represented by an ideal circular final surface with a diameter D measured in mm. The value of CAI is then obtained by multiplying D by 10. It usually ranges between 0.3 and 6.0.
			Abrasion value (AV) Abrasion Value Cutter Steel (AVS)	NTNU abrasion tests	The test apparatus (Lien, 1961; Bruland et al., 1995) consists of a rotating plate in which the sample of crushed rock (particle size <1 mm) is fed regularly on the plate through a vibrating feeder. A 10 kg load is applied on a tungsten carbide sample piece to be tested against abrasion of the crushed powder. The AVS is obtained in the same apparatus using samples from a steel cutter disc.
	Soil		Soil abrasion test (SAT)		The same procedure is adopted for assessing the abrasivity potential of soil in soil abrasion test (Nilsen et al., 2006). AV, AVS and SAT represent the weight loss in mg of the different pieces after a test time of 5 minutes (AV) or 1 minute (AVS and SAT), with a rotation speed of 20 rpm.
		Slurry abrasivity (Miller number)	Miller test		The Miller test (ASTM G75-15) is conducted in a tray filled with the test slurry. A standard steel block is dipped into the test slurry, loaded with a fixed weight (22.24 N) and driven in a reciprocating motion for 6 hours. The mass loss of the steel block gives the Miller number.

indicating the water content at which the soil passes from semi-solid to plastic state and from plastic to liquid state, respectively (Weh et al., 2010). A more extended and refined study is available from Hollmann and Thewes (2013), who proposed a universal classification chart for critical consistency changes regarding clogging and dispersing. A good review of the clogging potential has been provided by Alberto-Hernandez at al. (2017). More informations are provided in Chapter 2 of Volume 2.

4.10   IN SITU TESTS

In situ testing may be conducted in association with boreholes and sampling. Several standard techniques can be used, to obtain indirect measurements of soil properties. The derivation of mechanical parameters is in most cases based on both empirical and semi-empirical formulations, sometimes theoretically founded on the mechanical interpretation of the soil response during tests. Since the adopted relationships refer to tests that are performed according to well-defined procedures, it is important that appropriate standards to carry out tests and interpret their results are followed. In this section, the main in situ testing techniques and their applicability are summarized in the form of tables.

4.10.1   Soil mechanics tests

Table 4.5 summarizes the procedures and the main results of a few very common in situ tests for soils: *standard penetrometer test, cone penetrometer test* and its modifications, *pressuremeter tests, dilatometer test* and *vane test.* Such tests are widely adopted in ground investigations to build the geotechnical model. However, a few remarks should be made on the applicability of such field testing to tunnel projects in soils:

- Standard penetrometer test is easily performed whenever a vertical borehole is carried out; however, their use is generally limited to shallow tunnels in coarse-grained soils.
- Cone penetrometer test can be carried out also within tunnels, with special arrangements to gain reaction and test along multiple orientations.
- Pressuremeter and dilatometer tests are useful to get a reliable assessment of both soil and rock mass deformation characteristics; moreover, they can be used to get information on k_0.
- Vane tests are used at shallow depths; but in case of tunnelling at depths between 20 and 50 m and deeper, French practice is to use Menard pressuremeters.

4.10.2   Rock mechanics tests

Table 4.6 summarizes the main tests used in rock mechanics to define the in situ stresses and assess the deformation modulus of the rock mass. *Hydraulic fracturing, over-coring* and *flat jack tests* are routinely used to investigate the stress state in the rock mass. Loading tests, including *flat jack, radial jack* and *plate loading tests,* are carried out in tunnels to measure the deformation modulus. Although they are limited to a small

Table 4.5 Main in situ tests for soils (modified after FHWA (2002))

Test	Procedure	Applicable soil types	Soil properties	Limitations/remarks
Standard penetration test (SPT)	The *standard penetration test (SPT)* is performed (ASTM D1586) at the bottom of a borehole. A standard thick-wall sampling tube is driven into the soil by impacting a standard mass on it from a fixed height. The result of the test is the number of blows needed to penetrate the soil, N_{SPT}.	Mainly coarse-grained soils	Sand: φ_p' and D_r	Although very frequent, it needs special care in conducting and interpreting. Some correlations are provided with soil stiffness, but alternative measuring techniques should be preferred for it.
Cone penetration tests (CPT, CPTu and SCPTu)	The *cone penetration test (CPT)* is performed (ASTM D5778) from ground surface. A cylindrical probe is hydraulically pushed vertically through the soil measuring the resistance at the conical tip of the probe and along the steel shaft. In the case of CPTu pore water pressures are additionally measured using a transducer and porous filter element. In the case of SCPTu, shear waves generated at the surface are recorded by a geophone at 1 m intervals throughout the profile for the calculation of shear wave velocity.	Both sands and fine-grained soils	Estimation of soil type and stratigraphy Sand: φ_p', D_r and σ_{h0}' Clay: c_u and OCR additionally In clay: c_h and k_h (in CPTu) V_s and G_{max} (in SCPTu)	No soil sample is retrieved. Possible stuck/damage in gravel. Uncomplete saturation or wear of filters may alter pore pressure readings.
Flat dilatometer tests (DMT and SDMT)	The flat dilatometer test is carried out (ASTM D6635-15) by using a steel blade with a circular thin membrane on it. This is hydraulically pushed into the soil to a desired depth; at approximately 20–30 cm intervals, the pressure required to expand the thin membrane is recorded. Three intermediate parameters are measured: material index I_D, horizontal stress index K_D and dilatometer modulus E_D. The seismic DMT (SDMT) provides shear wave velocity measurements.	Sands, silts, clays and hard soils	Estimation of soil type and stratigraphy Sand: k_0, φ_p', ψ, D_r and E' Clay: k_0, OCR, c_u, E_u, c_h and k_h	Membranes may be overinflated and deformed, providing inaccurate readings. Leaks in connections may lead to wrong readings. Small strain stiffness can be measured directly.

(Continued)

Table 4.5 (Continued) Main in situ tests for soils (modified after FHWA (2002))

Test	Procedure	Applicable soil types	Soil properties	Limitations/remarks
Pressuremeter tests (PBP/ PMT, SBP and PIP)	The pressuremeter test (ASTM D4719) is performed with a cylindrical probe that applies a uniform pressure to the wall of a borehole through a flexible membrane. Depending on the installation method, three groups are defined: pre-bored pressuremeters (PBPs); self-boring pressuremeters (SBPs) and push-in pressuremeters (PIPs). The pressure required to expand the cylindrical membrane to a certain volume or radial strain is recorded.	Fine-grained soils mainly; may be used in sands	Clay: E_u, G and c_u Sand: φ_p', G, ψ and σ_{h0}'	The preparation of the borehole is the most important step in PMT. Possible disturbance during advancement in PIP. Stiffness can be measured directly.
Vane shear test (VST)	The vane test (ASTM D2573-01) is carried out using a four-blade vane. This is hydraulically pushed below the bottom of a borehole and then slowly rotated while the torque required to rotate the vane is recorded for the calculation of peak undrained shear strength. The vane is rapidly rotated for ten turns, and the torque required to fail the soil is recorded for the calculation of remoulded undrained shear strength.	Clays	c_u and OCR	Partial drainage may occur in fissured clays and silty materials, leading to errors in calculated strength.

Symbols used in the table: φ_p', effective peak friction angle; G_{max}, small strain shear modulus; D_r, relative density; G, shear modulus; ψ, dilatancy angle; E', drained Young's modulus; c_u, undrained shear strength; E_u, undrained Young's modulus; c_h, horizontal coefficient of consolidation; V_s, shear wave velocity; k_h, horizontal hydraulic conductivity; σ_{h0}', in situ horizontal effective stress; OCR, over-consolidation ratio; k_0, coefficient of at-rest earth pressure.

Table 4.6 Common in situ test methods for rocks (modified after USACE (1997))

Parameter	Test method	Procedure/limitations/remarks
In situ stress	Hydraulic fracturing	It is performed in vertical boreholes, after sealing a short segment between two inflatable packers. Water is then pumped in under increasing pressure until the surrounding rock undergoes tension failure forming a fissure (hydro-fracture): a sharp drop occurs in water pressure, pumping is stopped, and fracture closes at a 'shut-in pressure'; subsequent pumping will cause a reopening of the fissure. In a vertical hole, the hydro-fractures are expected to be vertical and perpendicular to the minimum horizontal stress; hence, the minimum and maximum horizontal stresses can be determined.
	Over-coring	Rock stresses are determined indirectly by measuring the size changes occurring in a rock volume at the bottom of the borehole that is isolated ('over-cored') from the stresses in the surrounding rock. The bottom of the borehole is instrumented with deformation meter. Over-coring releases the horizontal stresses, and the accompanying deformation is measured. Elastic theory is used to calculate the corresponding stresses.
	Flat jack test	A flat hydraulic jack is inserted into a notch that is created in the rock in the area where the stresses need to be measured and it is cemented in place. The creation of the notch results in a certain amount of deformation that is measured using strain gauges. The flat jack is therefore pressurized until such a deformation is compensated for, thus restoring the original stress. It is assumed that such a stress, applied by the flat jack, corresponds to that originally acting in perpendicular direction to the slot.
Modulus of deformation		From the deformation and stress measured in the flat jack test, the deformation modulus can be calculated.
	Plate loading test	Plate loading tests are frequently used in rock engineering to determine the deformability of the rock mass. In such tests, a load plate is pressed against a rock surface (e.g. the tunnel wall) using a hydraulic jack and the plate displacement is measured. In tunnels, it is possible to carry out double-plate loading test, where the opposite wall, or the crown or invert, is used to provide a reaction. In order to make the test results representative of jointed rock, large loading plates should be used to encompass several joints.

(Continued)

Table 4.6 (Continued) Common in situ test methods for rocks (modified after USACE (1997))

Parameter	Test method	Procedure/limitations/remarks
	Radial jack test	Flat jacks can be deployed between a reaction steel ring and the wall of a (pilot) tunnel, arranged to apply load in radial direction. The expansion of the cavity is measured. Hence, the deformation modulus can be back-calculated according to the theory of elasticity.
	Borehole expansion tests (pressuremeter/ dilatometer)	A borehole expansion test can be conducted by a rock dilatometer or pressuremeter probe. The probes used in such a test consist of a cylindrical rubber cell that is hydraulically expanded against the borehole wall. The borehole expansion is measured either by the flow of pressurizing fluid into the cell, or by built-in displacement transducers. Both monotonic and cyclic loading can be applied, thus allowing the deformation modulus to be estimated. By performing the test in boreholes with different directions, information on modulus anisotropy can be obtained. However, since the borehole test affects a relatively small volume of rock, its results may not be representative of the whole the fracture system in the rock mass. Furthermore, problems of borehole stability may occur in highly fractured rocks. In soft rock, as in soil, the expansion tests can also provide a measure of the radial stresses.
	Geophysical tests	These are described in Table 4.7.

volume of rock that may be not representative of the behaviour of the rock mass, *expansion tests* (either by *pressuremeter* or by *dilatometer* probes) can be performed in boreholes to measure the deformation modulus.

4.10.3   Geophysical testing methods

Geophysical tests are largely employed in tunnelling design and execution given the typically large ground volumes investigated. This characteristic makes these tests very valuable for the investigation of ground conditions ahead of the tunnel face during construction in combination with drilling (e.g. Li et al., 2017), especially for deep tunnels for which pre-excavation geotechnical characterization is inevitably rather limited. Nowadays, with the increasing demand of assessing the conditions of old existing tunnels, also for their seismic retrofitting, these methods are also employed to evaluate the lining geometry and integrity as well as the possible presence of voids or weakened zones around the tunnel boundary.

A summary of the principal methods is proposed in Table 4.7, distinguishing between invasive (but more reliable) and non-invasive (sometimes lacking a proper calibration) techniques.

Table 4.7 Geophysical testing methods (after AASTHO (1988))

Method	Procedure	Limitations/remarks
Invasive techniques		
Cross-hole	A seismic pulse is generated in a borehole and recorded by a geophone located at the same depth in another borehole. To eliminate the uncertainties related to the starting time of the test, additional receiver boreholes are used. Repeating the test at different depths, profiles of V_S and V_P with depth are obtained. They can be used to estimate the profiles of soil stiffness at very small strain.	Receivers must be properly oriented and securely in contact with the borehole wall. Boreholes deeper than about 9 m should be surveyed using an inclinometer or other devices to determine the correct travel distance between holes.
Down-hole	The seismic wave is generated at the surface, and one or more sensors are placed at different depths within the borehole. More accurate V_S and V_P profiles are obtained with the 'true-interval' technique (i.e. using sensors with double receivers). They can be used to estimate the profiles of soil stiffness at very small strain.	Data limited to the area very close to the borehole. Difficult interpretation in the presence of several reflected and refracted waves generated at the soil layer interfaces.
Suspension logging	It is one of the available methods for determining the shear and compression wave velocity profiles in both soil and rock. Measurements are made in single, uncased, fluid-filled boreholes. The average wave velocity (V_S or V_P) of a 1-metre-high segment of the soil column surrounding the borehole is determined by measuring the arrival times of a wave propagating upwards.	The data quality is heavily influenced by the borehole conditions. Mud rotary wash method should be used to achieve the best borehole conditions (Biringen & Davie, 2010).
Non-invasive techniques		
Seismic refraction	Detectors (geophones) are positioned on the ground surface at increasing distance from a seismic impulse source. The time required for the seismic impulse to reach each geophone is recorded and analysed. The test allows the identification of V_P value and thickness of the more superficial layers.	Distance between the closest and the furthest geophones must be 3–4 times the depth to be investigated. The presence of low impedance layer below a more rigid one cannot be identified.
Seismic reflection	The seismic wave recorded at the receiver is analysed in order to determine the first arrival time as well as the arrival times of reflected waves.	Reflection from a hard layer may prevent the identification of deeper layers.
Electrical resistivity	Four electrodes are placed partially in the soil, in line and equidistant from each other. A low-magnitude current is passed between the outer electrodes, and the resulting potential drop is measured at the inner electrodes. Useful to map fractured and karst zones ahead of the tunnel face during construction.	Results may be influenced by the presence of underground obstructions, such as pipelines and tanks.

(Continued)

Table 4.7 (Continued) Geophysical testing methods (after AASTHO (1988))

Method	Procedure	Limitations/remarks
Non-invasive techniques		
SASW/MASW	SASW ('spectral analysis of surface waves') and MASW ('multichannel analysis of surface waves') are geophysical investigation techniques that make use of spectral analysis of Rayleigh waves generated at the ground surface by vertical seismic sources, such as hammers (Nazarian & Stokoe, 1984; Park et al., 1999; Xia et al., 1999). In a layered medium, where seismic velocity changes with depth, Rayleigh waves have dispersion property (different wavelengths have different penetration depths and propagate with different velocities), that is indicative of elastic moduli of near-surface soil layers.	MASW makes use of multiple receivers (usually more than 12), which enables seismic data to be acquired relatively quickly when compared to the SASW method (which generally makes use of two receivers only). Both techniques can be used to estimate the profiles of soil stiffness at very small strain and as a quality assurance technique for ground improvement.
Ground penetrating radar	Repetitive electromagnetic impulses are generated at the ground surface or in boreholes, and the travel times of the reflected pulses to return to the transmitter are recorded. Useful to predict water bodies and fractured rock ahead of the tunnel face.	The presence of a clay layer may limit the investigations of layers below. Compared to other seismic methods, it is characterized by a higher resolution, but a lower penetration depth.

4.10.4   Tests for the determination of hydraulic parameters

A variety of in situ tests are available for determining the hydraulic parameters of soils and rock masses. They are performed in wells, boreholes or piezometers, according to the synthetic summary proposed in Table 4.8, and under the assumption that the flow is governed by Darcy's law.

4.11   THE GEOTECHNICAL MODEL

4.11.1   Purpose

The *geotechnical model* for an underground work is the reference for the definition of design solutions and for the evaluation of the performance of the tunnel during the construction phase and under operating conditions. It represents the synthesis of the process of *geotechnical investigation and characterization* of the volume involved. In the geotechnical model also *the hazards* that may occur during the construction phase and, in general, the interactions of the tunnel (during both tunnel construction and operation) with either the ground or the surface are identified and the uncertainties that persist after both the study and the investigation phases are highlighted.

Table 4.8 In situ tests for the determination of hydraulic parameters (modified after USACE (1997))

Method	Procedure	Limitations/remarks
Pumping tests	This test is generally carried out in soils. Water is pumped from a well, normally at a *constant rate* over a certain time period and the drawdown of the water table or piezometric head is measured in the well itself and in piezometers or observation wells in the vicinity.	Large volumes of soil are typically involved, thus providing the average horizontal hydraulic conductivity and storage parameters of the aquifer. A major disadvantage is the time required to reach stationary conditions (durations of I week or longer are not unusual).
Lefranc test	This type of test is carried out in boreholes to determine the permeability coefficient of a soil. It can be performed in two different ways: at *constant head* (in high-permeability soil) and at *variable head* (either *falling or rising head*).	This test is particularly suitable for soils with a coefficient of permeability greater than 10^{-7} m/s.
Lugeon test (packer test)	It is conducted by pumping water at a *constant pressure* into a borehole interval sealed off by packers and measuring the flow rate. This test is carried out in rocks only.	The test is rapid and simple to conduct and, if repeated at different depths, can provide a permeability profile. The limitation of the test is to affect a relatively small volume of the surrounding medium.

The geotechnical model is defined by starting from the *geological model,* which is the essential preparatory reference describing the nature and origin of the soil layers and rock masses, rock type and minerals, hydrogeology and spatial distribution of geological structures along the tunnel alignment.

The geotechnical model must be developed on the basis of specific *geotechnical surveys,* defined by the designer in relation to the characteristics of the project. For this reason, the *geotechnical model* is specific for a single project: it is necessary to know the route and overburden ground conditions, the geometrical characteristics, the main construction phases, the connections with other underground or surface works and any interference with pre-existing works.

4.11.2   Content

The geotechnical model identifies *homogeneous zones* along the tunnel route from the physical-mechanical point of view (i.e. unit weight, stiffness, strength and permeability) and in terms of overburden stresses (or depth); during the analysis, this classification may be associated with homogeneous sections with respect to the stress–strain ground response to the excavation.

The geotechnical model of a tunnel must:

- identify the *geotechnical units*, describing their geometrical layout and possible spatial variability; these may or may not coincide with the geological units identified and described in the geological model;
- define the *physical* and *mechanical parameters* of the geotechnical units, identifying their characteristic values for the specific limit states considered in the design;
- define the *structural characteristics of rock masses*, with reference to the discontinuities;
- define the specific characteristics of the *tectonic structural elements* (e.g. faults), in terms of *position* and *location* in relation to the tunnel axis, ground *conditions* (e.g. degree of fracturing and permeability) and *extent* of the affected zone;
- characterize the expected *hydraulic conditions* in terms of permeability of the crossed formations and piezometric/hydraulic head levels at the tunnel depth and their variation with time;
- provide the estimation of the *in situ stress state*.

The geotechnical model must also highlight all the *special conditions* that may occur along the tunnel route, that stem from the geological, geomorphological and hydrogeological studies, such as karst cavities, interference with water resources, landslide and slope deformation phenomena, possible presence of gas (harmful or flammable), possible presence of aggressive waters for the concrete structures, and mineralogical characteristics that are relevant to the definition of excavation method, such as the abrasiveness due to high quartz concentrations.

A specific detailed geotechnical model must be developed for the *portals*, also highlighting, in addition to the above, the thickness and geotechnical characteristics of the shallow layers and of the loose, weathered and altered soils/rock masses; specific characteristics (e.g. aperture and alteration) of the rock mass discontinuities near the surface; effect of weathering phenomena on the mass characteristics; and factors predisposing to phenomena of instability at the portals.

4.11.3   Representation

The geotechnical model is illustrated in the *geotechnical report* and synthetically represented in the *geotechnical profile* of the tunnel (longitudinal section parallel to the tunnel axis), accompanied by cross sections; the latter is necessary especially in low-coverage and parietal conditions, or in the presence of significant tectonic elements or in contexts characterized by a marked variability of lithotypes and geotechnical units. A detailed representation of ground conditions is required for the portals.

The geotechnical model must be accompanied by a document illustrating the location of the carried-out surveys, with the analysis and summary of their results, explaining on the one hand the methods of analysis and interpretation of data and commenting, on the other hand, the reliability and representativeness of the results.

4.11.4 Design stage

During the design stage, the geotechnical model makes possible the identification of the *design solutions* (such as excavation methods and ground stabilization) appropriate to the geotechnical context and the verification of their performance, also taking into account the *uncertainties* remaining after both the study and the investigation phases. Uncertainties can refer to various aspects:

- aspects pertaining to the *ground conditions* crossed by the tunnel, for example heterogeneity of soils/rock masses, the presence of boulders and definition of the structural characteristics of the rock mass at the tunnel depth;
- aspects of the project, first of all the *overburden* and the *accessibility conditions* of the site, which may make deep geotechnical investigations either difficult or not feasible; and
- aspects related to both the *analysis* and the *interpretation* of the results of site and laboratory *investigations* or the extrapolation of shallow data to tunnel depth.

Some uncertainties identified at the design stage can be overcome by employing *targeted investigations* and tests carried out *during the tunnel construction phase*. The uncertainties highlighted in the geotechnical model can be managed according to the *observational method approach* (Chapter 10) and by formulating different *scenarios* to which alternative solutions defined in the design correspond (Chapter 7).

The extremely variable nature of soils and rock masses implies significant uncertainties on the geotechnical parameters that are obtained from geotechnical investigation, empirical correlations and engineering judgement. Therefore, the definition of the design value of a geotechnical property must take these uncertainties into account, to manage the associated risks by assuming a target reliability level.

To account for this, EN 1997-1:2004 (CEN, 2004), as other national codes, requires the selection of a *characteristic value* for geotechnical parameters as 'a cautious estimate of the value affecting the occurrence of the limit state'. At the design level, this is a crucial aspect of the risk management: although in most cases the 'cautious' estimate is subjective, a proper methodology should be adopted taking several factors into account, such as the variability of material properties, the uncertainty in their assessment, the number of independent measurements and their spatial distribution, the extent of the ground zone governing the tunnel behaviour (as well as any other geotechnical structures) at the considered limit state, the type of failure mechanism and the influence of the material properties on the limit state (see IEG, 2008). It must also be noted at this point that a geotechnical parameter is a mere simplification of a rather complex non-linear relationship between stresses and strains.

Eurocode also introduces the *partial factor* method as a way to achieve the target reliability level, by affecting with partial factors the characteristic values of resistance and loads. Partial factors on soil properties are generally applied in design to account for 'the possibility of an unfavourable deviation of a material property from its characteristic value'. Their values are often prescribed by codes (i.e. Eurocode annexes), somehow limiting their potential to be calibrated to achieve a uniform reliability level in a realistic range of design scenarios (Phoon & Retief, 2015; Prästings et al., 2019). The use of partial factors will be addressed in Section 10.1.1 of Chapter 10.

4.11.5 Construction stage

During the construction stage, the geotechnical model is verified, refined and detailed through direct feedback from the tunnel face, the excavated material, in situ observations and measurements during excavation and investigations and tests carried out in progress, as already defined in the design. With reference to the observational method, investigations and tests carried out during construction must be foreseen in the design and specifically addressed to the verification/determination of significant parameters for the choice of alternative solutions already defined in the design stage. All the findings, investigations and measurements carried out during tunnel construction are incorporated and integrated in the drafting of the *as-built* geotechnical model.

AUTHORSHIP CONTRIBUTION STATEMENT

The chapter was developed as follows. Bilotta: chapter coordination and §4.1, 4.2, 4.3, 4.6, 4.7, 4.9, 4.10, 4.11; Barbero: §4.3, 4.4, 4.5, 4.9; Boldini: §4.2, 4.8, 4.10; Graziani §4.2, 4.3, 4.7; Martinelli §4.9.8. 4.9.9; Sciotti §4.11. All the authors contributed to chapter discussion and review. The editing was managed by Bilotta and Boldini.

REFERENCES

AASTHO (1988) *Manual on Subsurface Investigations.* American Association of State Highway and Transportation Officials, Washington, D.C.

AASHTO (2018) T 236–08, Standard method of test for direct shear test of soils under consolidated drained conditions. American Association of State Highway and Transportation Officials.

AGI (1977) Raccomandazioni sulla programmazione ed esecuzione delle indagini geotecniche. 94 pages.

Alberto-Hernandez Yolanda, Kang Chao, Yi Yaolin, Bayat Alireza, (2017) "Clogging potential of tunnel boring machine (TBM): a review" International Journal of Geotechnical Engineering. 12:3, 316–323.

ASTM (2007) D2573-01, *Standard Test Method for Field Vane Shear Test in Cohesive Soil.* ASTM International, West Conshohocken, PA.

ASTM (2011) D1586 / D1586M-18, *Standard Test Method for Standard Penetration Test (SPT) and Split-Barrel Sampling of Soils.* ASTM International, West Conshohocken, PA.

ASTM (2015) D2850-15, *Standard Test Method for Unconsolidated-Undrained Triaxial Compression Test on Cohesive Soils.* ASTM International, West Conshohocken, PA.

ASTM (2015) D6635-15, *Standard Test Method for Performing the Flat Plate Dilatometer.* ASTM International, West Conshohocken, PA.

ASTM (2015) G75-15, *Standard Test Method for Determination of Slurry Abrasivity (Miller Number) and Slurry Abrasion Response of Materials (SAR Number).* ASTM International, West Conshohocken, PA.

ASTM (2016) D2166 / D2166M-16, *Standard Test Method for Unconfined Compressive Strength of Cohesive Soil.* ASTM International, West Conshohocken, PA.

ASTM (2018) D3080 / D3080M-11, *Standard Test Method for Direct Shear Test of Soils under Consolidated Drained Conditions.* ASTM International, West Conshohocken, PA.

ASTM (2020) D4719–20, *Standard Test Methods for Pre-bored Pressuremeter Testing in Soils.* ASTM International, West Conshohocken, PA.

ASTM (2020) D4767-1, *Standard Test Method for Consolidated Undrained Triaxial Compression Test for Cohesive Soils*. ASTM International, West Conshohocken, PA.

ASTM (2020) D5778-20, *Standard Test Method for Electronic Friction Cone and Piezocone Penetration Testing of Soils*. ASTM International, West Conshohocken, PA.

Atkinson, J.H. & Sallfors, G. (1991) Experimental determination of soil properties. In: *Proceedings of the 10th ECSMFE*, vol. 3, Florence, 915–956.

Attewell, P.B. & Boden, J.B. (1971) Development of the stability ratios for tunnels driven in clay. *Tunnels and Tunnelling*, 3(3), 195–198.

Barton, N., Lien, R. & Lunde, J. (1974) Engineering classification of rock masses for the design of tunnel support. *Rock Mechanics*, 6, 189–236.

Bieniawski, Z.T. (1989) *Engineering Rock Mass Classifications: A Complete Manual for Engineers and Geologists in Mining, Civil, and Petroleum Engineering*. Wiley, New York.

Biot, M.A. & Willis, D.G. (1957) The elastic coefficients of the theory of consolidation. *Journal of Applied Mechanics*, 24, 594–601.

Biringen, E. & Davie, J. (2010) Suspension P-S logging for geophysical investigation of deep soil and bedrock. *GeoFlorida 2010: Advances in Analysis, Modeling & Design*. GSP 199.

Broms, B.B. & Bennermark, H. (1967) Stability of clay at vertical opening. *Journal of the Soil Mechanics and Foundations Division*, 93(1), 71–94.

Brox, D. (2017) *Practical Guide to Rock Tunneling*. CRC Press/Balkema, Leiden.

Bruland, A., Dahlø, T.S. & Nilsen, B. (1995) Tunnelling performance estimation based on drillability testing. In: *Proceedings 8th ISRM Congress*, Tokyo.

BTS/ICE (2004) *Tunnel Lining Design Guide*. British Tunnelling Society, Institution of Civil Engineers (Great Britain), Thomas Telford.

Carter, T.G. (1992) Prediction and uncertainties in geological engineering and rock mass characterization assessment. In: *Proceedings on 4th International Rock Mechanics and Rock Engineering Conference* – Paper 1, 1–23, MIR 92 Turin, Italy.

CEN (2004) *EN 1997-1:2004 Eurocode 7: Geotechnical Design – Part 1: General Rules*. CEN, Brussels, Belgium.

Cerchar—Centre d'Etudes et des Recherches des Charbonnages de France (1986) The Cerchar abrasivity index. Verneuil.

Clayton, C.R.I., Matthews, M.C. & Simons, N.E. (1995). *Site Investigation*. Blackwell Science Inc., Oxford, England; Cambridge, MA.

Coli, N., Berry, P. & Boldini, D. (2011). In situ non-conventional shear tests for the mechanical characterisation of a bimrock. *International Journal of Rock Mechanics and Mining Sciences*, 48(1), 95–102.

Dahl, F. (2003). *DRI, BWI, CLI Standards*. NTNU, Trondheim.

Dahl, F., Bruland, A., Grøv, E. & Nilsen, B. (2010) Trademarking the NTNU/SINTEF drillability test indices. *Tunnels & Tunnelling International*, June, 44–46.

Geertsma, J. (1957) The effect of fluid pressure decline on volumetric changes of porous rocks. *Petroleum Transaction of AIME*, 210(1), 331–340.

Graziani, A., Rossini, C. & Rotonda, T. (2002). Characterization and DEM modeling of shear zones at a large dam foundation. *International Journal of Geomechanics*, 12(6), 648–664.

Handin, J. (1963) Experimental deformation of sedimentary rock under confining pressure: pore pressure tests. *American Association of Petroleum Geologists*, 47, 717.

Hoek, E. (1983) Strength of jointed rock masses. The Rankine Lecture 1983. *Geotechnique*, 33(3), 187–223.

Hoek, E. (1994) Strength of rock and rock masses. *ISRM News Journal*, 2, 4–16.

Hoek, E. & Brown, E.T. (2019) The Hoek–Brown failure criterion and GSI – 2018 edition. *Journal of Rock Mechanics and Geotechnical Engineering*, 11(3), 445–463.

Hoek, E., Carranza-Torres, C. & Corkum, B. (2002) Hoek-Brown failure criterion – 2002 Edition. In: *5th North American Rock Mechanics Symposium & 17th Tunneling Association of Canada Conf.*, NARMS-RAC, Toronto, 1, 267–271.

Hoek, E. & Diederichs, M.S. (2006) Empirical estimation of rock mass modulus. *International Journal of Rock Mechanics & Mining Sciences*, 43, 203–215.

Hoek, E., Kaiser, P.K. & Bawden, W.F. (1995) *Support of Underground Excavations in Hard Rock*. Balkema, Rotterdam.

Hoek, E., Marinos, P. & Benissi, M. (1998) Applicability of the Geotechnical Strength Index (GSI) classification for very weak and sheared rock masses. The case of the Athens Schist Formation. *Bulletin of Engineering Geology and the Environment*, 57, 151–160.

Hollmann, F.S. & Thewes, M. (2013) Assessment method for clay clogging and disintegration of fines in mechanised tunnelling. *Tunnelling and Underground Space Technology*, 37, 96–106.

IEG (2008) *Swedish National Annex to Eurocode 7, Geotechnical Design – Part 1: General Rules (EN 1997-1)*. Swedish Geotechnical Society, Stockholm, Sweden.

ISRM (1978a) *Blue Book: The Complete ISRM Suggested Methods for Rock Characterization, Testing and Monitoring: 1974–2006*. Edited by R. Ulusay and J.A. Hudson. ISRM, Ankara.

ISRM (1978b) Suggested methods for the quantitative description of discontinuities in rock masses. *International Journal of Rock Mechanics and Mining Science*, 15, 319–368.

ISRM (1989). Suggested methods for laboratory testing of argillaceous swelling rocks. *International Journal of Rock Mechanics and Mining* Science, 26, 415–426.

ISSMGE (1998) *Recommendations of the ISSMGE for Geotechnical Laboratory Testing*. Beuth Verlag, Berlin.

ITA (2015) Strategy for Site Investigation for Tunnelling Project. ITA Working Group 2. Report number: 15.

Jager, J.C. (1960) Shear failure of anisotropic rocks. *Geological Magazine*, 97, 65–72.

Jaky, J. (1944) The coefficient of earth pressure at rest. *Journal of the Union of Hungarian Engineers and Architects*, 355–358 (in Hungarian).

Li, S., Liu, B., Xu, X., Nie, L., Liu, Z., Song, J., Sun, H., Chen, L. & Fan, K. (2017) An overview of ahead geological prospecting in tunneling. *Tunnelling and Underground Space Technology*, 63(2017), 69–94.

Lien, R. (1961) An indirect test method for estimating the drillability of rocks. PhD thesis, NTH Dept. of Geology.

Lord, C.J., Johlman, C.L. & Rhett, D.W. (1998) Is capillary suction a viable cohesive mechanics in chalk? In: *Proceedings of the Eurock'98, SPE/ISRM*. Trondheim, Norway, 479–485.

Lunardi, P. (2008) *Design and Construction of Tunnels*. Springer, Milano.

Marinos, P. & Hoek, E. (2000) GSI - A geologically friendly tool for rock mass strength estimation. In: *Proceedings of International Conference on Geotechnical and Geological Engineering (GeoEng 2000)*. Technomic Publishing Co. Inc., Melbourne, Australia, 1422–1440.

Marinos, V., Marinos, P. & Hoek, E. (2005) The geological strength index: applications and limitations. *Bulletin of Engineering Geology and the Environment*, 64, 55–65.

Mayne, P.W. & Kulhawy, F.M. (1982) K_0-OCR relationships for soils. *Journal of Geotechnical Engineering Division*, American Society of Civil Engineers, 108(6), 851–872.

Medley, E.W. (1994) The engineering characterization of mélanges and similar block-in-matrix rocks (bimrocks). PhD thesis. Department of Civil Engineering, University of California, Berkeley, California.

Nazarian, S. & Stokoe II, K.H. (1984) In-situ shear wave velocity from spectral analysis of surface waves. In: *Proceedings of 8th World Conference on Earthquake Engineering*, 3, 31–38.

Nicksiar, M. & Martin, C.D. (2012) Evaluation of methods for determining crack initiation in compression tests on low-porosity rocks. *Rock Mechanics and Rock Engineering*, 45(4), 607–617.

Martin D. (2014) The impact of brittle behaviour of rocks on tunnel excavation design. In: *Rock Engineering and Rock Mechanics: Structures in and on Rock Masses* (Alejano, Perucho, Olalla & Jiménez, Eds.) Taylor & Francis Group, London, 51-62. ISBN: 978-1-138-00149-7.

Nilsen, B., Dahl, F., Holzhauser, J. & Raleigh, P. (2006) SAT: NTNU's new soil abrasion test. *Tunnels and Tunnelling International*, 38, 43–45.

NTNU (1998) *Report 13A-98: Drillability—Test Methods*. Department of Civil and Transport Engineering, Trondheim.

Park, C., Miller, R. & Xia, J. (1999) Multichannel analysis of surface waves (MASW). *Geophysics*, 64.

Pesendorfer, M. & Loew, S. (2007) Detection efficiency of long horizontal pre-drillings to identify water conductive structures. *Felsbau*, 25(4), 18–27.

Phoon, K.K. & Retief, J.V. (2015) ISO 2394:2015 annex D (reliability of geotechnical structures). *Georisk: Assessment and Management of Risk for Engineered Systems and Geohazards*, 9(3), 125–127.

Prästings, A., Spross, J. & Larsson, S. (2019) Characteristic values of geotechnical parameters in Eurocode 7. *Proceedings of the Institution of Civil Engineers – Geotechnical Engineering*, 172(4), 301–311.

Ribacchi, R. (2000) Mechanical tests on pervasively jointed rock material: Insight into rock mass behaviour. *Rock Mechanics and Rock Engineering*, 33(4), 243–266.

Singh, J., Ramamurthy, T. & Rao, G.V. (1989) Strength anisotropies in rocks. *Indian Geotechnical Journal*, 19, 147–166.

Skempton, A. (1961) Effective stress in soils, concrete and rocks. In: *Proceedings of Conference on Pore Pressure and Suction in Soils*, London, 4–16.

Stille, H. & Palmström, A. (2008) Ground behaviour and rock mass composition in underground excavations. *Tunnelling and Underground Space Technology*, 23(1), 46–64.

Sulem, J. & Ouffroukh, H. (2006) Hydromechanical behaviour of Fontainbleau sandstone. *Rock Mechanics and Rock Engineering*, 39, 185–213.

Terzaghi, K. (1923) Die berechnung der durchlassigkeitzifer des tones aus dem verlauf der hydrodynamischen spannungserscheinungen, *Mathematish-naturwissenschaftliche, Klasse*. Akademie der Wissenschaften, Vienna, 125–138.

Thomas, A.H. (2009) *Sprayed Concrete Lined Tunnels*. Taylor & Francis, Abingdon.

USACE (1997) Engineering and design, tunnels and shafts in rock, EM 1 1 10-2-2901. U.S. Army Corps of Engineers.

Vermeer, P.A. & de Borst, R. (1984) Non-associated plasticity for soils, concrete and rock. *Heron*, 29(3), 3–64.

Weh, M., Ziegler, M., Zwick, O. (2010) Verklebungen bei EPB-Vortrieben in wechselndem Baugrund: eintrittsbedingungen und Gegenmassnahmen, *STUVA-Tagung 2009*, 185–189, Hamburg, Bauverlag, Gütersloh.

Xia, J., Miller, R.D. & Park, C.B. (1999) Estimation of near-surface shear-wave velocity by inversion of Rayleigh waves. *Geophysics*, 64(3), 691–700.

Xu, W.J. (2008) Study on meso-structural mechanics (M-SM) characteristics and stability of slope of soil-rock mixtures (S-RM). PhD thesis. Institute of Geology and Geophysics, Chinese Academy of Science, Beijing, China (in Chinese).

Zhang, H., Xu, X., Boldini, D., He, C., Liu, C. & Ai, C. (2020) Evaluation of the shear strength parameters of a compacted S-RM fill using improved 2-D and 3-D limit equilibrium methods. *Engineering Geology*, Article 105550.

Chapter 5

Environmental aspects

D. Peila, C. Todaro, and A. Carigi
Politecnico di Torino

S. Padulosi, F. Martelli, and N. Antonias
Italferr SpA

CONTENTS

5.1 INTRODUCTION

The construction phases of major transport infrastructures must be environmentally managed starting from the early stages of the design in order to be incisive and effective. In fact, the environmental component plays a fundamental role in the setting up of the project and strongly influences the operative and design choices, being one of the key constraints that describe the frame within the design that must be developed (Cotecchia, 1993). The environmental impact (i.e. hazard), in a broad sense, plays a role that is certainly not less than other technical and functional aspects. A good design inserts the construction in the land while minimizing the irreversible environmental damage with a proper development of mitigation measures (ITA-AITES WG 14 & WG 19, 2016). Regarding the interaction between the underground works and the environment, the ITA WG 15 (1998) recommends considering the following key topics at the design stage: the management and organization of the job sites, architectural and landscaping considerations, water and air issues, ground contamination, the impact of noise, dust and vibrations on the population living near the tunnel entrance (Selleri, 2019), the natural biotypes (flora and fauna) and a correct natural resources management.

Generally speaking, the environmental impact of a tunnel can be permanent when it is related to the existence of the tunnel or temporary when it occurs during the

DOI: 10.1201/9781003256175-5

construction, but disappears when the tunnel has been excavated and properly lined (Pelizza et al., 2002). The permanent environmental impacts can be positive (compared to those of other infrastructure works) since the tunnel does not create physical barriers across the territory, it is not visible, and it is not subject to rockfalls, landslides and avalanches (except at the portals). Furthermore, in road tunnels the traffic is not conditioned by atmospheric factors and it is possible to concentrate the traffic emissions (noise, exhaust gases and dust) only at the entrances or at the outlet of ventilation shafts, where it is easier to evacuate and purify them. Considering the permanent interaction with the environment, it should be considered that the presence of tunnel portals can affect the landscape. In this case, Peila and Pelizza (2002) presented two design options that can be developed: to hide the portal with earthworks and a suitable use of vegetation in order to mitigate its landscape impact (i.e. portal integration), or to select a design that is able to give architectural importance to the tunnel portal. Another important permanent impact is related to the disposal of the excavated muck (or mucking, or spoil), which usually requires a new landfill to be inserted in the landscape. Other permanent impacts are polluting materials in the muck and drainage of the underground water.

The temporary environmental impact can be either positive (compared to those of other infrastructural works) since the construction site is not visible and part of the required plant and infrastructure can be placed underground (for example, the plant for concrete production of the Brenner Base Tunnel on the Italian side), or negative. Examples of effects that have to be properly mitigated are interference of the transport of muck, from the tunnel entrance to the landfill, with the ordinary traffic; interference of the construction in terms of the emission of dust, noise, vibration, potential drainage and pollution of the underground or surface waters ('dirty' water coming from the job site has to be treated before sending it to the surface receptors). Considering the temporary hazards, special care must be taken over the environment conditions inside the tunnel, both during the construction phase and during the operational one. These aspects are mainly linked to water ingress and egress, and to the presence of polluting material and gases in the ground and their ingress into the tunnel. The permeation of both water and gases through the final lining (and/or the waterproofing system) can affect the tunnel operations, and hence great care is required in tunnel design. The leakage of groundwater can affect the humidity in the tunnel and result in the misting of rail tunnels, causing an increase in the ventilation loading or condensation on the lining and other tunnel components. In conclusion, it increases the cost of waste water treatment and tunnel maintenance.

In the following, the most important environmental constraints are presented and briefly discussed.

5.2 PORTAL TUNNEL DESIGN

According to Peila and Pelizza (2002), the most important aspects to be considered for the environmental and static design of tunnel portals are as follows:

- Landscape and environmental analysis
  - evaluation of the landscape and the architectural environment;
  - special constraints and social aspects;

- technological needs during use (i.e. tunnel ventilation);
- architectural choice between portal inscription or portal integration.
- Site investigations
  - geological and geotechnical investigations (special attention should be paid to landslides and rockfall-prone areas);
  - seismic condition of the area.
- Surface and underground constraints
  - influence of portal building on the slopes and on nearby structures;
  - influence of the portal on the surface waters and on groundwater both during construction and in operation.
- Analysis of ground behaviour and design of mitigation measures with reference to landslides, avalanches and rockfalls
- Architectural and environmental design of the portal
  - design of the technological buildings;
  - design of final rehabilitation if portal insertion is considered;
  - evaluation of environmental and landscape impacts of rockfall protection structures;
  - evaluation of ancillary works (access roads, working areas, etc.).
- Structural design and calculations
  - portal construction method and required excavation stabilization devices;
  - choice and design of the excavation techniques.
- Structural design of the portal
  - foundation design;
  - structural design taking into account the influence of the reinforcing techniques;
  - structural safety factor evaluation during each backfilling phase.

A good example of the rendering of the impact of a portal on a slope is presented in Figure 5.1.

Figure 5.1 Rendering examples of different hypotheses of Delle Grazie Tunnel portal of the Gronda Project (Italy). The pictures show the different tunnel portal impacts and excavation at different stages of the project development (from left to right). The design choices were focused to minimize the visual impact of the portal on the hill slope, to reduce the extent of the side roadways (Francesconi & Degni, 2019).

5.3   INTERFERENCE BETWEEN THE TUNNEL AND THE UNDERGROUND WATER

The interference between underground constructions and underground water is an important problem due to the need for water resources protection, for the effectiveness of the construction techniques and for tunnel stability control against filtration forces (Pelizza, 1997; Pazzagli, 2001; Lunardi et al., 2016). In this context, the correct prediction of the water inflow and the relative disturbances to aquifers represents one of the substantial hazards, without forgetting that water is increasingly becoming a fundamental resource for humanity and must therefore be protected and rationally used. However, this consideration must not lead only to a technical solution to avoid water drainage caused by excavations: that is too simplistic and sometimes impossible to apply. In fact, this problem is complex because on the one hand it involves the territory, and on the other, it has an impact on the construction works (Figure 5.2). In cohesionless soils, the hydraulic conductivity usually allows the drainage of an aquifer as a consequence of the tunnelling construction. When this occurs, subsidence may potentially affect the top soil due to consolidation of the ground, having an impact also on buildings or infrastructure on the surface. This hazard is managed by foreseeing the use of shield machines when it is possible (the face of the tunnel is stabilized by pressure and aquifer drainage is avoided, thanks to linings) or by performing grouting in a conventional excavation.

In the case of deep tunnels realized below the water table, the issue of water management is more complex mainly due to the difficult identification of the water paths.

(a)   (b)

Figure 5.2 Example of water impact on the tunnel excavation. (a) Picture of water drainage during construction on the Genoa railway junction project. In the photograph, it is possible to see the water ingress during the construction process that must then be controlled. (Courtesy of Italferr S.p.A.) (b) Example of large ingress of water during the Pont Ventoux Tunnel excavation through a rock joint.

For these types of tunnels, the main problem is the management of the potentially high water pressure, which is sometimes difficult to control structurally with waterproof linings. The mitigation measures can resort to the reduction in the hydraulic conductivity of the rock masses by permeation grouting (this procedure is by far the most frequently used) and the use of a definitive waterproof lining. When the interference is permanent, it is necessary to mitigate the potential disturbances caused on the territory by compensating aquifer resources reduced by the tunnel drainage with clean water. This type of intervention must be prepared before the harm occurs and must be promptly implemented.

Studies focused on the hydrogeological balance concerning the disturbed water catchment areas of the tunnel and studies of the water needs of the various users (fauna and vegetation present in the areas) are undoubtedly essential, as is updating them based on the results of hydrogeological monitoring both on the surface and inside the tunnel. Pertaining to the aspect of safeguarding wooded and naturally vegetated areas, the experience of many Italian tunnels that drained the aquifer in the last century shows that sometimes the feared environmental problems did not materialize and alterations of the vegetation were not appreciable.

5.4 EXCAVATED MUCK MANAGEMENT

The correct management of muck is an important environmental constraint since it needs to be disposed of somewhere after the excavation (Pigorini et al., 2014; AFTES, 2016). Today, muck management is based on the concept not only that the muck is a landfill material, but that a properly designed reuse process should be preliminarily identified to minimize the impact on the environment. ITA WG 15 (1998), ITA WG 14 & WG 15 (2019) and ITA WG 14 & WG 19 (2016) are the most important documents related to mitigation measures and provide a complete discussion of the topic of muck management (Figure 5.3).

When the material cannot be reused, an appropriate landfill should be foreseen, designed and used for the minimization of the landscape impact. Finally, it should not be forgotten that in the case of tunnelling excavation by using full-face machines, chemical products that are able to change the behaviour of the ground (for example, during the conditioning process for excavation with EPB) are added to the soil. Depending on the uses for which the muck is intended, it can be used as it is excavated (without treatment) or easily mechanically treated (granulometric classification, comminution and, sometimes, washing). Due to the method of excavation, there is a great difference between the grain size curves of the excavated rock mass obtained from conventional drilling and blasting and rock TBM. The latter is smaller than that obtained from drill and blast (D&B) excavation, and it is characterized by a typical elongated and flattened shape. For the various possible reuses, this aspect is important and must be carefully considered (Gertsch et al., 2001; CETU, 2016; Barrel & Salot, 2019; Perugini, 2019). For reuse as an inert material for cement agglomerates, the standard and well-known lithological requirements for aggregates must be respected. Among others, the absence of altered, soft and soluble minerals that could react with the mortars or hinder their setting, the absence of lamellar crystalline individuals and the requirement for average morphometric coefficients that guarantee the mechanical isotropy of the agglomeration are fundamental requirements for obtaining a good concrete. The muck can be used as a raw material for industrial production, through

Figure 5.3 Hinterrigger site plan view; area for the spoil management of Brenner Base Tunnel, Italian side. Part of the excavated is used in the concrete for the production of the segment lining. (Courtesy of BBT.)

physical–chemical transformation processes (Bellopede et al., 2011; Galler, 2019; Haas et al., 2020), but when reuse is not possible, the excavated muck must be put in a landfill. The storage in landfill, even if water-tightly performed, constitutes an inevitable environmental disturbance that, in principle, causes the following types of damage: modification of morphology, change in stability of the existing slopes, modification of surface hydrology, vegetation loss, mobilization of mineral particles, road modification and alteration in the landscape. The construction of a landfill always changes the original morphology of the location and normally persists indefinitely. Furthermore, for its arrangement it is necessary to pay attention to the stability both of the substrate and of the heap, as well as to the induced modification of the surface and underground hydrology and, finally, to the possible pollution of the filtered water, due to the washing away of solid fractions or soluble minerals, or due to the presence of chemical products added to the soil during the landfill construction. However, the use of landfills is the simplest method available for muck management and it consists in the specific transport with a low environmental impact of the waste materials to landfills in safe and not visible places. After listing the drawbacks, it should also be noted that landfills can be easily regulated, kept under control, elevated with stability criteria (Oggeri et al. (2014) suggested using lime to stabilize and improve the mechanical properties of clayey soils) and adequately masked and definitively set in the new landscape. Sometimes, the excavated material is successfully adopted to reclaim areas that have already been compromised from the landscape point of view, for example exhausted quarry sites. The construction of landfills follows the usual schemes of earthworks.

Figure 5.4 Plattner storage of the Brenner Base Tunnel, Italian side. Photograph of the site ante-operam and reconstruction post-operam with the environmental mitigation and plan of the deposit. (Courtesy of Italferr.)

Special attention should be provided to the re-vegetation design (Figure 5.4). The criterion of using native species is widely adopted in environmental restoration and mitigation works. Local species, due to their coherence with the sites, are more adaptable to the climatic condition of the area. Thus, they ensure an easier success of the intervention. They are also more resistant to external elements (unexpected frosts, drought and parasitoids), and they need less maintenance, allowing the use of chemical fertilizers or pesticides to be reduced.

5.4.1   Management of soil excavated by EPB-TBM

With the aim of the sustainable management of EPB-TBM muck, it is important to verify the environmental compatibility of the conditioned excavated soil and its possible reuse as a by-product. In fact, the conditioned excavated soil could represent a critical issue for the project and it could potentially require a large landfill area. As known, the most common types of soil conditioners are water, foam (consisting of air and aqueous solution of surfactant also called foaming agent) and solutions of polymers, if needed, depending on the soil type (Firouzei et al., 2020) as summarized in Figure 5.5.

Foaming agents are provided by chemical suppliers and are usually used in EPB in water solution, at a concentration by volume ranging from a minimum of 0.5% to a maximum of 6%. The foaming agents are, in turn, a solution of water and anionic surfactants among which sodium lauryl ether sulphate (SLES) is one of the main compounds of most commercial products used in the tunnelling industries. Commonly, foaming agents have an SLES concentration ranging from 5% to 30%. Furthermore, some specific polymers that are able to adsorb free water increase the plasticity of the spoil; high dispersing capacity polymers (anti-clogging polymers) or lubricant agents can be used with a different chemical formulation. A wide range of products can be found on the market, ranging from natural polymers, such as starches, guars and modified natural polymers, including carboxymethyl cellulose and polyanionic cellulose, to synthetic polymers, particularly derivatives of polyacrylamides. Continuous

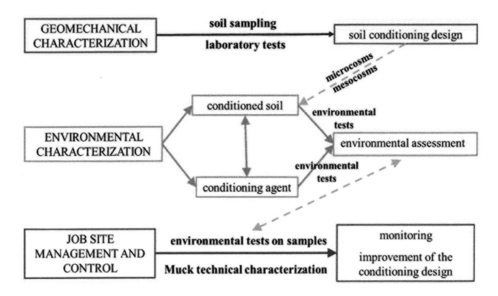

Figure 5.5 Flow chart of the three steps of soil conditioning design, highlighting the role of the environmental assessment of the soil excavated by EPB-TBM (Martinelli et al., 2019).

improvement and development of these products are taking place, and more environmentally friendly products are and will be developed.

Starting from the design phase, the environmental compatibility of the conditioned excavated soil should be assessed and subsequently verified during the construction phase, as summarized in Figure 5.4. Environmental tests should be carried out in order to assess the toxicity of the conditioned soil and the speed of biodegradation of substances (Grenni et al., 2018; Tommasi et al., 2019; Firouzei et al., 2020; Patrolecco et al., 2020). The ecotoxicity of an additive or of a conditioned soil consists in its capacity to produce toxic effects on target organisms and microorganisms (plants and aquatic and non-aquatic organisms) with which it comes into contact. The level of ecotoxicity is assessed via ecotoxicological tests, observing the direct effects on the microorganisms in terms of their mortality or anomalous behaviour. The biodegradability of a product is a measure of the aptitude of microorganisms that are naturally present in the environment to decompose chemical substances present in the conditioned soil. The two concepts, ecotoxicity and biodegradability, are strictly connected one to the other. In the design phase, ecotoxicological tests at laboratory scale are carried out both on the commercial conditioning agents and on the conditioned soil (Martelli et al., 2017; Padulosi et al., 2019). The ecotoxicological features of the commercial products depend on the chemical formula, i.e. on the main components and on the quantities thereof. The ecotoxicological test results on conditioned soil depend upon the quantity of the product and water used for soil conditioning and the grain size distribution, as well as on the soil mineralogy, pH and chemical properties. The environmental assessment of conditioned excavated soil also depends upon its destination site and its possible interaction with the ground and with underground/surface water.

The amount of conditioning agent to be used, usually expressed as the treatment ratio (i.e. litres of product per cubic metre of bulk soil), should be determined by a specific soil conditioning assessment at the design phase. The environmental risk assessment should focus on lithotypes that require a higher treatment ratio, i.e. the 'worst-case scenarios' to be taken as reference in the ecotoxicological study. As a precautionary measure, the ecotoxicological study should consider a further increase in the treatment ratio compared to the value calculated by geo-engineering conditioning testing. This precaution should be foreseen in order to take into account any potential dosage rising due to unexpected geological or operational conditions encountered during the works.

In addition to the study of environmental and ecotoxicology behaviour with time, it must be considered if and when the conditioned soil recovers the original geotechnical properties. This phenomenon is fundamental, since the whole jobsite performance depends on the required recovery time. If the soil has to remain at the jobsite for a long time, the following may occur: slowing down of the excavation, need for a large surface dedicated to soil storage or a high cost due to storage in a temporary landfill. A laboratory procedure for the evaluation of the recovery of the geotechnical parameters for a conditioned soil has been presented by Carigi et al. (2020).

5.5 MANAGEMENT OF ASBESTOS-BEARING ROCK MASSES

Asbestos embraces a series of minerals with a fibrous morphology, belonging to the mineralogical class of silicates, whose health risk is linked to the inhalation of airborne fibres. In 1986, the World Health Organization indicated as dangerous all asbestos fibres with length > 5 µm, diameter < 3 µm and length/diameter ratio greater than 3:1 that are defined as 'breathable' (i.e. able to reach the alveolar area of the human respiratory system). The National Institute for Occupational Safety and Health (NIOSH) has established that exposure to asbestos fibres causes cancer and asbestosis in humans. NIOSH recommends reducing exposure to the lowest concentration possible. For these reasons, the presence of these minerals is a critical hazard for both the OH&S of the workers and the management of the polluted excavated muck. The OH&S topics are beyond the scope of the present chapter and are developed in Chapter 6, while the topics of the environmental management and controls on the excavated muck when asbestos is present in the rock mass are discussed in the following.

Tunnel design where geological formations with asbestos are potentially present requires the definition of the reference geological model, where their occurrence and the probability of finding them during the works are clearly assessed. Furthermore, technical arrangements and operating procedures, aimed to avoid the dispersion of asbestos fibres in the environment (through soil debris, air and water), must be defined and designed also considering proper threshold values. For example, in the large tunnelling project of the Terzo Valico dei Giovi (Italy) in the monitoring plan for airborne asbestos in the living environment, the value of 1 fibre/L in the air measured in SEM over an 8-hour measurement period was adopted as the threshold reference following the Air Quality Guidelines for Europe (WHO, 2000). During the works, the geological model must be verified by in situ surveys: in conventional excavation, this can be done through direct observation of the excavation face, while in the case of excavation

Figure 5.6 Example of fan system and filters in the case of the Cravasco tunnel of the
Terzo Valico dei Giovi (Italy). (Courtesy of Italferr.)

with full-face TBMs, the survey can be performed on the extracted material since the
excavation face is not accessible (for example on the conveyor belts, in the case of EPB-
TBM, or in the pipeline in the case of hydraulic transport). It is important to highlight
that the sampling methods at the face affect the final result, as asbestos is not ubiqui-
tously distributed within the rock; therefore, an adequate scheme of sampling should
be planned as shown in Figure 5.6.

The analytical methods used to check for the presence of asbestos are important,
too, since adequately precise measurements are needed: scanning electron microscopy
with energy-dispersive spectroscopy (SEM/EDS) and polarized light optical microscopy (MOCF/MOLP) are suggested, while X-ray diffractometry (XRD) or Fourier
transform infrared spectroscopy (FTIR) has a very high detection limit and does not
distinguish the morphology of the fibres.

5.5.1   Mitigation measures for the management of excavation material containing asbestos

Various legislations can have different threshold values for the admissible content of
asbestos in the muck. For its potential reuse (irrespective of the presence of these minerals), even if it complies with the legal limits, a health risk assessment related to its
reuse is required. Commonly, this health risk assessment results in the identification
of any prevention measures that must be adopted during the entire management process of the material. In Italy, for example, the reference legislation defines a threshold
limit for asbestos content (equal to 1,000 mg/kg) below which the excavated material

can be reused, as a by-product, for backfilling, filling, etc. Instead, above this value, the excavated material must be handled as waste and delivered to suitable authorized treatment plants. Alternatively, in common cases, in the case of muck with an asbestos concentration over the limit, the transport has to be organized by packaging the material itself in big bags.

To minimize the risk of dispersing asbestos fibres into the environment, it is necessary to provide for technical preparations and specific operating procedures (Mancarella, 2017). These measures must be adopted throughout the whole of the muck excavation and management process and in the related areas (i.e. from the production phase to the final disposal), paying particular attention to the intermediate phases of temporary storage and transport (by road, by conveyor belt, etc.). The main mitigation measures are summarized in the following, based on the experiences in the tunnels of Terzo Valico dei Giovi (Italy) (Meistro et al., 2019a, b).

During the excavation phase, it is necessary to provide a compartmentalization of the tunnel, identifying physically separated zones, each associated with the potential exposure to airborne asbestos fibres. This subdivision, the extent of which depends on the method adopted for the excavation (conventional or full-face mechanized excavation), can be simply summarized as zones A, B and C. Zone A is the contamination area. This is the section of the tunnel close to the excavation face where the rock containing asbestos is excavated. In this zone, all the equipment needed for the primary containment and abatement of the asbestos fibres released during the excavation activities must be concentrated. Zone B is the decontamination area. This is the tunnel area adjacent to the contamination zone A, where the decontamination operations of personnel, vehicles and equipment used for the excavation are carried out. Finally, zone C is the uncontaminated area. This is the portion of the tunnel between zone B and the portal where activities not directly connected with the excavation of materials containing asbestos are carried out. As an example, in the Finestra Cravasco tunnel of the Terzo Valico dei Giovi (Italy) the separation of the three areas was obtained by 'physical compartments' consisting of quickly removable metal structures, equipped with automated doors and suitable nozzles expressly located in order to create a blade of water when opening the doors. In each of these areas, the circulating air must then be adequately managed in order to avoid the dispersion of asbestos fibres. The measures to be designed are aimed at confining the source of potential contamination (the excavation face, in this context) from the rest of the tunnel and depend on the excavation methods and construction site logistics. Particularly important is the use of an air circulation and filtration system that prevents the spread of fibres from the inside to the outside of the tunnel (Figure 5.6). It is necessary to use suction ventilation and verify its effectiveness by monitoring the airborne fibres both in the tunnel (in the various zones) and in the outside environment. For the outside living environment, monitoring can be foreseen at: points inside the construction site at the tunnel portal, points located close to the construction site fence located externally (first belt points) and finally points outside the construction site corresponding to sensitive receptors (second belt points), such as schools and meeting places.

Regarding the excavation techniques when conventional excavation is carried out, it is preferable to use high-energy impact hammers rather than D&B. Hammers are more easily humidified, as are the excavated material and the face during the excavation process. In the excavation with EPB-TBM or with slurry-shield TBM, the excavation area is physically separated from the machine environment and the use of conditioning in EPB-TBMs or bentonite mud in SS-TBMs limits the dispersion of fibres. However, specific measures must be adopted at the points where dispersion of the excavated material could potentially occur, which, once dried, could release fibres into the air. For instance, the conveyor belts should work in an encapsulating way or possibly maintain the material in depression through air suction. Furthermore, de-dusting systems should be foreseen, together with the use of water atomizers, sprinklers and washing stations. Concerning the use of full-face machines, special attention must be paid during access to the excavation chamber for inspections and tool changes. Workers must wear safety devices to ensure adequate protection for working in environments that are potentially characterized by a high concentration of fibres. In addition, specific operating procedures must be defined for accessing and exiting from the excavation chamber, in order to isolate the work area from the remaining part of the TBM. Finally, a proper decontamination system expressly for the work clothes of the personnel entering the chamber must be provided. The tools used must also be properly decontaminated by thorough washing inside the excavation chamber, before taking them out.

Transport by conveyor belt, characterized by all the above-mentioned precautions and by mitigation procedures, should always be preferred for storage of the extracted material in specific sites close to the tunnel. However, this solution is not always achievable, so it may be necessary to use classic road transport. In this case, all precautions must be taken to minimize dust emissions, such as always transporting thoroughly wet material, using lorries equipped with tarpaulin covering that must be used for both loaded and unloaded vehicles, and using boxes with sealing gaskets. Furthermore, the washing of wheels and vehicles must be planned, as well as the identification of the shortest possible routes and limiting stops as much as possible.

The water resulting from the washing in the various steps of the process should be conveyed to a contaminated water circuit where a physical–chemical treatment plant eliminates asbestos fibres (for example with specific ultrafiltration units and filter presses that may be contained in a vacuum environment). The residual sludge from the treatment must be disposed of as a polluted material.

5.5.2   Mitigation measures at the storage sites (intermediate or final)

During the final repository of the material (both in the intermediate and in the final sites), all precautions must be taken to minimize dust emissions. It should be considered that dust emissions are also related to the movement of vehicles. First, it is fundamental to minimize the handling of vehicles containing asbestos soil, and second, the speed limit of these vehicles should be reduced. Consequently, the whole excavation process speed is reduced, taking into account the need to perform the washing of the wheels with closed-circuit systems and the washing of the vehicles with suction inside the cabin at the end of each work shift. Special care should be taken with the management of the muck. In the case of wind, it is necessary to suspend the unloading activity, to carry out an immediate reduction in the discharged material by limiting the movements and falls from great heights that can lead to the fragmentation of the rock, to humidify the muck using cannon fog (or equivalent systems) and to organize the work phases in order to facilitate the vertical development of the soil coverage with respect to the horizontal one, so as to have less extensive exposed surfaces. Furthermore, the rainwater must be properly regulated; in the case of prolonged stops, humidification of the material or covering with sheets or a thin film of agglomerating product and/or film-forming product; at the end of the works, covering the already folded surface with the excavated material (1 m) free of asbestos must be done and appropriate signage must be placed.

Another interesting example of the management of the muck coming from the excavation of an asbestos-bearing rock mass is the Cesana tunnel (Figure 5.7). In this example, an underground deposit in a non-asbestos-bearing rock mass was excavated starting from the tunnel. Inside it, the polluted excavated material was stored directly underground in a safe condition, creating the first underground waste disposal site in Italy (Testa et al., 2008; Alessio et al., 2009).

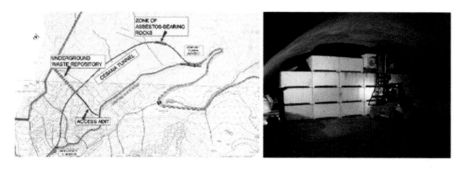

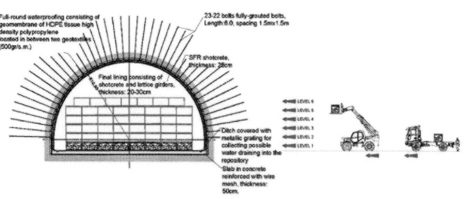

Figure 5.7 Plan of the underground deposit constructed in the Cesana tunnel (Turin, Italy). It has been constructed in a non-asbestos-bearing rock mass, and it is orthogonal to the already constructed tunnel. In the cross section and in the picture, the position of the concrete container of the asbestos rock mass is highlighted. The lining of the deposit has been designed to guarantee the stability and the isolation of the asbestos from the environment. (Courtesy of AK Company.)

5.6   LIFE CYCLE ASSESSMENT AND CARBON FOOTPRINT OF A TUNNEL CONSTRUCTION

In the modern world, there is a great attention from the public towards the development of an economy that needs to be more careful about the environmental impact, the climate changes and the carbon footprint of each activity. Therefore, public acceptance of an infrastructure is a possible constraint that could affect the design and construction of the infrastructure as a whole, including tunnels. For these reasons, it becomes fundamental to measure the value of an infrastructure project in terms of its sustainability to communicate the sense and the opportunities of the project for the community and the territory of reference, as discussed in another chapter of this book. In this perspective, sustainability assessment methods and protocols are effective tools to promote an innovative engineering concept that interprets the design as an opportunity for dialogue with the communities involved. The dialogue should be focused on local needs and the environmental context, enhancing the reference territory and communicating to the public, in a clear and transparent way, the benefits deriving from the realization of the works.

An evaluation of the life cycle of the work aimed at integrating the paradigms of the circular economy, which by its nature is regenerative, in the feasibility analysis of a specific project could be a useful instrument for the purpose of communication with locals. In particular, the reuse of materials excavated during underground construction, as has already been discussed, becomes a strategic means to reduce the CO_2 emissions linked to the transportation of the material outside the building site, promoting also a reduction in the total amount of materials to be supplied during the construction phase. Besides the CO_2 reduction, the reuse also allows an overall reduction in traffic flows for off-site transportation (reducing the carbon footprint following the standard UNI ISO 14064).

These studies should allow the measurement and reporting of emissions produced in infrastructure construction activities and a preventive energy assessment of the works to be performed, facilitating the project manager's interventions right from the first planning phases, which is necessary for the modification of any possible form of irrational consumption of resources. It could be useful that specific contractual regulations will oblige the construction companies to procure their construction materials from environmentally aware suppliers and to steer towards environmentally friendly means of transport, thus rewarding firms that actively collaborate and contribute to reducing CO_2 emissions into the atmosphere. In this way, contractors are assessed based on the environmental improvements originating from their 'environment-friendly' choices when procuring materials and selecting the transportation methods for materials. They can choose low environmental impact products by buying them from suppliers who have gained Environmental Product Declaration (EPD) certification. As regards transportation, contractors can privilege suppliers using environmentally compatible means of transport, such as trains instead of trucks. In a broader perspective, life cycle assessment (LCA) is the modern reference method for evaluating the interaction of an infrastructure project with its territorial context in a long-term vision, and it becomes an operational tool for integrating sustainability into the development of the project and for measuring the environmental and energy loads of the entire 'infrastructure – including the tunnels – as a system' (Antonias, 2019). The objective of

LCA is to analyse the 'environmental profile' of an infrastructure project and its construction process, taking account of all the stages of its useful life (from the extraction and acquisition of raw materials up to final decommissioning), while also assessing the opportunities for mitigating or reducing its environmental impact. The designer should develop an LCA study (in conformity with the standard ISO 14040 and ISO 14044). The conscious management of complex information referring to the sustainability of the design choices made highlights the opportunities for a synergy between the LCA and the BIM approach, which brings undoubted advantages in terms of project efficiency and quality, allowing an organic and integrated management process through the broadest sharing of information concerning the entire life cycle of the infrastructure. Specific tools and methodologies to measure the carbon footprint and life cycle of infrastructure projects are explicitly required by the Envision protocol (Gigli et al., 2017) (the first rating system to design and build sustainable infrastructure), which provides criteria to improve design solutions in a new sustainable approach to Quality of Life, Leadership, Resource Allocation, Natural World, and Climate and Resilience.

ACKNOWLEDGEMENTS

D. Putzu, G. Dajelli and F. Nigro of Italferr S.p.A.; L. Captini of Webuild Company; G. Parisi and C. Zippo of COCIV are gratefully acknowledged for their help and assistance with the preparation of this chapter.

AUTHORSHIP CONTRIBUTION STATEMENT

The chapter was developed as follows. Peila: chapter coordination and §5.1,5.2,5.3,5.4; Todaro and Carigi: §5.1,5.2,5.3,5.4; Carigi: §3,4; Martelli (with the help of Nigro and Dajelli) §5.4, Padulosi (with the help of Nigro, Dajelli, Captini, Parisi, Potzu and Zippo) §5.5, Antonias §5.6. All the authors contributed to chapter review. The editing was managed by Peila, Todaro and Carigi.

REFERENCES

AFTES (2016) Recommandation sur la gestion et l'emploi des matériaux excavés, *Recommandation de l'AFTES*, GT35R1F2, AFTES (Paris).

Alessio C., Kalamaras G., Pelizza S., Peila D., Testa C. & Carlucci L. (2009) Asbestos isolation. *Tunnels and Tunneling International*, 4, 21–23.

Antonias N. (2019) Designing sustainable railway infrastructure: envision protocol and the carbon footprint. In: Peila, D., Viggiani, G. and Celestino, T. (eds.) *ITA-AITES 2019 World Tunnel Congress: Tunnels and Underground Cities: Engineering and Innovation Meet Archaeology, Architecture and Art: Proceeding of WTC 2019*, 3–9 May 2019, Naples, Italy. London, Taylor & Francis. pp. 4311–4317.

Barrel A. & Salot C. (2019) Online identification of the excavated materials on the Saint Martin La Porte site. In: Peila, D., Viggiani, G. and Celestino, T. (eds.) *ITA-AITES 2019 World Tunnel Congress: Tunnels and Underground Cities: Engineering and Innovation Meet Archaeology, Architecture and Art: Proceeding of WTC 2019*, 3–9 May 2019, Naples, Italy. London, Taylor & Francis. pp. 251–259.

Bellopede R., Brusco F., Oreste P. & Pepino M. (2011) Main aspects of tunnel muck recycling. *American Journal of Environmental Sciences*, 7(4), 338–347.

Carigi A., Todaro C., Martienlli D., Amoroso C. & Peila D. (2020) Evaluation of the geo-mechanical properties property recovery in time of conditioned soil for EPB-TBM tunneling. *Geosciences*, 10(11), 438. https://doi.org/10.3390/geosciences10110438.

CETU (2016) Materiaux géologiques naturels excavés en travaux souterrains – Spécificités, scenarios de gestion ed role des acteurs. *Document d'information CETU*, Bron.

Cotecchia V. (1993) Opere in sotterraneo: rapporto con l'ambiente. In: *Atti del XVIII Convegno Nazionale di Geotecnica AGI, Rimini, Italy*. pp. 145–190.

Firouzei Y., Grenni P., Barra Caracciolo A., Patrolecco L., Todaro C., Martinelli D., Carigi A., Hajipour G., Hassanpour J. & Peila D. (2020) The most common laboratory procedures for the evaluation of EPB TBMs excavated material ecotoxicity in Italy: a review. *Geoingeneria Ambientale e Mineraria*, 160(2), 44–56.

Francesconi E. & Degni R. (2019) Tunnel entrances integrated into a landscape: the example of the Gronda motorway bypass. In: Peila, D., Viggiani, G. and Celestino, T. (eds.) *ITA-AITES 2019 World Tunnel Congress: Tunnels and Underground Cities: Engineering and Innovation Meet Archaeology, Architecture and Art: Proceeding of WTC 2019*, 3–9 May 2019, Naples, Italy. London, Taylor & Francis. pp. 311–320.

Galler R. (2019) Tunnel excavation material – waste or valuable mineral resource? – European research results on resource efficient tunnelling. In: Peila, D., Viggiani, G. and Celestino, T. (eds.) *ITA-AITES 2019 World Tunnel Congress: Tunnels and Underground Cities: Engineering and Innovation Meet Archaeology, Architecture and Art: Proceeding of WTC 2019*, 3–9 May 2019, Naples, Italy. London, Taylor & Francis. pp. 342–349.

Gertsch L., Fjeld A., Nielsen B. & Gertsch R. (2001) Use of TBM much as construction material. *Tunnelling and Underground Space Technology*, 15(4), 374–402.

Gigli P., Orsenigo L. & Ciraci S. (2017) Il sistema di rating Envision^{TM} per progettare infrastrutture sostenibili. *Gallerie e grandi opere sotterranee*, 121, 31–36.

Grenni P., Barra Caracciolo A., Patrolecco L., Ademollo N., Rauseo J., Saccà M.L., Mingazzini M., Palumbo M.T., Galli E., Muzzini V.G., Polcaro C.M., Donati E., Lacchetti I., Di Giulio A., Gucci P.M.B., Beccaloni E. & Mininni G. (2018) A bioassay battery for the ecotoxicity assessment of soils conditioned with two different commercial foaming products. *Ecotoxicology and Environmental Safety*, 148, 1067–1077.

Haas M., Galler R., Scibile L. & Benedikt M. (2020) Waste or valuable resource – a critical European review on reusing and managing tunnel excavation material. *Resources Conservation & Recycling*, 162. https://doi.org/10.1016/j.resconrec.2020.105048.

ITA-AITES WG 14 & WG 19 (2016) Recommendations on the development process for mined tunnel. [Online] Available from: https://about.ita-aites.org/publications/wg-publications/content/19-working-group-14-mechanized-tunnelling [Accessed 10th October 2020].

ITA-AITES WG 14 & WG 15 (2019) Handling, treatment and disposal of tunnel spoil materials. [Online] Available from: https://about.ita-aites.org/publications/wg-publications/content/20-working-group-15-underground-and-environement [Accessed 10th October 2020].

ITA WG 15 (1998) Report on the analysis of relations between tunnelling and the environment. [Online] Available from: https://about.ita-aites.org/publications/wg-publications/content/20-working-group-15-underground-and-environement [Accessed 10th October 2020].

Lunardi G., Cassani G., Bellocchio A., Pennino F. & Perello P. (2016) Studi Idrogeologici per la progettazione delle gallerie AV/AC Milano/Genova. Verifica e mitigazione degli impatti dello scavo sugli acquiferi esistenti. *Gallerie e Grandi Opere Sotterranee*, 117, 17–24.

Mancarella A. (2017) The underground works of the new Milan-Genoa high speed/capacity railway line: the project and work progress in rock masses with possible asbestos presence. *Gallerie e grandi opere sotterranee*, 122, 29–40.

Martelli F., Pigorini A., Sciotti A., Martino A. & Padulosi S. (2017) Main issues related to EPB soil conditioning and excavated soil. In: *Congrès International AFTES: "L'espace souterrain notre richesse"*: Proceeding of AFTES 2017, 13–15 November 2017, Paris, France. C3–7, pp. 1–9.

Martinelli D., Todaro C., Luciani L., Peila L., Carigi A. & Peila D. (2019) Moderno approccio per lo studio del condizionamento dei terreni granulari non coesivi per lo scavo con macchine EPB. *Gallerie ed opere sotterranee*, 128, 19–28.

Meistro N., Parisi G., Zippo & Captini L. (2019a) Management of tunneling machines excavation material. In: Peila, D., Viggiani, G. and Celestino, T. (eds.) *ITA-AITES 2019 World Tunnel Congress: Tunnels and Underground Cities: Engineering and Innovation Meet Archaeology, Architecture and Art, WTC 2019*, 3–9 May 2019, Naples, Italy. London, Taylor & Francis. pp. 453–463.

Meistro N., Poma F., Russo U., Ruggiero F. & D'Auria C. (2019b) Excavation with traditional methods through geological formations containing asbestos. In: Peila, D., Viggiani, G. and Celestino, T. (eds.) *ITA-AITES 2019 World Tunnel Congress: Tunnels and Underground Cities: Engineering and Innovation Meet Archaeology, Architecture and Art: Proceeding of WTC 2019*, 3–9 May 2019, Naples, Italy. London, Taylor & Francis. pp. 4932–4941.

Oggeri C., Fenoglio T.M. & Vinai R. (2014) Tunnel spoil classification and applicability of lime addition in weak formations for muck reuse. *Tunnelling and Underground Space Technology*, 44, 97–107.

Padulosi S., Martelli F., Mininni G., Sciotti A., Putzu D.F. & Filippone M. (2019) Environmental risk assessment of conditioned soil: some Italian case studies. In: Peila, D., Viggiani, G. and Celestino, T. (eds.) *ITA-AITES 2019 World Tunnel Congress: Tunnels and Underground Cities: Engineering and Innovation Meet Archaeology, Architecture and Art: Proceeding of WTC 2019*, 3–9 May 2019, Naples, Italy. London, Taylor & Francis. pp. 505–514.

Patrolecco L., Pescatore T., Mariani L., Rolando L., Grenni P., Finizio A., Spataro F., Rauseo J., Ademollo N., Muzzini V., Donati E., Lacchetti I., Padulosi S. & Barra Caracciolo A. (2020) Environmental fate and effects of foaming agents containing sodium lauryl ether sulphate in soil debris from mechanized tunneling. *Water*, 12, 2074. doi: 10.3390/w12082074.

Pazzagli G. (2001) L'interazione tra opere in sotterraneo e falde idriche. Introduzione al tema, *Geologia dell'Ambiente*, IX(3), 2–5.

Peila D. & Pelizza S. (2002) Criteria for technical and environmental design of tunnel portals. *Tunnelling and Underground Space Technology*, 17(4), 335–340.

Pelizza S. (1997) L'utilizzo di opere in sotterraneo per interventi di sistemazione idrogeologica. In: Atti del IX Congresso Nazionale dei Geologi, Roma. pp. 410–415.

Pelizza S., Peila D., Oreste P.P., Oggeri C. & Vinai R. (2002) Environmental aspects of underground construction and exploration for underground structures. State of the art, Congress ACUUS, Torino, Italy.

Perugini V. (2019) Low energy nobilitation of clay waste from tunnelling. In: Peila, D., Viggiani, G. and Celestino, T. (eds.) *ITA-AITES 2019 World Tunnel Congress: Tunnels and Underground Cities: Engineering and Innovation Meet Archaeology, Architecture and Art: Proceeding of WTC 2019*, 3–9 May 2019, Naples, Italy. London, Taylor & Francis. pp. 515–522.

Pigorini A., Martino A., Martelli F., Padulosi S. & Putzu D. (2014) Gestione terre e rocce da scavo: nuovi orizzonti o nuovi limiti? In: *Convegno SIG: Terre e rocce da scavo nelle opere in sotterraneo: un problema o una opportunità?* 7–9 May, 2014, Verona, Italy. pp. 89–102.

Selleri A. (2019) The Genoa Bypass project; the new highway system to overcome the congestion in the Genoa area. The geological and engineering challenges. In: Peila, D., Viggiani, G. and Celestino, T. (eds.) *ITA-AITES 2019 World Tunnel Congress: Tunnels and Underground Cities: Engineering and Innovation Meet Archaeology, Architecture and Art: Proceeding of WTC 2019*, 3–9 May 2019, Naples, Italy. London, Taylor & Francis. pp. 5180–5190.

Testa C., Carlucci P., Pelizza S., Peila D., Kalamaras G. & Alessio C. (2008). Tunnel design in the asbestos-bearing rocks and design of an underground cavern for storing the contaminated muck material – the case of Cesana Tunnel. In: *Congrès International AFTES: "Le souterrain : espace d'avenir", AFTES 2008*, 6–8 October 2008, Monaco, France.

Tommasi P., Lollino P., Di Giulio A. & Belardi G. (2019) Investigation on the geotechnical properties of a chemically conditioned spoil from EPB excavation, a case study. In: Peila, D., Viggiani, G. and Celestino, T. (eds.) *ITA-AITES 2019 World Tunnel Congress: Tunnels and Underground Cities: Engineering and Innovation meet Archaeology, Architecture and Art: Proceeding of WTC 2019*, 3–9 May 2019, Naples, Italy. London, Taylor & Francis. pp. 3235–3244.

World Health Organization (2000) *The world health report 2000- Health systems: improving performance, World Health Report*. Geneva, World Health Organization.

Chapter 6

Occupational Health and Safety aspects

A. Sorlini
Tunnel Euralpin Lyon Turin

M. Patrucco and S. Pentimalli
Universita' degli Studi di Torino

R. Nebbia and C. Todaro
Politecnico di Torino

CONTENTS

6.1 Introduction...119
6.2 Occupational safety and health key points.. 120
    6.2.1 OS&H RAM principles 124
    6.2.2 Use of the proposed approach 126
6.3 OS&H hazard identification 127
    6.3.1 Context-related hazards... 127
    6.3.2 Work-related hazards... 128
6.4 Some insights 128
    6.4.1 Emergency and rescue design criteria... 128
        6.4.1.1 Safety concepts 129
        6.4.1.2 Survival shelters 130
        6.4.1.3 Escaping from underground 131
    6.4.2 Ventilation design criteria..131
        6.4.2.1 Goal and design principle of the ventilation systems.............. 132
Acknowledgments... 136
Authorship contribution statement 136
References.. 136

6.1 INTRODUCTION

Consideration of occupational safety and health (OS&H) of the workers developed with the advance of the industrial revolution in the 19th century, when the social revolution that followed requested to improve the working conditions in all industrial sectors, albeit with slow progress. Over the last 200 years, a clear improvement in working conditions has been seen, due to technological and social innovations, discoveries in medicine and occupational hygiene, and an awareness of the importance of life and human health. Tunneling and mining industry was included in this process, which is clearly described in Figures 6.1 and 6.2 that show the decrease in fatalities in American

DOI: 10.1201/9781003256175-6

mines from 1930 to 2000 and in many important tunnels in the world. Figure 6.3 shows the trend of the OS&H performance index, defined as the relationship between the number of serious or fatal events and the technological, social, and cultural evolution of work, during the 19th and 20th centuries. The goal to be reached is a performance index of zero that can be obtained only if properly programmed in tunnel design phase to create a solid basis for construction phase.

6.2　OCCUPATIONAL SAFETY AND HEALTH KEY POINTS

Tunnel construction can nowadays refer to organizational models, advanced techniques and technologies, specialized personnel, surveillance and rescue systems, and widespread use of the information technology that globally make underground construction sites safer and healthier than in the past, but the pivotal step to achieve further improvement in the OS&H is the creation of the "culture of safety" that must be integrated in the design and operations management activities (Eeckelaert et al., 2011; Hollnagel, 2014; Reason, 1997; De Cillis et al., 2017).

Within the frame of this chapter, "project" identifies the set of activities correlated among themselves, directed toward the attainment of an aim, that establishes the characteristics and performance of the goal itself, at a given time and with given resources, as a tunnel is. In this context quality, in the sense of response to the expected requirements and performance, plays a very important role, to guarantee correct OS&H conditions for the workers involved in the operational and maintenance phases, as well as for the dwellers of the area around the site and final users and environmental aspects (discussed in Chapter 5 of this volume), together with compliance with the laws in force.

In the case of tunneling operations, the need to minimize the risks from the very beginning of the design process is of paramount importance, considering that the tunnel excavation and construction is a very complex system in terms of techniques and technologies used, volumes, pollutant dynamics, and evolution of the job sites that could interfere one with the other (Council Directive 89/391/EEC, 1989; Council Directive 92/57/EEC, 1992; Labagnara et al., 2011).

Furthermore, the usual very tight schedule for a tunnel construction phase imposes high predictive and response capabilities to minimize the need for decision-making steps during the excavation process and the tunnel design, where the OS&H is considered, requires a wide multidisciplinary knowledge, not only based on the usual engineering aspects, but also considering specialists with expertise in safety and health and environmental protection, in industrial medicine and in metrology.

It is important to remember that to guarantee the achievement of the project aims within the expected budget, it is necessary to consider three aspects affecting the efficiency of the operations and directly linked with the OS&H:

- injury costs: an injury involves major economic costs that are both direct (refund, penalties, and medical costs) and indirect (stoppage of activity, personnel management, equipment restoration);
- legal aspects;
- stoppages of the excavation that can cause a direct economic loss.

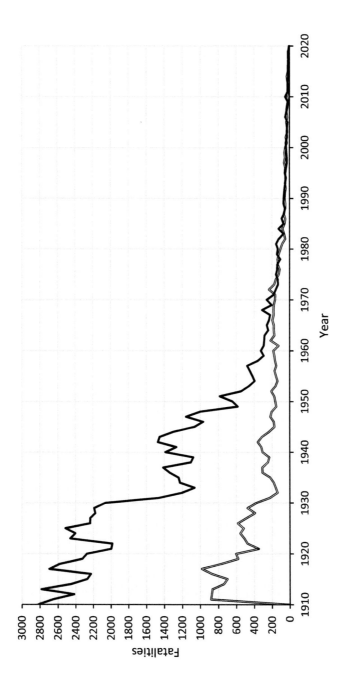

Figure 6.1 U. S. mine fatalities, 1910–1999 (based on data taken from Katen 1992 and Ramani and Mutmansky 2000).

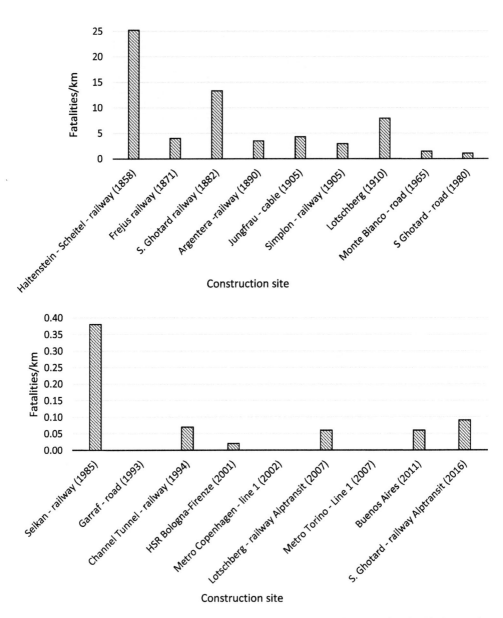

Figure 6.2 Comparison between historical and recent tunneling works, highlighting the important decrease in fatalities. The data were taken from historical documents and the literature.

The development of a tunnel design by including OS&H aspects from the initial phase of the activities assures relevant advantages in terms of general effectiveness of the tunneling design and construction process. The amount of potential errors compromising OS&H of the workers is directly linked with the degree of involvement of designers and OS&H experts. If the safety and health experts work within the tunnel

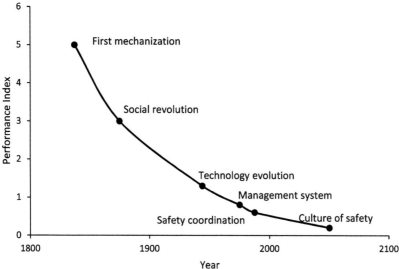

Figure 6.3 Historical trend of OH&S improvement performance versus technological, social, and cultural evolution.

design team from the very beginning (i.e., from the preliminary design stage) and interact in all the design phases, there is the minimization of the potential errors that can't be solved later.

These concepts are also clearly defined in the "Prevention through Design (PtD)" approach, that was introduced by the National Institute for Occupational Safety and Health (ASSE, 2009; CCPS, 2011; NIOSH, 2013; Labagnara et al., 2013), and that is coherent with what has been stated since 1989 in the 89/391/EEC Directive, stating what are wise choices: "choices based on an all-encompassing Risk Analysis developed from the very first tunnel design phase, basically contribute to a safe tunnel use, an effective emergency management, and to effective and safe maintenance operations." Furthermore, it should not be forgotten that all the people involved in OS&H should be univocally identified in a well-defined chain of responsibilities and roles, should cooperate closely: this leads to the definition of the general OS&H policy, based on synergy between management, operating line, and staff organizations. Moreover, an unbiased documental and technical information transfer, based on input data and systematically updated data sharing, is of pivotal importance.

It is not to forget what the 89/391/EEC Directive (1989) states: "whereas the improvement of workers' safety, hygiene and health at work is an objective which should not be subordinated to economic considerations" and that:

whereas employers shall be obliged to keep themselves informed of the latest advances in technology and scientific findings concerning work-place design, account being taken of the inherent dangers in their undertaking, and to inform accordingly the workers' representatives exercising participation rights under this Directive, so as to be able to guarantee a better level of protection of workers.

Furthermore, there are general obligations on employers in the choice of the workers' OS&H protection measures: "the choice should take into account the technical progress and the development of a coherent overall prevention policy covering technology, organization of work, working conditions, social relationships and the influence of factors related to the working environment."

Starting from the above-mentioned concepts, a basic general approach to OH&S Risk Assessment and Management requires an interaction between the following key phases:

- the assessment of the workplace safety with the evaluation of workplace conditions in terms of structure, materials, systems, etc., associated with the activation of General Support Services (GSA) that is the technical and organizational countermeasures to criticalities and emergencies according to general and specific regulations (i.e., fire protection systems, accident management and first aid, and communication and alarm systems).
- the evaluation, for each worker, of his exposure to the total number of hazard factors as systematically identified through:
  - the detailed analysis of the operations performed by each worker and logical breaking up of every complex operation into basic ones;
  - the determination of the average duration of each basic operation and throughout hazard identification;
  - the analysis of each equipment, that should make reference to up-to-date regulations and standards such as the 2006/42 European Directive (Directive 2006/42/EC of the European Parliament and of the Council, 2006), and to the techniques and technologies progress (e.g. anti-collision systems (Patrucco et al., 2021) which at present cannot be considered an optional choice).

The process should then be driven by quality management, whose intent is to establish, document, implement, and continually improve the OS&H policy, in compliance with the law in force and the best techniques requirements (International Organization for Standardization, 2018a).

6.2.1   OS&H RAM principles

In this frame, starting from the basic definition of hazard (the feature or ability of something, the hazard factor, to threaten life, health, property, or environment), it is possible to say that it can be classified as: dormant, when the situation presents a potential hazard, but no people, property, or environment are currently affected; armed, when people, property, or environment are in potential harm's way; and, finally, active when a harmful incident, deviation, involving the hazard has actually occurred.

The risk is defined as the probability of losing something of value: life, health, property, or environment, which can be compromised by a given action, activity, or inaction involving a hazard factor. Risk implies that an exposure value (contact factor) greater than zero is present, that's to say that the hazard is armed. To attribute a numerical value to the risks involved in a specific situation so that a hierarchical order approach can be defined, the following relationship can then be written: $\text{Risk} = M \times P = \text{DE} \times \text{CF} \times P$, where M is the potential loss consequent to the unwanted event occurrence, expressed in general terms (e.g., m^2 of polluted areas); DE is the detailed value of the loss (e.g., lost days from temporary or impairing injuries and fatalities); CF is the exposure conditions of the affected parties (e.g., the percentage of a work shift); and P is the likelihood of occurrence of unwanted events (i.e., the possibility of deviation from the correct work organization/development), expressed in terms of probability.

Based on the occupational risk assessment and management approach developed since 1994 (Faina et al., 1996), and adopted as official reference by Italian agencies, of proven effectiveness in different NACE—Nomenclature statistique des Activités économiques dans la Communauté Européenne—sectors (Eurostat—European Commission, 2008), including tunneling, the following steps should be followed to reach a numerical risk evaluation, unbiased by subjective estimation:

- the damage entity (DE) is expressed in terms of lost days according to work-related accident statistics and injury frequency/severity rates (Eurostat—European Commission, 2013);
- the contact factor (CF) can be estimated in terms of percentage of the work shift involving the exposure to hazard;
- the likelihood of occurrence of unwanted events (P) can be numerically evaluated in a simplified way substituting P with PR (Equation 6.1): expected frequency of occurrence level. A PR value > 1 points out an unacceptable situation, while $\text{PR} \leq 1$ represents a correct situation, in which the DE value can be expressed in terms of worst credible case:

$$\text{PR} = \frac{expected\ frequency\ of\ occurrence\ of\ the\ event\,(present\ situation)}{maximum\ expected\ frequency\ of\ occurrence\ in\ compliance\ with\ up\text{-}to\text{-}date\ safety\ standards} \quad (6.1)$$

Finally, an effective identification of the hazard factors can be based on the following approach:

- preliminary general risk analysis and control of site characteristics in terms of intended use, fittings, general support service (fire and accident management, emergency organization, etc.) by means of suitable hazard identification techniques (e.g., Preliminary Hazard Analysis - PHA);
- identification and management of interferences (for example using Project Evaluation Review Technique – PERT, Gantt & Functional Volumes Analysis) (Mohan et al., 2007; Labagnara et al., 2016a; Tender et al., 2017);
- a safety analysis and control of every working activity (for example through the use of a computer image generation for job simulation) (Lainema & Nurmi, 2006; Zheng & Liu, 2009; Bersano et al., 2010);

- failure analysis and control by means of hazard evaluation techniques (International Organization for Standardization, 2018b; International Electrotechnical Commission, 2019).

6.2.2   Use of the proposed approach

The decision-making process leading to a project definition often involves the discussion of a number of technical/technological alternatives. In spite of the fundamentals of OS&H, the selection of the modus operandi is even today sometimes carried out neglecting the risk assessment phase, considering solely the technical feasibility and economic evaluations. Apart from the OS&H implications, this approach can induce poor quality of the output, and the expected economic advantage is in the long run deceptive (Borchiellini et al., 2016).

A correct approach, coherent with Equation 6.1, involves following the below steps:

- Step 1: comparative analysis of the feasible options, to identify the one entailing Minimum Risk (MR)—technically achievable with the present-day techniques and technologies (the reasonable fault scenarios taken into account);
- Step 2: additional prevention measures, if any, should be defined for the alternative solutions, to ensure they also reach the minimum risk technically achievable and the related cost estimated;
- Step 3: economic decisions will be applied to the results, leading to the cheapest modus operandi ensuring the safest operating conditions.

A project is based on a correct OS&H risk assessment approach only if it includes the proof of the attainment of minimized risk technically feasible in relation to the latest advances in technology and scientific findings. Moreover, also some constraints are necessary to ensure an expected frequency of occurrence level ≤ 1, i.e., a minimized risk result:

- confidence limits to inform contractors and users about the boundaries in which the design remains adequate; they also permit selecting the most suitable techniques for in situ verifications;
- a dynamic design approach, essential to redefine the operational safety parameters, if an overcoming of the design-planned condition occurs, making possible an alternative (safer) modus operandi.

However, things often go differently. The blame often falls on "simplification," which should be recommended in the bureaucracy of OS&H, but misunderstanding often occurs concerning the term, with serious consequences of (Borchiellini et al., 2018):

- excess of optimism: arbitrary guess from statistics to the extent of not including information in the violations of safety law, or subjective assumptions of the occurred accidents leading to incorrect forecasting of expectable accident rates, and unsubstantiated audit scheduling;

- subjectivity: from the use of qualitative or subjective approaches, typically within risk matrices (Duijm, 2015);
- incompleteness: general-purpose checklists on a limited number of hazard factors.

In conclusion, the following aspects lie at the very base of an effective OS&H risk assessment, which is essential to ensure the success of an operation:

- a thorough hazard identification, essential to ensure that no armed hazard is neglected or misclassified as dormant. A number of hazard identification techniques are available, among which the most suitable should be selected for each special problem;
- a careful evaluation of the probability of occurrence of the unwanted events: to start a project, input data (technical, economic, environmental, etc.) are essential. The level of representativeness of these data can affect the results in many ways, determining deviations from the expected goal.

6.3 OS&H HAZARD IDENTIFICATION

As already discussed in Chapter 2, the first step in a correct risk analysis is the identification of all the related hazards that also affect the health and safety levels that can be achieved. The information collected in the survey makes it possible to catalog the hazards related to OH&S risks into two categories: the context-related hazards and the work-related hazards.

Both categories have contexts which, on the basis of information and technical possibilities, can be eliminated early in the design phase. An example is the choice between two routes: a shorter one that has to cross asbestos-bearing rocks; a longer one that goes around them. The design has to be developed through a multi-criteria analysis, where variables are many, with different weights, and where the hazards will be linked back to the risks, in order to compare the possible alternatives. Then there are inevitable contexts, which derive from structural and technological choices; for example, if a TBM technology is chosen for the excavation, all the hazards related to the use of explosives are not present, but those related to the mechanized excavation are unavoidable.

6.3.1 Context-related hazards

Without claiming to be exhaustive, in the following the most relevant OH&S hazards that may arm themselves or be activated by a tunneling excavation related to geology, hydrogeology, and geomechanics can be summarized: minerals present in the rocks crossed by the project; high levels of SiO_2 releasable in the air; the presence of minerals of the asbestos group, amphiboles, pyroxenes, or other categories able to release fibres in air and water; radioactive minerals; the presence of radon gas, also transported by underground water; toxic or explosive gases such as CH_4, CO, CO_2, H_2S, and SO_x; rock burst and tensional releases; seismicity; geothermal gradient and thermal springs; underground water in the fractured rock mass or in permeable soils; karst; and rock mass joint patterns.

With reference to the hazards that can affect the tunnel portals and the working yards, the key issues can be related to landslides and rockfall events; avalanches; flooding due to the nearby presence of rivers and streams; severe weather conditions: thunderstorms, hurricanes, snowfalls, and exceptional frost; and tropical or desert areas.

The most important anthropogenic hazards to be considered are chemical or biological pollution of the surface area that can reach the underground through rock joint systems or soil permeability, mainly transported by underground water; active or abandoned mining activities, structures, foundations, preexisting pipelines, buried dumps, wells, reservoirs, underground power lines, and gas and oil pipelines; the presence of unexploded bombs or other dangerous objects; interferences with existing networks such as roads, railways, and airports; and, finally, health conditions of the area or the nation and adequacy of care and rescue facilities.

6.3.2   Work-related hazards

The list of the work-related hazards for a specific project derives from the nature of the work activities that we have to carry out, and from the technologies chosen for these activities (defined as already discussed after the risk analysis process). Therefore, they must be discussed case by case, without claiming to be exhaustive, in the following. The most relevant work-related hazards are access and egress to the construction or worksite; diesel exhaust emissions; excavation methodology; explosives; fatigue and shift patterns; heat/cold; hyperbaric working; lifting and lowering operations; lighting conditions; materials and material handling; mechanical and electrical equipment used, including TBMs; mental health and well-being; natural and artificial ventilation; occupational noise; off-site spoil transport and the associated hazards with heavy vehicle safety; the presence of pedestrians in the tunnel; placement of concrete and concrete formwork; tunnel transport; underfoot conditions ("slips, trips, and falls"); the use of chemicals, resins, and grouts; vibrations; welding fumes and gases; and working at heights.

6.4   SOME INSIGHTS

Some important topics, are very important in reducing and managing OS&H risks, need to be tackled in the design phase: emergency and rescue, and ventilation.

6.4.1   Emergency and rescue design criteria

The priority goal of correct emergency planning for underground scenarios is to allow that the workers are able to exit safely from the tunnel and, for this task, the structural choices for the construction process and operation planning play a decisive role since even a small accident, which would not cause any damage on the surface, can have very serious consequences for the safety of people and rescuers underground.

The technological/structural choices can have a decisive influence on the management of the emergency during the construction phase (e.g. with reference to ventilation, in the presence of asbestos (Eskesen et al., 2004; CCPS, 2008; Špackova et al., 2013; Labagnara et al., 2016b)).

The major emergencies in workplaces underground consist of:

- events that prevent survival in the working environment, requiring total or partial underground evacuation, or waiting for rescue in safe places;
- serious accidents requiring fast rescue of one or more persons, even on complex sites, far from the portals;
- possible combination of the first two points.

The measures applied in design phase and during the construction phase are therefore different:

- construction work is made up of a series of transients that constitute the construction sequence, so a scenario analysis is essential every time the structural setup below ground changes;
- the priority of "self-rescue" in the management of emergencies: need for trained, educated, and specially equipped personnel;
- difficulty of evacuation, relying on the availability of equipped safe places.

For the various scenarios, the designer will therefore have to analyze the availability of the necessary escape routes to reach safe places within the same tunnel or other interconnected tunnels, if not directly to the surface. Similarly, safe access routes must be available for rescuers, who themselves must not risk their own safety.

The scenario analyses can follow the indications of EU Regulation 1303/2014 (Commission Regulation (EU) No. 1303/2014), which distinguishes between hot scenarios (fire, explosion, smoke, or gas emission) and cold scenarios (collision between vehicles, etc.). These have to be analyzed following the development of the construction site in order to design the construction planning phase minimizing the risks.

Project and construction planning are therefore the first weapon for dealing with emergencies and accidents below ground.

6.4.1.1   Safety concepts

Self-rescue: today's complex construction sites rarely allow personnel to go outdoors immediately during an emergency. Rather than face long transfers, it is preferable to gather people in well-equipped safe places, which allow for safe conditions while the system is restored and they can be picked up by rescuers. It is the tunnel staff themselves who have to reach these places, having been trained and instructed in the procedures. Similarly, first aid will have to be provided by internal staff, taking into account the arrival times of external rescuers.

The key tools to be considered for the design of the self-rescue tasks are personal breathing apparatus, illuminated path markers, survival shelters, training, communications, and the definition of proper procedures.

Search and rescue: outside rescuers must be able to access the underground without jeopardizing their own safety and have the means to find their way around, reaching shelters and collecting waiting personnel. It is essential to know the number and location of people underground with the utmost precision. In the mining sector, there are many computer applications available today, which are still too little used on civil engineering sites. Exercises carried out in conjunction with the local rescue services are essential.

The key tools to be considered for the design of the search and rescue task are special vehicles, radar/IR vision systems, GPS or similar tracking and navigation systems, training, communications, and the definition of proper procedures.

6.4.1.2   Survival shelters

The safe place should be as close as possible and wide enough to accommodate all those present in its area. The chosen distance is not defined by a standard rule, and it has to be verified in each scenario and in each of our projects, taking into account excavation methods, the number of faces, interconnections, slope, and other parameters that may slow down or hinder the flow of people. The guiding parameter should therefore be the travel time, not distance.

As an example, and with reference to the large-scale projects developed in the last decade, a survival shelter should be at a distance of a maximum of 500 m for anyone below ground at the time. Shelters are therefore located every 1,000 m along the tunnels in order to respect this parameter. The distance of 500 m was chosen conventionally, based on the concept that this distance can be covered by well-trained personnel in 10–15 minutes, even in conditions of poor visibility and smoke, wearing special Personal Protective Equipment (PPE) such as survival masks, which are now standard equipment for workers underground. This time must be measured during exercises that are part of the workers' training. The route can be facilitated by high-visibility light signals or other installations.

On TBMs, it is now the rule to have a shelter to accommodate the machine crew. The survival shelter is easily accessible and is useful in cases where the direct access route to the TBM is impeded or if there are breakdowns in vital equipment to the working environment. In this case, the distance is well under 500 m. The shelters behind will be installed in sequence, in accordance with the rationale provided by the scenario analysis of the site.

In traditional tunnel excavation, the placement of shelters close to the face must reconcile proximity to workers with sufficient distance so that the shelter is not damaged by excavation operations.

In a complex environment with many interconnections, it is possible to optimize the position of shelters, so as to reduce their number, by making interconnections in advance (bypasses between two tunnels, for example).

So far, prefabricated mobile shelters, which can be easily installed and moved along tunnels, have been the most frequently used even if on complex and very extensive construction sites it is recommended to set up larger fixed safe places at nodal points of the system.

For example, the compartmentalization of a bypass, equipped with survival and communication systems, offers large waiting areas, especially if close to caverns where several faces converge or where there are installations manned by technical personnel. Such installations can be used also during the operation life of the project.

Since visitors are a common sight at large construction sites, their number must be taken into account in the scenario analyses and in the design of the shelters (Sorlini & Gilli, 2015).

6.4.1.3   Escaping from underground

If the emergency is located at the tunnel face, in some cases the use of the next safe room may not be the correct choice. It will therefore be necessary to leave the face more quickly and reach another safe place or the surface directly. For this reason, it is practical to have an equipped evacuation vehicle available at the front. Many solutions have been observed: if the transport is by rail, an emergency train always present behind the TBM backup and on a different track may be a good solution, and in other cases, one or more vehicles available a few hundred meters from the tunnel face (they may be the same ones as used for the shift change).

The key tools to be considered for the design of escaping from underground task are escape vehicles and the definition of proper procedures.

6.4.2   Ventilation design criteria

The ventilation systems (Figure 6.4a and b) should be considered only after the demonstrated completion of three preliminary design steps, i.e., (i) minimization of the hazard factor (e.g., route modifications), (ii) control of the pollutant emissions (e.g., possibly focusing on the influence of the abatement methods and techniques), and (iii) control of the pollutant dispersion into the working environment (e.g., discarding a priori abatement techniques whose emissions cannot be managed in the proximity of source).

The design of the tunnel ventilation is important in the frame of OS&H, since there are risks arising from the possible presence of environmental pollutants in underground environment. The design of the ventilation must start from the very first design stages and is based on a specific risk assessment requiring the knowledge of the pollutants, induced both by the tunneling operations and by the characteristics of the rock formations themselves.

The latter can be classified as follows:

- mobile: firedamp and other gases present in the rock mass can reach the tunnel area from great distances, directly through fractures or, if soluble, can be present in the underground water systems;
- localized: hazardous pollutants or agents present in the excavated rock formations (e.g., asbestos, silica, and radioactive minerals).

Moreover, the criticality should be considered in terms of possible consequences of the exposure of the workers to the pollutants. The main parameters to be considered for an effective management of these criticalities in the workplaces in relation to the presence of pollutants are the follows:

- harmfulness of the pollutant:
  - potential impact on exposed workers vs. concentration or level, possibility of early diagnosis and treatment (e.g., moderate exposures to asbestos and ionizing radiation can, respectively, have a latency time of more than 30 and from 5 to 10 years, and associated pathologies are critical (mesothelioma, lung cancer, leukemia, etc.);
  - other possible associated hazards (e.g., explosion and abnormal response of engines or numerical control devices);
- specific characteristics of the pollutant:
  - capacity of dispersion in more or less extensive portions of the environment, and dispersion velocity;
  - persistence characteristics in the environment;
  - tendency to temporary or permanent segregation (e.g., as function of the density, and the miscibility in air);
  - tendency to evolve in terms of chemical nature (e.g., oxidation);
  - tendency to aggregation, flocculation, etc.;
  - surface tension (wettability) and hygroscopicity;
- characteristics of the source:
  - stationary or mobile;
  - spatial characteristics (from localized to ubiquitous);
  - spatial characteristics of emission (e.g., directivity);
  - temporal characteristics of emission (continuous, cyclic, intermittent, and occasional);
- characteristics of the environment:
  - the presence of air movements (droughts);
  - other parameters conditioning the dispersion or propagation of the pollutant;
  - seasonal variations;
- characteristics of the exposed workers:
  - exposure characteristics (in terms of exposure factor ↔ aspects of work organization);
  - subjective characteristics of the workers (sensitivity to the pollutant);
- other hazard factors:
  - the presence of other pollutants and possible synergistic effects;
  - conveyance of other pollutants (e.g., combusted carbon particles and polycyclic aromatic hydrocarbons).

6.4.2.1   Goal and design principle of the ventilation systems

The purpose of the underground ventilation is as follows (CARSAT Rhône-Alpe & CETU, 2013; CARSAT Rhône-Alpe, 2017):

- to supply the workplaces below ground with sufficient clean air to ensure adequate environmental hygiene conditions as regards the percentage of oxygen and

concentrations of pollutants; of course, due account must also be taken of the comfort aspects linked to microclimatic conditions;
* to remove the polluted air and transfer it to the outside, after pretreatment coherent with the pollutant emissions into the atmosphere official regulations.

The design of a ventilation layout can be described by the following steps:

* definition of the general layout of the system. Figure 6.4a and b show the possible alternatives and examples (auxiliary operations along the tunnel (opening of niches, chambers, etc.) should be fed with clean air derived from the main flow, and exhaust connected with the main exhaust airways). The selection should be based on the approach to MR discussed in Chapter 2.2, to guarantee correct OS&H conditions for workers involved in the operational and maintenance phases, the safety and health of dwellers of the area around the site, and general environmental protection;
* once the general layout has been defined, the essential parameter for the design of a ventilation system is the definition of the flowrate necessary to manage the environmental pollutants, as a function of the pollutant emission of the source.

In conclusion, the following considerations can be drawn:

* ventilation systems are an essential response (although not the first in hierarchical terms in the context of prevention measures) to the management of hygiene and environmental conditions in underground workplaces, and emissions from the portals;
* the design must necessarily be based on accurate risk assessments—in conformity with the PtD approach—whereas, however, in the case of underground excavation activities, there is often a need for over-sizing because it is impossible to obtain detailed information on the local characteristics of the rock formations, in particular in the presence of highly harmful mineral contents distributed, for geological reasons, according to models that cannot be represented in geo-statistical terms (Labagnara et al., 2016b);
* the design must make the whole ventilation system (engines, actuators, ducts, and monitoring, control and alarm systems) capable of responding to emergency situations, and contain solutions to ensure the desired degree of availability (Copur et al., 2012; Labagnara et al., 2015).

Further improvements to the effectiveness of the underground environment management systems in an OS&H point of view may in future be based (Figure 6.4a and b) (Patrucco et al., 2018) on:

* refinement in the selection criteria of techniques, technologies, and organization for the excavation activities, increasingly consistent with an approach in PtD, based on a more rigorous RAM;
* the generalization of the use of analysis techniques, such as Hazard and Operability analysis (HazOp) and Fault Tree Analysis (FTA) / Event Tree Analysis (ETA), for a thorough assessment of the actual availability of ventilation systems;

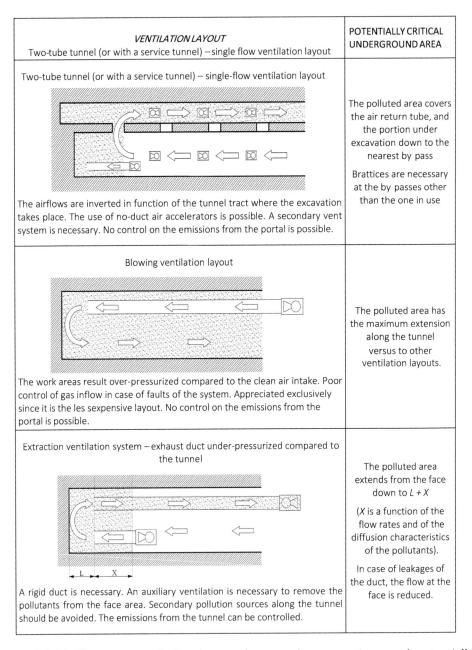

VENTILATION LAYOUT Two-tube tunnel (or with a service tunnel) – single flow ventilation layout	POTENTIALLY CRITICAL UNDERGROUND AREA
Two-tube tunnel (or with a service tunnel) – single-flow ventilation layout	The polluted area covers the air return tube, and the portion under excavation down to the nearest by pass
The airflows are inverted in function of the tunnel tract where the excavation takes place. The use of no-duct air accelerators is possible. A secondary vent system is necessary. No control on the emissions from the portal is possible.	Brattices are necessary at the by passes other than the one in use
Blowing ventilation layout	The polluted area has the maximum extension along the tunnel versus to other ventilation layouts.
The work areas result over-pressurized compared to the clean air intake. Poor control of gas inflow in case of faults of the system. Appreciated exclusively since it is the les sexpensive layout. No control on the emissions from the portal is possible.	
Extraction ventilation system – exhaust duct under-pressurized compared to the tunnel	The polluted area extends from the face down to $L + X$
	(X is a function of the flow rates and of the diffusion characteristics of the pollutants).
A rigid duct is necessary. An auxiliary ventilation is necessary to remove the pollutants from the face area. Secondary pollution sources along the tunnel should be avoided. The emissions from the tunnel can be controlled.	In case of leakages of the duct, the flow at the face is reduced.

Figure 6.4 (a) The main ventilation layouts for tunneling operations, and potentially critical aspects.

VENTILATION LAYOUT	POTENTIALLY CRITICAL UNDERGROUND AREA
Blowing ventilation system – ventilation duct over-pressurized compared to the tunnel 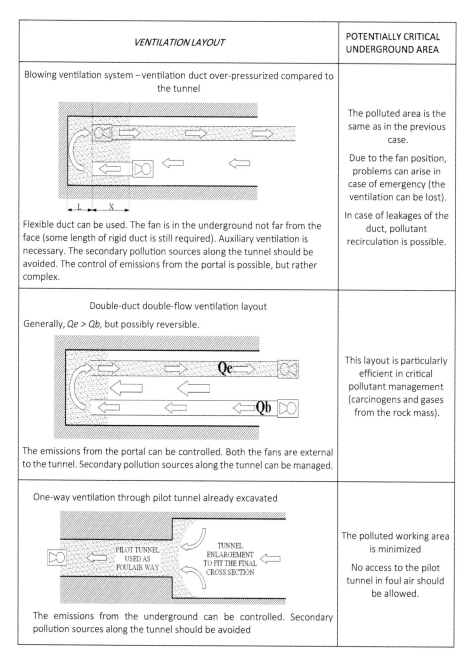 Flexible duct can be used. The fan is in the underground not far from the face (some length of rigid duct is still required). Auxiliary ventilation is necessary. The secondary pollution sources along the tunnel should be avoided. The control of emissions from the portal is possible, but rather complex.	The polluted area is the same as in the previous case. Due to the fan position, problems can arise in case of emergency (the ventilation can be lost). In case of leakages of the duct, pollutant recirculation is possible.
Double-duct double-flow ventilation layout Generally, $Qe > Qb$, but possibly reversible. The emissions from the portal can be controlled. Both the fans are external to the tunnel. Secondary pollution sources along the tunnel can be managed.	This layout is particularly efficient in critical pollutant management (carcinogens and gases from the rock mass).
One-way ventilation through pilot tunnel already excavated The emissions from the underground can be controlled. Secondary pollution sources along the tunnel should be avoided	The polluted working area is minimized No access to the pilot tunnel in foul air should be allowed.

Figure 6.4 (b) The main ventilation layouts for tunneling operations and potentially critical aspects.

- verification of actual pollution conditions in workplaces below ground based on a strict representativeness approach;
- field tests to verify the effectiveness of the detection techniques for typically predictable pollutants in underground excavation activities, taking into account the difficulties arising from often demanding microclimatic and logistical conditions;
- field surveys for the evaluation of the effectiveness of containment and segregation systems, and polluted air recirculation and in-line separators, to analyze the possibility of use even in the presence of pollutants with high toxicity;
- careful investigation on the possibility of increasing the integrated design approaches of air flow management during tunneling operations and future tunnel uses, to improve both the underground environmental quality, and the general effectiveness and return of the investment.

ACKNOWLEDGMENTS

Tunnel Euralpin Lyon Turin is gratefully acknowledged for cooperation and making available technical data and documents. D. Peila is acknowledged for the help in preparing and revising this text.

AUTHORSHIP CONTRIBUTION STATEMENT

The chapter was developed as follows: Sorlini and Patrucco: chapter coordination and jointly developed the chapter Todaro, Nebbia and Pentimalli: figures and final editing.

REFERENCES

ASSE (2009) *Guidelines for Addressing Occupational Risks in Design and Redesign Processes 1st Edn.* American Society of Safety Engineers Press, New York.

Bersano, D., Cigna, C., Fissore, F. & Patrucco, M. (2010) Computer image generation for job simulation: an effective approach for occupational Risk Analysis. *Safety Science*, 48, 508–516.

Borchiellini, R., De Cillis, E., Fargione, P., Maida, L., Nebbia, R. & Patrucco, M. (2018) The possible contribution of a well-tested Occupational Risk Assessment and Management technique to counter the recent unexpected rise in the work-related accidents. In: *Proceedings of the International Symposium on Occupational Safety and Hygiene – SHO 2018*, 26–27 March, Guimaraes, Portugal.

Borchiellini, R., De Cillis, E., Fargione, P. & Patrucco, M. (2016) Risk Assessment and Management: easier said, perhaps with too many words, than done - the importance of the culture of prevention. In: *Proceedings of the – The Occupational Safety and Hygiene Symposium – SHO 2016*, 23–24 March, Guimarães, Portugal, ISBN: 978-989-98203-6-4.

CARSAT Rhône-Alpe (2017) Mise en œuvre de dispositifs de ventilation en travaux souterrains linéaires. *Recommandation Du Comité Technique National Des Industries Du Bâtiment Et Des Travaux Publics – R.494.*

CARSAT Rhône-Alpe & CETU (2013) Guide de Bonnes Pratiques pour la sécurité et la protection de la santé lors de travaux en souterrain, pp. 61–62.

CCPS (2008) *Guidelines for Hazard Evaluation Procedures 3rd Edn.* Wiley, Hoboken, NJ, ISBN: 10-0471978159, p. 576.

CCPS (2011) *Guidelines for Hazard Evaluation Procedures 3rd Edn.* John Wiley & Sons, New York, ISBN: 10-1118211669, p. 576.

Commission Regulation (EU) No 1303/2014 of 18 November 2014 concerning the technical specification for interoperability relating to 'safety in railway tunnels' of the rail system of the European Union Text with EEA relevance.

Copur, H., Cinar, M., Okten, G., Bilgin, N. (2012) A case study on the methane explosion in the excavation chamber of an EPB-TBM and lessons learnt including some recent accidents. *Tunneling and Underground Space Technology*, 27, 159–167.

Council Directive 89/391/EEC of 12 June 1989 on the introduction of measures to encourage improvements in the safety and health of workers at work.

Council Directive 92/57/EEC of 24 June 1992 on the implementation of minimum safety and health requirements at temporary or mobile construction sites (eighth individual directive within the meaning of Article 16 (1) of Directive 89/391/EEC).

De Cillis, E., Fargione, P. & Maida, L. (2017) The dissemination of the culture of safety: Innovative experiences from important infrastructures and construction sites. *Geoingegneria Ambientale e Mineraria*, 151(2), 118–127.

Directive 2006/42/EC of the European Parliament and of the Council of 17 May 2006 on machinery, and amending Directive 95/16/EC.

Duijm, N.J. (2015) Recommendations on the use and design of risk matrices. *Safety Science*, 76, 21–31.

Eeckelaert, L., Starren, A., Van Scheppingen, A., Fox, D. & Brück, C. (2011) Occupational Safety and Health culture assessment – a review of main approaches and selected tools. European Agency for Safety and Health at Work (EU-OSHA). ISBN: 978-92-9191-662-7.

Eskesen, S.D., Tengborg, P., Kampmann, J. & Veicherts, T.H. (2004) Guidelines for tunnelling risk management: International tunnelling association, working group No. 2. *Tunneling and Underground Space Technology*, 19, 217–237.

Eurostat – European Commission (2008) Statistical classification of economic activities in the European Community. Available from: https://ec.europa.eu/eurostat/documents/3859598/ 5902521/KS-RA-07-015-EN.PDF [accessed on 3rd January 2021].

Eurostat–EuropeanCommission(2013)EuropeanStatisticsonAccidentsatWork(ESAW).Available from: https://ec.europa.eu/eurostat/documents/3859598/5902521/KS-RA-07-015-EN.PDF [accessed on 3rd January 2021].

Faina, L., Patrucco, M. & Savoca, D. (1996) Guidelines for risk assessment in Italian mines. Doc. No 5619/96 EN - S.H.C.M.O.E.I., Luxembourg, 17 July 1996. pp. 47–71, and Doc. No 5619/1/96 EN - S.H.C.M.O.E.I., Luxembourg, 17 January 1997, 46–47.

Hollnagel, E. (2014) Is safety a subject for science? Safety Science, 67, pp. 21 – 24.

International Electrotechnical Commission (2019) IEC 31000:2019. Risk management – risk assessment techniques.

International Organization for Standardization (2018a) ISO 45001:2018. Standard occupational health and safety management systems – requirements with guidance for use.

International Organization for Standardization (2018b) ISO 31000:2018. Risk management – Guidelines.

Katen, K.P. (1992) Health and safety standards. In: H.L. Hartman (eds.), *SME Mining Engineering Handbook*. Society for Mining, Metallurgy, and Exploration, Littleton, CO, pp. 162–173.

Labagnara, D., Maida, L. & Patrucco, M. (2015) Firedamp explosion during tunneling operations: suggestions for a prevention through design approach from case histories. *Chemical Engineering Transactions*, 43, 2077–2082.

Labagnara, D., Maida, L., Patrucco, M. & Sorlini, A. (2016a) Analysis and management of spatial interferences: a valuable tool for operations efficiency and safety. *Geoingegneria Ambientale e Mineraria*, 16(3), 35–43.

Labagnara, D., Martinetti, A. & Patrucco, M. (2013) Tunneling operations, occupational S&H and environmental protection: a prevention through design approach. *American Journal of Applied Sciences*, 10(11), 1371–1377.

Labagnara, D., Patrucco, M. & Sorlini, A. (2011) Aspetti Tecnologici e di Valutazione e Gestione del rischio occupazionale nei cantieri per la realizzazione di opere infrastrutturali in sotterraneo. In: D. Flaccovio (eds.), *Ingegneria forense: metodologie, protocolli e casi di studio*. Palermo, Italy, pp. 247–272. ISBN: 9788857901015.

Labagnara, D., Patrucco, M. & Sorlini, A. (2016b) Occupational safety and health in tunnelling in rocks formations potentially containing asbestos: good practices for risk assessment and management. *American Journal of Applied Sciences*, 13(5), 646–656.

Lainema, T. & Nurmi, S. (2006) Applying an authentic, dynamic learning environment in real world business. *Computers & Education*, 47(1), 94–115.

Mine Safety and Health Administration (2021) MSHA: coal fatalities – metal-nonmetal fatalities. [Online] Available from: https://arlweb.msha.gov/stats/centurystats/coalstats.asp; https://arlweb.msha.gov/stats/centurystats/mnmstats.asp [Accessed 3rd May 2021].

Mohan, S., Gopalakrishnan, M., Balasubramanian & Chandrashekar, A. (2007) A lognormal approximation of activity duration in PERT using two-time estimates. *Journal of the Operational Research Society*, 58(6), 827–831.

Patrucco, M., Fargione, P., Maida, L. & Pregnolato, M.G. (2018) Sistemi di ventilazione nelle attività cantieristiche in sotterraneo: aspetti di sicurezza e salute del lavoro. *PinC – Prevenzione in Corso*, 3, 20–32.

Patrucco, M., Pira, E., Pentimalli, S., Nebbia, R. & Sorlini, A. (2021) Anti-collision systems in tunneling to improve effectiveness and safety in a system quality approach: a review on the state of art. *Infrastructures*, 6(3), 42.

Ramani, R.V. & Mutmansky, J.M. (2000) Mine Health and Safety at the turn of the millennium. *Mining Engineering*, 51(9), 25–30.

Reason, J.T. (1997) *Managing the Risks of Organizational Accidents*. Ashgate Publishing Co, Aldershot.

Sorlini, A. & Gilli, P. (2015) Emergency and rescue: a new safety concept for the construction of la Maddalena exploratory adit-high capacity rail line between Turin (Italy) and Lyon (France). In: *Proceedings: Rapid Excavation and Tunneling Conference, RETC* 2015, January 2015, New Orleans, LO, pp. 1040–1044.

Špackova, O., Novotna, E., Šejnoha, M. & Šejnoha, J. (2013) Probabilistic models for tunnel construction risk assessment. *Advances in Engineering Software*, 62(63), 72–84.

Tender, M.L., Martins, F.F., Couto, J.P. & Pérez, A.C. (2017) Study on prevention implementation in tunnels construction: Marão Tunnel's (Portugal) singularities. *Revista de la Construcción*, 16, 262–273.

The National Institute for Occupational Safety and Health (2013) Prevention through design. Available from: https://www.cdc.gov/niosh/topics/ptd/default.html [accessed on 3rd January 2021].

Zheng, X. & Liu, M. (2009) An overview of accident forecasting methodologies. *Journal or Loss Prevention in the Process Industry*, 22, 48–491.

Chapter 7

Preliminary risk assessment

L. Soldo and E.M. Pizzarotti
Pro Iter Srl

G. Russo
Geodata Engineering SpA

CONTENTS

7.1 INTRODUCTION

Technical risks inherent in a certain tunnelling project are specific and different throughout its life. Political and Country, market and financial risks (exchange, inflation, interest and credit/counterparty), and regulatory and legal risks are here not specifically considered here; nonetheless, they can influence, directly or indirectly, the technical aspects of a project. Based on a chronological classification, it can be classified into the following categories:

DOI: 10.1201/9781003256175-7

- Pre-completion phase risks: they include those related to the activities such as planning, design and technological solutions, several occurring during construction (both because inappropriate selection of construction techniques and because procedures or bad application). Because of these risks, the project can remain uncompleted, be completed with cost overruns and/or delayed, or be completed with performance deficiency. Apart from a merely economic point of view, in the worst cases, it must be considered the risk of injury or catastrophic failure with the potential for loss of life, and personal injury, extensive material and economic damage.
- Post-completion phase risks (or operational risks): they can be a consequence of performance deficiency (not meeting functional design, operational, maintainability and quality standards) or the occurrence of some incoming events, including accidents and external hazards (e.g. natural hazards).
- Risks common to pre-completion and post-completion phases: they can be a consequence of performance deficiency or the occurrence of some incoming events, including accidents, environmental risks and external hazards (e.g. natural hazards).

Risk Management in tunnelling has become progressively important, also considering the increasing requests in terms of safety and environmental and socio-economic sustainability coming from citizens, owners, lenders and insurers (Grasso et al., 2016). Risk Management is intended not only for risk avoidance and mitigation, but also as a means to value creation, ameliorating the overall project. The 'Guidelines for Tunnelling Risk Management' (International Tunnelling Association, 2004) emphasize as the Risk Management processes are significantly improved by using an early-activated, systematic approach, starting from the very beginning design phase ('preliminary') throughout the entire tunnel project development: 'the use of risk management from the early stages of a project, where major decisions such as choice of alignment and selection of construction methods can be influenced, is essential'.

7.2   UNCERTAINTIES AND RISKS

A source of misunderstanding stands into the concept of 'uncertainty', often intended as a direct synonymous of risk. Because uncertainty can result in not only risks, but also opportunities, worst or better conditions are expected. It is also necessary to differentiate among risks of an event with a known probability, and true uncertainty, which is a known event with an unknown probability or, at worst, a completely unknown event. It must be noted that an engineering project can be negatively affected also in case of unpredicted favourable conditions, e.g. using a TBM tailored for expected adverse geotechnical conditions, while, on the contrary, they are favourable.

The field of deterministic knowledge remains confined to identified subjects that can be analysed with an (acceptable) certainty (something identified and certain, 'known known', with possible risks coming from an event with a known probability). The event can be faced with a rational design, based on sufficient and good data input and robust processing methodologies.

Out of this field, the project must be faced with uncertainties that can be grouped in:

- true uncertainty, a known event with an unknown probability ('known unknowns', should be mitigated by a flexible contractual architecture, leaving room for adequate design and construction procedure to the actually faced conditions); poor characterization (qualitative and/or quantitative) of the identified/foreseen element/property; and uncertain location of the identified/foreseen element (boundary and extension) (something identified/foreseen with a limited level of knowledge). This classification labels also the 'identification' of uncertainties together with the 'level of knowledge' (including the level of knowledge about the 'impact' when considering the risk assessment);

- something not identified/foreseen, but knowable (sometimes called untapped knowledge or 'unknown known');
- complete ignorance, that is when a certain aspect remains completely unknown, not identified/predicted ('unknown unknowns'). It refers to the worst scenario, when 'we do not know what we do not know' and the project solutions can be largely inadequate to the actual conditions. This last case also includes the so-called 'black swans' (Nassim, 2007) that are high-impact, hard-to-predict, rare events.

In this scenario, if the nature of an event is certain, uncertainties can derive from the limited knowledge affecting both what we already know (i.e. known known) and what we don't yet know (i.e. unknown known), while if the nature of an event is uncertain, its occurrence can be uncertain (i.e. with an occurrence probability less than one) and its impact can be uncertain as well and more severe if the uncertainties are due to our limited knowledge. It is not trivial to highlight that the real conditions vary gradually between these two extremes.

7.3   THE RISK MANAGEMENT PLAN

A Risk Management Plan (RMP) is based on a four-step procedure: hazard identification, risk assessment/quantification, risk response (mitigation and control, also including the process of risk allocation and transfer) and monitoring of risk response during construction. The framework on which the RMP is built is the Risk Register. A RMP includes the following parts:

A.   families of hazards list and, within each family, the list of hazards and their causes,
B.   quantification of the likelihood and impact of the hazards and, hence, of the initial risks,
C.   identification of a specific strategy to reduce each initial risk (mitigation measures),
D.   quantification of residual risks (after the mitigation measures) (Figure 7.1).

The basis of the analysis starts from the recognition of the underground project environment and conditions, with their potential hazards and related risks.

As already discussed, the construction of a tunnel affects geological materials, some part of which will be excavated, leaving space for the tunnel structure, and that during both work construction and life, surround the void, i.e. the tunnel. The knowledge of this context with its spatial and, sometime, time variability influences all the steps of the RMP. Therefore, a correct design process must be based and integrated with a correct Risk Management. The design and the risk management must therefore be integrated in one unique rational process that can be defined as "'Risk Analysis-driven

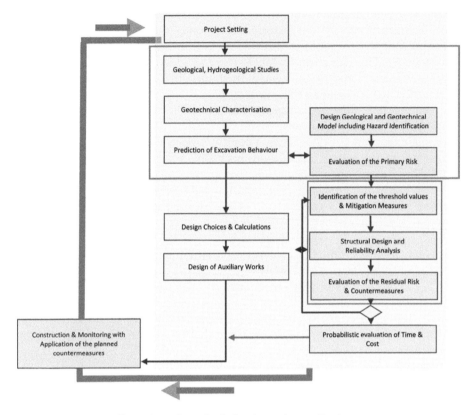

Figure 7.1 Illustrative flow chart for a Risk Analysis-driven Design.

Design' (RAdD, as discussed by Russo (2014)). An example of this design approach is also presented at the beginning of Chapter 8.

To assess the acceptability of the risks to all the involved actors, after all the hazards have been identified, should follow both a qualitative and quantitative evaluation and, after this step, it is possible to plan, to design or to process the countermeasures and to mitigate unacceptable residual risks, thereby making them acceptable.

The combination of hazards and their effects is generally represented by a two-dimensional matrix, called the 'risk space' that is usually divided into regions of different levels of risk. Frequently, the risk space is divided into three zones: a region where the combination of hazard and severity of the event is acceptable since the probability of occurrence or severity or both are low, an unacceptable rock region for which the probability of occurrence or severity or both are high, and the so-called 'as low as reasonably practicable' region in between. The classic representation of risk, using the bidimensional green–yellow–red chart, usually comes from multiplying probability and impact/severity ratings. The final risks can be quantified visualizing the hazards and their effects along the course, chain, of events occurring during tunnel construction. It must be noted that the analysis must follow along further steps, also continuing during the construction phase, based on additional, factual data.

7.4   UNDERGROUND WORKS, GEOLOGY, HYDROGEOLOGY AND GEOTECHNICS

A relevant source of risks comes from an inadequate – before construction – knowledge of the geological and geotechnical conditions (Design Geological and Geotechnical Reference Model – DGGM): several geological, hydrogeological and geotechnical aspects can remain unknown, partially or completely, before actual construction. Quoting the ITA WG17 reports on Long Tunnels (2010),

> at great depth … very little is known about the geological, hydrogeological and geotechnical conditions: the deeper the tunnel, the larger the uncertainties; the higher the probability of encountering adverse or unforeseen conditions for tunnelling, the greater the effort and the cost for site investigations to reduce the uncertainties.

And in the subsequent revision (2017), 'The timely identification of the geotechnical hazards and the understanding of their consequences are essential … to minimize the risks during construction'.

The DGGM must be focused on the analysis of the engineering needs of the project. Because of the unavoidable limits rising from accuracy and completeness with which subsurface conditions may be known, it is also a means for identifying their related variability and uncertainties and the related hazards and risks, providing the basis for planning possible additional site investigation and a correct design procedure as discussed by Soldo et al. (2019). The DGGM can be then described as the framework in which the expected risks are recognized and characterized, and it should include the assessment of its effectiveness and reliability and an estimated reliability degree and evaluation of the related uncertainties and risks. Some approaches also consider the quality of the investigation procedures, while others consider only the quality of the input data and others the quality of the complete model of the underground (Soldo et al., 2019). Generally speaking, the DGGM visualizes, describes and quantifies for a certain volume around the project work (whose extension is a property of the DGGM itself, establishing the extension of the 'influence zone', meaning both what is influenced by the works and symmetrically what is capable of influencing the works) the following features and properties (with their time variation if they could change significantly during the construction and operation life):

- lithological and petrographical characteristics,
- stratigraphical and sedimentological description for sedimentary bodies, lithified or not,
- geomorphological characterization with the prediction of the evolution of active processes or forms (landslide, weathering, erosion, karst and subsidence),
- tectonic and structural features: fault zones (properties and evaluation of their eventual activity), joints at the regional and local scales, folds, natural stress field and seismicity,
- geological bodies' structure (geometry) at depth, with their boundaries,
- hydrogeological context (underground water and gases), and hydrodynamical and hydrochemical characteristics,

- geotechnical (soils) or geomechanical (intact rock and rock masses) properties, at different scales,
- geotechnical or geomechanical behaviour along the project (i.e. for certain stress conditions).

7.5   UNCERTAINTIES AND RISKS

The largest source of uncertainty, as previously said, refers to imperfect and inexact knowledge of the tunnel-related environment and geological context that could arise from the need for simplifications, heterogeneity, inherent randomness, imperfect interpretative concepts and hypotheses, measurement inaccuracies, sampling limitations, insufficient sample numbers and others. At the end, from these limits a limited capability derives to predict the geometry of the geological bodies and structures and their features underground also at various scales (Soldo et al., 2019). These prediction limits may reflect in possible mistakes of the reference model (e.g. the complete missing of a geological formation, an overestimation or underestimation of the size of a geological formation or the erroneous prediction of the contact between the various geological formations).

We have not to forget that all geological data are subject to several sources of uncertainty (as discussed by Wellmann et al. (2010)). This source of uncertainty includes:

- imprecision and measurement error;
- stochasticity and insufficient and/or imprecise knowledge;
- model inadequacy.

Typical examples include uncertainty in raw data, used for modelling (e.g. the position of a formation boundary or the orientation of a structural feature), stochasticity, inherent randomness (commonly the uncertainty in interpolation between and extrapolation from known data) and imprecise knowledge of ore conceptual ambiguities.

Most part of the DGGM (Soldo et al., 2019) derives from speculations, and only a limited amount of it can be based on direct factual data (Soldo et al., 2019; Dematteis & Soldo, 2015; AFTES, 2012). Its reliability is deeply influenced, among others, by:

- the intrinsic complexity of the natural context,
- depth and dimensions of the investigation zone (with them directly increase the limits and the costs of the investigation methodologies),
- the technical limits of the investigation methodologies,
- the time and money budget for the investigation,
- the competence and adequate multidisciplinary of the geologist and geotechnical team.

In addition to the data coming from the factual and interpretative models, the DGGM, matched with the adopted design model, allows the study of possible alternative construction scenarios (managing uncertainty by providing alternative views of the expected conditions, considering combinations of uncertainties and their significant interactions) with a complete risk-based design allowing a complete risk analysis-based design (identifying the possible hazards, the failure modes, their probability and consequences as discussed by Grasso et al. 2016).

Considering the alternative scenarios, the designer should describe a range of future operating environments (a condition of 'possible situations awareness') and identify relevant failure triggers and corresponding tolerance ranges for each scenario. This triggering phenomenon then should be monitored on an ongoing basis, to provide management with advance warning of material changes in the operating environment.

Whenever the predefined triggers and their tolerances are exceeded (meaning that a new scenario becomes effective), tactics using the previously developed plans will be adjusted, generating new forecasts reflecting this change.

7.6   PRELIMINARY ASSESSMENT OF GEOLOGICAL AND GEOTECHNICAL SOURCES OF UNCERTAINTIES AND RISKS

The DGGM establishes a division of the project-influenced space into 'homogeneous' sectors/volumes, associated with different geotechnical behaviours, then also including:

*   geotechnical (soils) or geomechanical (intact rock and rock masses) properties, at different scales,
*   geotechnical or geomechanical behaviour along the project (i.e. for certain stress conditions) with which specific engineering assumptions and solutions are finally associated.

One main component of the design procedure turns around the prediction of the geotechnical behaviour (stress and strain relationship) of the geological materials which host an underground structure. The geotechnical response to the excavation must be treated, as mentioned before, in terms of potential hazards and risks, analysing the potential probability of failure and its impacts, sometimes extended at distance.

Geomechanical properties include the strength and deformability of the intact rock and of the rock mass as a whole. The risk analysis must consider all the relevant aspects related to the intact rock, the rock mass and the context (engineering geological situation, e.g. gas presence) together with the influence of engineering works. It should be noted that some geological aspects can be considered as direct hazards, as pointed out, for example, for toxic or noxious gases. On the other hand, most of the geomechanical hazards must be considered derived hazards, e.g. the presence of weak rocks as a reason for squeezing behaviour.

The instability of rock masses surrounding the tunnel may be divided into three main categories:

1.   Block failure or 'structurally controlled failures', where pre-existing blocks in the roof and side walls become free to move because of the excavation, according to different failure modes (loosening, ravelling, block falls, etc.). Discontinuities (stratification, foliation and joints) govern the mode of rock failure and collapse.

2. Failures induced from overstressing, i.e. the stresses developed in the ground exceed the local strength of the material, occurring in two main forms:
    *   overstressing of massive or intact, medium to hard, rock, occurring in the mode of spalling, popping, rock burst, etc.
    *   overstressing of medium to poor materials, i.e. heavy jointed rocks, where squeezing and creep may take place.
3. Failures induced because of the presence of evaporitic material, causing swelling and/or solubilization.

Squeezing can be defined as a time-dependent shearing of the ground, leading to inward movement of the tunnel perimeter. Squeezing is associated with plastic deformation, governed by both material properties and overstress conditions around the underground excavation. Some effects of squeezing are evident immediately after excavation (i.e. progressive convergences independent of excavation advance), but normally the long-term effect is prevalent, including progressive ground deformation and support loading. Squeezing ground commonly refers to materials displacing into a tunnel, due to the action of the surrounding stress gradient. The effects of squeezing immediately become evident during an excavation as closure starts to take place at the tunnel face.

Fragile, brittle rapid failure is widely described in the specialist literature as 'spalling' or 'bursting'. The definition of spalling is associated with rock damage (with the development of visible fractures) and consequent failure processes near the excavation boundaries under high stress. Spalling can be either violent or non-violent and, in some cases, can be time dependent. With rock bursting and strain bursting, it is usually described as the violent rupture of a volume of rock walls under high stresses.

Swelling behaviour implies the response of soils to stress due to change in the presence of water due to a significant swelling of some minerals (clays, such as montmorillonite and, to a lower extent, hillite and kaolinite; evaporitic minerals, such as anhydrite). Both squeezing and swelling are stress- and strain-related phenomena, with a time-dependent volume increase in the case of swelling.

Running ground and liquefaction refer to the case of soil particles freely moving; for example, in the presence of loose sands running ground phenomena can occur, both under dry conditions and more frequently under saturated conditions, when liquefaction takes place.

A specific attention must be paid to those risks related to the hydrogeological context.

Within rock masses, in relation to the variations of water conductivity and flow entity with depth three main zones can be schematically distinguished, with the depth of the transition boundaries extremely variable (because of several complex reasons, e.g. the rock mass types and structural characteristics, and the natural stress field), both gradual or sharp: (i) a deep zone, almost impervious, (ii) an intermediate zone, with most of the flow moving along fracture zones, and (iii) a near-surface zone, unsaturated, drained because of the intense network of open (because of decompression) fractures. Along the deep zone, the main joints and fractures/fault sectors, if un-filled with clays (related to alteration/weathering, e.g. because of hydrothermal alteration, or intense crushing along shear zones, e.g. 'gouge'-type fault rocks), contain water under pressure (the pressure is a fraction of the total hydrostatic overburden). Most of the deep waters stand without flowing, then reaching a substantial equilibrium (thermic and chemical) with the surrounding. The Void Index is low (<1%), and the overall conductivity, far away from the fracture zones, is near to be null. At large depths, tectonic stresses became significant, exceeding the hydrostatic pressure. In tectonic compressive areas, this means that most of the fractures tend to be closed. Nonetheless, some main fracture zones, particularly in areas undergoing tectonic distention, can exhibit significant conductivity, with a certain amount of water flow. In the intermediate zone, at lower depths (the depth must be considered in terms of distance from the surface, and then with complex variation considering the slope topography), the tectonic stresses decrease, with a variable degree of decompression, making the existence of networks of open fractures possible, with possible water flow, at variable velocities, along large distances. The water circulation varies, increasing at lower depths. The water pressure is progressively more related to the hydrostatic load (then with possible relevant variations related to topography). Near the topographic surface, there exists a decompressed (then with open fractures because of the stress release related to both the exhumation and erosion and, somewhere, with the load rebound after the Holocene glaciers melting) unsaturated zone, drained because of the fracture opening. A variable, discontinuous piezometric surface, also changing with seasons, limits at depth this zone. The waters coming from rainfall, lake and rivers or from snow and glaciers melting only flow through this zone, then reaching the stable aquifer limited by the piezometric surface. This zone disappears where the piezometric surface emerges. In each of the described main zones, the water flow can be classified into some main reference models from a dominant single fracture, from fracture networks and from fault zones, each of them varying considering the presence/ratio of pervious/impervious materials.

Soils deposits can be classified with respect to the type of the constituent particles: clastic sediments are those derived directly as particles broken from a parent rock source, and non-clastic sediments are from the newly created mineral matter precipitated from chemical solutions or from organic activity.

Because several of the largest historical cities worldwide were built along rivers, where the most common grounds in temperate regions are alluvial deposits, most of the existing risk analysis for urban tunnelling revolves around the understanding of subsidence effects. Nonetheless, also in these environments the geological complexity is not negligible. Apparently, simple Quaternary deposits can show abrupt horizontal and vertical changes (in lithology, structure and texture as well as in hydrogeology) because of the rapidly changing sedimentary processes. Coastal sedimentary deposition is, for example, influenced by tidal phenomena, changing river runoff and long-term cycles of global sea level changes all of them influenced by glaciations. Glacial-related deposits similarly show a complicated sequence, bearing boulders, that varies from unconsolidated, fine, lacustrine sediments with peat to concrete-like tills. But today this already significant complexity is increasing; urban tunnels are now driven into strongly varying, differing geological conditions, from weathered tropical soils and bedrock to karstified rock masses, or inside piedmont coastal alluvial fan with a related high degree of geotechnical and hydrogeological complexity. Most of the cities in Asia, because of the past and present climate conditions, lie on soils covering weathered bedrock. The unpredictable mixed conditions, often abruptly varying from soils, eventually with embedded rock blocks, or weathered or sound bedrock are a major challenge to both the tunnel designers and the builders. Fault sectors (that usually exhibit poor geomechanical properties and frequently represent a preferred path for groundwater) in urban areas are difficult to predict because most or all the territory is hidden under asphalt and buildings. Discontinuities (stratification, foliation and joints) influence the mode of rock failure and collapse. Tunnelling inside karstified rock masses (mostly in carbonate rocks) must face with cavities of different sizes, filled or empty, sinkholes (over time the roof of cavities may dissolve or collapse triggering sinkholes or depressions at the ground surface) and frequently heavy water inflows, in particular in coastal cities. Tunnels could cross the soil–bedrock contact, running among sharply varying pinnacles, cliffs and ravines (karstified rockhead). Criticisms could, for example, arise for tunnels in karstified limestone, with possible sudden, huge water and debris inrush or uncontrolled loss of pressure when TBMs are employed.

Risks can arise from the hydrogeological context, in terms of hydrodynamic parameters (governing the groundwater flow), geometric characteristics (the hydro-structure) and water chemistry. Aggressive waters (carbon dioxide, sulphates and chlorides) influence the design of the concrete mix (pre-cast segments and cast-in-place concrete), foams and additives for TBMs (for torque reduction, soil plasticization, reduction in permeability, protection from stickiness, consolidation and void filling, anti-clogging agents, tail sealants, anti-wear and dust suppression). From an environmental point of view, considering the impacts on water resources is mandatory.

The occurrence of swelling or soluble rocks (as it is the case inside evaporitic sequences or expansive clay-bearing deposits) is of significance during both construction and operation of the tunnel. The modern TBMs offer today reliable solutions for undersea (and river) urban tunnels. Nevertheless, their excavation could require special attention, for example, where tunnels pass from sound rocks to loose, young unconsolidated deposits (crossing, for example, a river), with possible phenomena of settlement, sinking or flotation of the TBM and tunnel itself. The presence of rocks bearing noxious or radioactive minerals must be carefully predicted for the adoption of the necessary countermeasures for both workers and the surrounding environment, with an increased level of risks in densely populated urban areas.

Settlement phenomena can arise due to the loss of ground at a tunnel face, ineffective filling of the tail voids, water inflow with the soil inrush and poor ground control at the shield: damage to already existing structures (buildings and utilities) can occur after settlement phenomena. In the presence of cyclic loads (usually in the case of earthquakes or humanly induced vibrations), some saturated soils, with prevalence of sand, with a low overburden can be subject to liquefaction (annulment of the effective forces and instantaneous collapse of the shear strength) with the relative collapse of the ground and propagation of the strain phenomena to the surface.

Quartz-rich soils with sharp and angular grains may be highly abrasive for the TBM cutter head, the cutting tools, screw conveyor and some other parts of the machine in contact with the soil or the outcoming muck.

Adhesion and clogging, in soils with high prevalence of fine grains, containing clayey minerals, can affect (the so-called 'sticky behaviour') the tools, the walls of the excavation chamber and the mucking plants. Furthermore, clay clogging is a source of problems to TBM steering, mucking and cutter head rotation.

Within several types of soils (glacial deposits, mega-fan, landslide, etc.) or along the contact with weathered bedrocks, TBMs must frequently handle the presence of embedded boulders of various sizes and hardness values, often with relevant problems for their crushing and mucking off the excavation chamber.

In many over-consolidated or extremely dense or cemented soils, it is not uncommon to find voids (empty or filled, with or without water) of different dimensions (from little voids or conduits, sometimes cavities). They may originate from man-made caves or wells, but they may be also natural piping conduits (progressive grains wash out along drainage circuits) or dissolution (from weathering) voids.

7.7   PRELIMINARY ASSESSMENT OF OTHER SOURCES OF UNCERTAINTIES AND RISKS

Several risks can be related to large-scale geodynamic phenomena, particularly because of their potential capability of damaging the underground structures during their operation life and, sometimes, during the construction phases. Hereinafter, a synthetic list of some of the most common of these risks is given.

Earthquakes are among the most important geodynamic perils, even if underground structures are considered less vulnerable than surface structures. They cause transient ground motion and permanent ground deformation, both of which affect tunnels and underground structures. Transient ground motion (travelling body P and S waves and surface waves, especially Rayleigh waves), especially velocity pulses of strong motion, can develop shear and tensile strains.

The principal causes of permanent ground displacement are faulting, where the tunnel intercepts active structures, tectonic uplift and subsidence, liquefaction, post-liquefaction consolidation, buoyancy effects, landslides (lateral spread, flow failure, etc.) and densification of loose granular deposits. Tunnels and underground structures are mostly affected by ground settlement due to liquefaction. Earthquake-induced subsidence can result from the densification of granular soils, consolidation of clays and liquefied soils.

Regional subsidence related to earthquakes, pumping of water, oil or gases, or even due to subsurface mining can cause major damages to underground structures, in terms of direct deformation or increasing indirect risks, such as flooding. If this phenomenon is caused by earthquakes, the subsidence can occur rapidly, with large damages also during construction.

Tsunamis are most often generated by earthquake-induced movement of the ocean floor, sometimes by landslides and volcanic eruptions. Associated risks for underground facilities include extreme flooding, water contamination, fires from ruptured tanks or gas lines and induced landslides, both during tunnel life and eventually during the construction phase.

Flooding and landslides (including lahar) also can damage underground structures (and the connected surface structures) during both the operation period and the construction phase.

During next years, climate change could have wide-reaching effects on infrastructures. Beyond the physical threats from climate changes, underground infrastructures stand to face an array of additional risks (increased risk of extreme flooding phenomena)

related to the use of underground spaces for the provision of basic services and public goods (water supply, physical infrastructure, transport, energy, etc.). Many coastal
cities will suffer extensive problems because of sea level rising. This phenomenon is
affecting many areas worldwide, and it is expected to accelerate. Increasing rates of
sea level rise caused by global warming are expected to lead to permanent inundation,
episodic flooding, beach erosion and saline intrusion in low-lying coastal areas. Impacts
include intrusion of salt, aggressive, waters; increase in the water table level; reduction
in capacity of flood control structures; and landward migration of freshwater wetlands.
For cities with bedrock of karstified limestone, impacts could be significantly worst. All
these aspects will impact progressively the existing tunnels, and also they are strongly
and rapidly changing in some areas in the construction conditions.

Volcanic eruption could represent an uncommon, rare, but potentially destructive
event, again during both the operation period and the construction phase. Impacts
could be different for the different types of eruption (with a broad distinction between
effusive and explosive phenomena).

7.8   A SPECIAL FOCUS ON ROCK BURST AND SQUEEZING HAZARDS

In the following pages, a detailed description of two specific types of geomechanical
hazards: rock burst and squeezing, is given. After their description, a description of
some possible risk analysis approaches follows. Rock burst and squeezing are the two
main modes of underground instability caused by the overstressing of the ground, the
first occurring in brittle hard rocks and the second in ductile soft rocks.

7.8.1   Rock burst

Rock burst, strain burst and *spalling* classification has been detailed in Chapter 3.

7.8.1.1   Spalling susceptibility: shear vs spalling (brittle) failure

According to Diederichs et al. (2010), *spalling* susceptibility mainly depends on the
geostructural rock mass conditions and the intact rock brittleness, as, respectively,
represented in Figure 7.2 and Table 7.1 by the Geological Strength Index (GSI) (Hoek
et al., 1995) and by the ratio index UCS/T (where UCS = intact rock strength and
T = intact rock tensile strength).

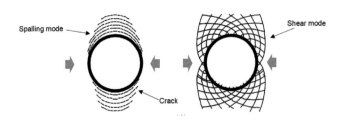

Figure 7.2 Comparison between spalling and shear failure mode involved in plasticity.

Table 7.1 Transition from brittle spalling to rock mass shear failure (from Diederichs (2017))

BI (Brittle Index)	GSI < 55	GSI = 55–65	GSI = 65–80	GSI > 80
UCS/T < 8	Shear	Shear	Shear	Shear
UCS/T = 9–15	Shear	Shear	Shear/spall	Spall/shear
UCS/T = 15–20	Shear	Shear/spall	Spall/shear	Spall
UCS/T > 20	Shear	Shear/spall	Spall	Spall

GSI, geological strength index (Hoek et al., 1995); UCS, intact rock strength; T, intact rock tensile strength.

On the basis of the combination of GSI and Brittle Index, different failure mechanisms are expected to occur and, consequently, appropriate geomechanical modelling needs to be applied in numerical simulation for design. As later remarked in Figure 7.5, for massive or moderately jointed rock (→high GSI) with a high UCS/T ratio, brittle spalling damage initiates at a wall stress of around 40%–60% of UCS, following the 'spalling criterion' in Figure 7.5.

7.8.1.2   Strain burst Potential (SBP) and Strainburst Severity (SBS)

According to Cai and Kaiser (2018), two conditions must be met for a strain burst to occur, largely defining its potential (SBP):

1. The rock mass must fail in a brittle manner, i.e. it must display a high intrinsic brittleness (Figures 7.5 and 7.6).
2. A high level of tangential stress must build up in the skin of the excavation.

The same authors remark that four additional factors contribute to the Strain burst Severity (SBS):

1. The volume of rock that actually bursts (which is called the burst volume).
2. The energy imposed by the surrounding rock mass through forces and deformations acting on the burst volume.
3. The energy consumed during the failure, i.e. the deformation of the reinforced volume of the fractured rock.
4. The volume increase (bulking) due to stress fracturing of the burst volume.

Diederichs (2017) proposed to estimate the Rock burst Potential by the assessment of Intact Rock brittleness and UCS (Figure 7.3) by introducing the Dynamic Rupture Potential (DRP). As it is possible to observe, the graph allows us also to derive a (rough) indication of the potential rock block ejection velocity.

Consistently with previous definitions, Cai and Kaiser (2018) remarked that the UCS (Y-axis) in Figure 7.3 is an indicator of the Strain burst Severity (SBS), while brittleness (X-axis) relates to spalling/strain burst Potential (SBP). Moreover, these authors observed that simple relations as in Figure 7.3 are suitable to assess in a preliminary manner SBP and SBS, but only for conditions that are not affected by mining

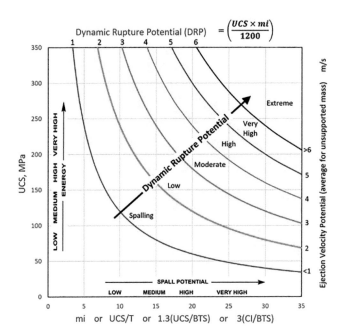

Figure 7.3 Dynamic rupture potential (DRP) indicator for massive rock (Diederichs, 2017). The horizontal axis proposes several indicators for rock brittleness.

and other factors that lower the loading system stiffness (LSS), that is the ratio of an induced stress change to the related strain increment in a rock volume. The lower the LSS or the softer the mine stiffness, the higher the energy input from the surrounding rock mass. Accordingly, the figure therefore is only quantitatively applicable for relatively high LSS values.

7.8.1.3   Rock burst damage severity

According to the Canadian Rockburst Support Handbook (CRSH Kaiser et al., 1996), the rock burst severity damage level is classified by the thickness of fractured rock as illustrated in Figure 7.4.

Moreover, Potvin (2009), as reported and modified by Cai and Kaiser (2018), proposed the Rock burst Damage reported in Table 7.2, as a function of the involved rock mass and surface.

Based on this rock damage definition shown in the figure above, the Canadian Rockburst Research Handbook (CRSH, 1996) provided indications for rock burst damage mechanisms and classification as in Table 7.3.

7.8.1.4   Damage Index (DI) and Depth of Failure (DOF)

In Figure 7.5, the basic empirical relationship between the Damage Index or Stress Level (s_{max}/UCS or s_{max}/UCS*) and Depth of Spalling (r/a) is shown (Diederichs et al., 2010).

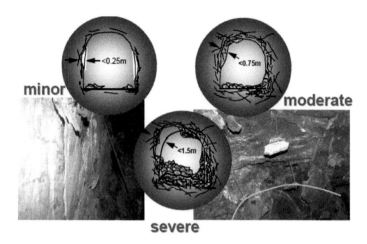

Figure 7.4 Rock burst damage severity level (as per CRSH).

Table 7.2 Rock burst damage scale proposed by Potvin (2009); Modified by Cai and Kaiser (2018)

Rock burst damage scale	Rock mass damage	Damaged surface area	Rock support damage
R1	No damage, minor loss	0	No damage
R2	Minor damage, less than 1 t displaced	$<1\,m^2$	Support system is loaded; loss in mesh; and plates deformed
R3	1–10 t displaced	$<10\,m^2$	Some broken bolts
R4	10–100 t displaced	$10–50\,m^2$	Major damage to support system
R5	100+ t displaced	$>50\,m^2$	Complete failure of support system

The reference classification for kinetic energy is reported in Table 7.4 (CRSH, 1996).

It is important to note that the fitting curve in Figure 7.5 is based on the maximum depth of failure recorded; moreover, the prevalent cases collected by Martin et al. (1999) refer to non-violent events in unsupported tunnels. From this figure, Table 7.5 can be derived. Recently, Kaiser and Cai (2018, with reference to Perras & Diederichs, 2016) have indicated that the mean depth of failure (DOF_{mean}) is about 20%÷30% of the maximum depth (DOF_{max}) provided by Figure 7.5, or in other terms (Figure 7.6):

$$DOF_{mean} \approx DOF_{max}/(3.5 \div 4.5)$$

7.8.1.5 Dynamic Rock burst Hazard

As it is possible to derive from previous items, the rock burst hazard is mainly related to the estimated likely depth of damage or yield in the rock (Figure 7.5) and the potential for that failure to be brittle and with high energy release (DRP in Figure 7.3).

Table 7.3 Rock burst damage mechanism (CRSH – Kaiser et al., 1996)

Damage mechanism	Damage severity	Cause of rock burst damage	Thickness [m]	Weight [kN/m2]	Closure^a [mm]	ve [m/s]	Energy [kJ/m2]
Bulking	Minor	Highly stressed rock	<0.25	<7	15	<1.5	Not critical
Without	Moderate	With little excess	<0.75	<20	30	<1.5	Not critical
Ejection	Major	Stored strain energy	<1.5	<50	60	<1.5	Not critical
Bulking	Minor	Highly stressed rock	<0.25	<7	50	1.5–3	Not critical
Causing	Moderate	With significant	<0.75	<20	150	1.5–3	2–10
Ejection	Major	Excess strain energy	<1.5	<50	300	1.5–3	5–25
Ejection by	Minor	Seismic energy	<0.25	<7	<150	>3	3–10
Remote	Moderate	Transfer to	<0.75	<20	<300	>3	10–20
Seismic event	Major	Jointed or broken rock	<1.5	<50	>300	>3	20–50
	Minor	Inadequate strength	<0.25	<7g/(a+g)	NA	NA	NA
Rockfall	Moderate	Forces increased	<0.75	<20g/(a+g)	NA	NA	NA
	Major	By seismic acceleration	<1.5	<50g/(a+g)	NA	NA	NA

ve is the velocity of displaced or ejected rock; a and g are seismic and gravitational accelerations.
^a Closure expected with an effective support system.

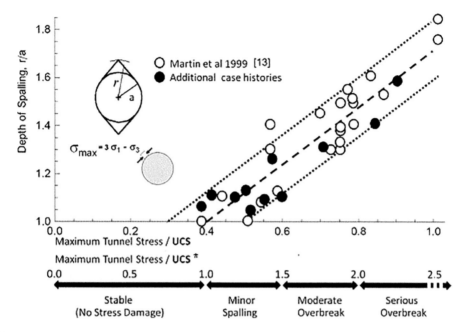

Figure 7.5 Empirical chart for estimating the normalized maximum Depth of Damage around a tunnel (Diederichs et al., 2010). Note: UCS* = CI is the Crack Damage Initiation threshold (in the figure UCS* = 0.4 UCS). The linear fitting is expressed by the equation r/a = 0.49 + 1.25(s_{max}/UCS) with a = tunnel radius. The normalized Depth of Failure DOF/a = 1.25(s_{max}/UCS)−0.51 with DOF = r−a.

Table 7.4 Damage intensity based on kinetic energy

Kinetic energy (kJ/m^2)	Damage intensity
<5	Low
5–10	Moderate
10–25	High
25–50	Very high
>50	Extreme

Diederichs (2017) proposed a method to take into account both these conditions for estimating the rock burst hazard.

The approach is based on the following basic estimations:

1. Damage Depth Potential (DDP) = a^*(DOF)$^{0.5}$, where DOF is the estimated Depth of Failure (m).
2. Burst Hazard Potential (BHP) = β^*(DRP*DDP)$^{0.5}$.

Here, α (= 4.5) and β (= 1) are empirical constants (in parentheses are the suggested values).

Table 7.5 Classification of damage intensity

Hazard for brittle rock mass	Minor spalling	Moderate overbreak	Serious overbreak	Very severe overbreak[a]
s_{max}/UCS	0.4–0.6	0.6–0.8	0.8–1	>1
s_{max}/UCS*(=CI)	1.0–1.5	1.5–2.0	2.0–2.5	>2.5
DOF/a (max) ($\approx$)	0.25	0.5	0.75	>0.75

[a]   Very severe overbreak class is added with respect to the indications in Figure 7.5.

Figure 7.6 Example of large overbreak resulting from overstressed zone in the crown of an adit tunnel for hydroelectric plant in Andean region (Russo, 2019).

Mainly on the basis of CRSH classification and other practical experiences, the hazard levels and related indicative conditions in Table 7.6 are proposed for a preliminary assessment.

Indications provided are useful for preliminary setting of design strategy, by considering limitations on DRP assessment previously remarked. A more comprehensive analysis can be developed with main reference to the in-progress Rockburst Support – Reference Book (Cai & Kaiser (2018) and following), in which, among others, the design rationale published in the CRSH is strongly revised, by shifting from a 'ground-motion-centric' and then energy-based to a deformation-based approach.

The previous description of the *rock burst*'s phenomena suggests a necessary parenthesis. As discussed in the previous chapters, natural phenomena are intrinsically complex, with the related prediction difficulties and uncertainties. Always, after the preliminary assessment of the risks, as a specific and foundational part of the Risk Management procedure, the design and construction phases must include a detailed and comprehensive monitoring programme, with attention and alarm levels, with

Table 7.6 Classes of dynamic rupture (burst) hazard potential (BHP)

Class	Hazard level	Issues
0	Low stress	No indications.
0.1	Local stress concentration	Minor local and intermittent stress noise.
0.5	Consistent stress	Consistent stress popping for several hours after blast.
1	Spalling	Visible slab formation developing after blast and scaling.
2	Minor dynamic rupture (burst)	Slabbing with noise, minor ejection (5–20 cm thick). Less than 2 kJ/m2 kinetic energy of release. Less than 5 cm closure potential.
3	Moderate dynamic rupture (burst)	Constant or strong noise with high frequency of ejection 20–80 cm thick @ > 1.5 m/s velocity for hours after blasting. Kinetic energy (2–10 kJ/m2) or 5–15 cm closure potential.
4	Major dynamic rupture (burst)	Large volumes of ejection or dynamic heave (floor) with constant and large noise events, release possible well after blast and back into tunnel away from face >80 cm thick @>3 m/s or 10–30 kJ/m2 ejection or 15–30 cm closure potential.
5+	Extreme bursting	Very large events >120 cm thick, >4 m/s velocity, kinetic energy >30 kJ/m2, >30 cm closure potential.

countermeasure. For example, for rock burst, monitoring includes the detection of Acoustic Emissions.

7.8.2 Squeezing

As previously described, with rock squeezing a condition of large time-dependent convergence during underground excavation is generally identified. It takes place when – because of the material properties and the induced stresses – in a certain zone around the tunnel, the limiting shear value is exceeded, at which the creep deformation of the rock mass starts. The deformation may terminate during construction or continue over time.

Various empirical or analytical–empirical methods have been proposed to predict high deformations and squeezing, some of them connected to the prediction of face stability.

The preliminary estimation of the risk of squeezing (Table 7.2) has been based on more approaches, following Jehtwa and Singh (1984), Bhasin (1994) and Marinos and Hoek (2000). The tunnel stability has also been analysed based on Panet (1995).

Similarly, the qualitative estimation of the risk of brittle, fragile rapid failure (Table 7.3) has been based on Zhen-Yu (1988) and Hoek et al. (1995) (Tables 7.7 and 7.8).

Table 7.7 Empirical–analytical methods for the estimation of the squeezing intensity and location

Squeezing indexes and face stability (Panet)

Jehtwa and Singh (2000)	No squeezing	$N_c = \dfrac{q_{cm}}{P}$	<0.4
	Mildly squeezing		0.4–0.8
	Moderately squeezing		0.8–2
	Highly squeezing		>2
Bhasin and Grimstad (1996)	No squeezing	$N_t = 2P/q_{cm}$	$N_t t < 1$
	Mild to moderate squeezing		$1 < N_t < 5$
	Highly squeezing		$N_t > 5$
Hoek and Brown (1980)	Few stability problems	p_z/σ_{ci}	0.1
	Minor squeezing		0.2
	Severe squeezing		0.3
	Very severe squeezing and face stability		0.4
	Extreme squeezing		0.5
Panet (1995)	Elastic	$N = 2P/q_{cm}$	$N < 1$
	Partially elastic		$1 < N < 2$
	Plastic	$\lambda_E = \dfrac{1}{4N}\left(\sqrt{m_b^2 + 8m_b N + 16s} - m_b\right)$	$2 < N < 5;$ $0.6 < \lambda e < 1.0$
	Stable – stable only in the short term		$N > 5; \ 0.3 < \lambda e < 0.6$
	Unstable		$N > 5; \ \lambda e < 0.3$

Note: p_z, vertical stress; σ_{ci}, uniaxial compressive strength of intact rock; P, vertical load stress; q_{cm}, rock mass compressive strength; m_b and s rock mass Hoek and Brown constants.

Table 7.8 Empirical methods for the estimation of the fragile, brittle rapid failure intensity and location

Rock burst indexes

Zehn-Yu (1988)	No rock bursting	σ_{ci}/σ_i	<2.5
	Low rock bursting activity		2.5–5.5
	Moderate rock bursting activity		5.5–13.5
	High rock bursting activity		>13.5
Hoek (2000)	Stability	p_z/σ_{ci}	0.1
	Spalling		0.2
	Severe spalling – slabbing		0.3
	Need important stabilization measures		0.4
	Cavity collapse (rock burst)		0.5

Note: p_z, vertical stress; σ_{ci}, uniaxial compressive strength of intact rock; σ_i, major stress component (geostatic field).

7.8.3 Illustrative approaches for a preliminary analysis of rock burst- and squeezing-related risks

7.8.3.1 Example 1 for a deep alpine tunnel

It is first considered the risk analysis of squeezing and rapid brittle failure phenomena along a deep alpine tunnel. The Geological and Geotechnical Model (refer Section 7.4) settled during the design phase visualizes, describes and quantifies for a certain volume around the project works the geological and hydrogeological features and properties. Then, considering these aspects, at the end, based on the model, a division of the project into 'homogeneous' sectors/volumes has been established, associated with different geotechnical behaviours, and then with specific design assumption and solutions, also including:

•   geotechnical (soils) or geomechanical (intact rock and rock masses) properties, at different scales, with their statistical variability;
•   geotechnical or geomechanical behaviour along the project (i.e. for certain stress conditions).

Some of the properties identified for the rock masses along the tunnel has then been analysed within the perspective of a Risk Analysis and Management-driven Design, considering the potential hazards, the potential consequent risks (reference scenarios) and a set of applicable countermeasures (risk mitigation) (Figure 7.7). Here, as said, it is considered, for example, the specific risks associated with squeezing and brittle, fragile failure.

Each rock mass type (i.e. geological unit, with its lithologies and structural features) has been classified into geomechanical classes using, e.g., the Bieniawski (1989) Rock Mass Rating classification and/or the GSI (Hoek, 1997 and mod.), and a set of characteristic geomechanical properties has been associated with each of them. For each of the identified rock mass classes and vertical stresses along the tunnel (p_z vertical stress, varying with the overburden), based on the above-described empirical–analytical methods, the potential risk of squeezing, face instability and rock burst along the tunnel (Table 7.9) has been estimated and then divided into zones, each with an

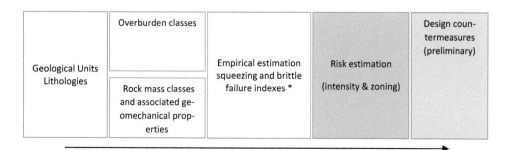

Figure 7.7 Conceptual scheme of the Risk Analysis-based Design (geotechnical risks).

Table 7.9 Exemplifying theoretical estimation of the potential risk of squeezing and fragile, brittle rapid failure (preliminary Risk Analysis) along the tunnel (tunnel zonation of risk), for each rock mass type (geological unit, lithology and structure), geomechanical class (Bieniawski's (1989) Rock Mass Rating) and vertical stress (overburden range) (Brenner Base Tunnel) (Marini et al., 2019)

Rock mass name	Class	Max. overburden (m)	Squeezing	Face stability	Rock burst
Rock mass I	I	1,060			
	II	900			
	II	700			
	III	1,200			
	IV	1,300			
	IV	600			
	V	1,300			

Green, no risk; pale yellow, low risk; yellow, moderate risk; red, high risk.

expected, specific level of risk and suggested countermeasures in terms of excavation procedures, temporary lining and final lining.

7.8.3.2 Example II for a deep Andean tunnel

As a further example of the preliminary geomechanical risk analysis, the "Multiple graph" (Figure 7.5) approach proposed by Russo (2014) is applied. The analysis refers to a deep Andean tunnel where both squeezing and brittle failure are expected to occur.

The Multiple Graph is divided into four sectors, based on the logical sequence of the following engineering steps, by proceeding clockwise from the first (bottom-right) to the final graph (top-right):

- Graph 1: Rock block volume (Vb) + Joint conditions (J_c) = Rock mass fabric (GSI).
- Graph 2: Rock mass fabric (GSI) + Strength of intact rock (σ_c) = Rock mass strength (σ_{cm}).
- Graph 3: Rock mass strength (σ_{cm}) + In situ stress (H) = Competency (CI).
- Graph 4: Competency (CI) + Self-supporting capacity (RMR) = Excavation behaviour → Potential geomechanical hazards.

The caving field identifies generic gravitational collapse of highly fractured volumes of rock mass from the cavity and/or tunnel face. Therefore, given their very poor self-supporting capacity, the highest risk of caving is associated with the most unfavourable RMR classes. Squeezing describes pronounced time-dependent deformations and is generally associated with rocks with low strength and high deformability, such as phyllites, schists, serpentines, mudstones, tuffs, certain kinds of flysch and chemically altered igneous rocks. The spalling and rock burst region identify the risk of brittle phenomenon. It is important to observe that the depth of failure (DOF) does not necessarily imply (or not only) a violent phenomenon (rock burst), which mainly depends on the rock strength and its related capacity to store energy. The potential of rock wedge failure is mainly associated with good (or fair) rock masses subjected to relatively low stress condition, i.e. when the response at excavation is dominated by the shear strength of discontinuities and a 'translational' failure should occur. Two 'improbable' zones have also been marked in the graph corresponding to unrealistic combinations between GSI and RMR: the first below the 'spalling/rock burst' region and the other in the upper right part ('caving' zone), where RMR class V and elastic behaviour theoretically overlap.

The graph in Figure 7.8 refers to the preliminary analysis for a tunnel zone 740 m long, crossing igneous rock masses with an overburden of about 1,000 m, with anisotropic in situ stress ($k = 1.5$) (Russo et al., 2014).

The hypothesized variability (based on a probabilistic analysis, Monte Carlo method) of the geomechanical parameters GSI (Geological Strength Index, Hoek et al. (1998)) and rock mass strength is plotted in the graph, giving a cloud of points of 'possible conditions'. Given the high variability of the rock mass quality (GSI~20÷80) and generalized overstress conditions, the analysis highlights a certain level of moderate to severe spalling/rock burst risk, with a dominant risk of severe to very severe squeezing. Note that a fictitious overburden has been considered to reproduce the same tangential stress in isotropic conditions.

It must be observed that the Multiple Graph approach is also useful in the construction phase, to select at the tunnel face the support section type to be applied in function of the encountered geomechanical conditions. Consequently, in the fourth quadrant the predefined field of application of the support section types must be remarked according to the design criteria of reference.

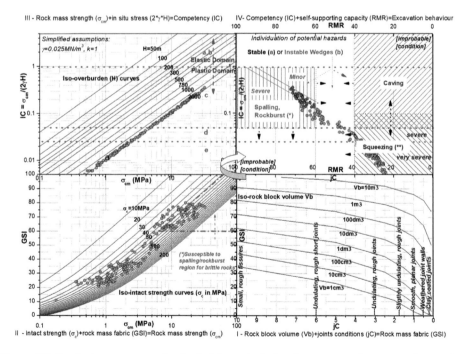

Figure 7.8 Preliminary geomechanical risk analysis for igneous rock masses along a base tunnel (overburden of about 1,000 m), with anisotropic in situ stress, based on the Multiple Graph approach (Russo, 2014).

AUTHORSHIP CONTRIBUTION STATEMENT

The chapter was developed as follows. Soldo: chapter coordination and § 7.1, 7.2, 7.3, 7.4, 7.5, 7.6, 7.7, 7.8.2, 7.8.3.1; Pizzarotti: § 7.1, 7.2, 7.3, 7.4, 7.6, 7.7, 7.8.2; Russo: §7.8.1, 7.8.3.2. The editing was managed by Soldo.

REFERENCES

AFTES Recommendation GT32.R2A1 (2012) Recommendations on the characterisation of geological, hydrogeological and geotechnical uncertainties and risks. *Tunnels & Espace Souterrain*, 232, 315–355

Bhasin, R., Grimstad, E., (1996). The use of stress-strength relationships in the assessment of tunnel stability. *Tunneling and Underground Space Technology*, 11(1), 93–98.

Bieniawski, Z.T. (1989) Engineering rock mass classifications: a complete manual for engineers and geologists in mining, civil, and petroleum engineering.

Cai, M., Kaiser, P.K. (2018) *Rockburst Support Reference Book*. Available from: http://www.imseismology.org/imsdownloadsapp/webresources/filedownload/Cai-Kaiser.

Dematteis, A., Soldo, L. (2015) The geological and geotechnical design model in tunnel design: estimation of its reliability through the R-Index, *Georisk: Assessment and Management of Risk for Engineered Systems and Geohazards*, 9(4), 250–260.

Diederichs, M.S. (2017) Early assessment of dynamic rupture and rockburst hazard potential in deep tunnelling. SAIMM 2017 in Capetown (Afrirock).

Diederichs, M.S., Carter, T., Martin, C.D. (2010) Practical rock spall prediction in tunnel. *Proceedings of World Tunnelling Congress'10 Vancouver*.

Grasso, P.G., Pescara, M., Soldo, L. (2016) Risk management in tunneling: a review of current practices and needs for future development from the designer's perspective. *International Tunnelling Association World Congress*, San Francisco.

Hoek, E., Brown, E.T. (1980) *Underground Excavations in Rock*. London: Institution of Mining and Metallurgy.

Hoek, E., Kaiser, P.K., Bawden, W.F. (1995) *Support of Underground Excavations in Hard Rock*. Rotterdam: Balkema, p. 215.

Hoek, E. and Brown, E.T. 1997. Practical estimates or rock mass strength. Intnl. J. Rock Mech. & Mining Sci. & Geomechanics Abstracts. 34(8), 1165–1186.

Hoek, E., Marinos, P., Benissi, M. (1998) Applicability of the GSI classification for very weak and sheared rock masses: the case of the Athens schist formation. *Bulletin of Engineering Geology and the Environment*, 57(2), 151–160.

ITA Working Group no. 2 (2004) Guidelines for tunnelling risk management. *Tunnelling and Underground Space Technology*, 19, 217–237.

ITA Working Group n°17 Long Tunnels at Great Depth (2010) Report n.4.

ITA Working Group n°17 Long Tunnels at Great Depth (2017) TBM excavation of long and deep tunnels under difficult rock conditions. Report n.19.

Jehtwa, J.L., Singh, B. (1984) Estimation of ultimate rock pressure for tunnel linings under squeezing rock conditions – a new approach. *Proceedings of the ISRM Symposium on Design and Performance of Underground Excavations*, Cambridge, UK.

Kaiser, P.K., Cai, M. (2018) Rockburst Support volume 1 – Rockburst phenomenon and support characteristics MIRARCO Mining Innovation, Laurentian University, Sudbury, Canada.

Kaiser, P.K., Tannant, D.D., McCreath, D.R. (1996) *Canadian Rockburst Support Handbook*. Sudbury, Ontario: Geomechanics Research Centre, Laurentian University.

Marini, D., Pizzarotti, E.M., Rivoltini, M., Zurlo, R. (2019) Brenner base tunnel, Italian side: mining methods stretches – design procedures. *Proceedings of World Tunnel Congress*, Naples.

Marinos, P., Hoek, E. (2000) GSI: A geologically friendly tool for rock mass strength estimation. *Proceedings GeoEng2000 Conference*, Melbourne, 1422–1442.

Martin, C.D., Kaiser, P.K., McCreath, D.R. (1999) Hoek–Brown parameters for predicting the depth of brittle failure around tunnels. *Canadian Geotechnical Journal*, 36, 136–151.

Nassim, T. (2007) *The Black Swan*. New York: Random House.

Panet, M. (1995) Calcul des tunnels par la méthode convergence-confinement. Presses de l'école nationale des ponts et chausses, Paris, France.

Potvin, Y. (2009). Strategies and tactics to control seismic risks in mines. *Journal of the Southern African Institute of Mining and Metallurgy* 109 (March), 177–186.

Russo, G. (2014) An update of the "multiple graph" approach for the preliminary assessment of the excavation behaviour in rock tunnelling. *Tunnelling and Underground Space Technology*, 41, 74–81.

Russo, G. (2019) Severe rockburst occurrence during construction of a complex hydroelectric plant. *Lecture for the 29th Mine Seismology Seminar- IMS Institute of Mine Seismology*, 5–7 May 2019, Phoenix, AZ.

Russo, G., Grasso, P.G., Verzani, L.P., Cabañas, A. (2014) On the concept of "Risk Analysis-driven Design". *Proceedings of the World Tunnel Congress - Tunnels for a better Life*, Foz do Iguaçu, Brazil.

Soldo, L., Vendramini, M., Eusebio, A. (2019) Tunnels design and geological studies. *Tunnelling and Underground Space Technology*, 84, 82–98.

Wellmann, F.J., Horowitz, F.G., Schill, E., Regenauer-Lieb, K. (2010) Towards incorporating uncertainty of structural data in 3D geological inversion. *Tectonophysics*, 490(3–4), 141–151.

Zhen-Yu, T. (1988) *Support Design of Tunnels Subjected to Rockbursting: Rock Mechanics and Power Plants*. ISRM International Symposium, Madrid, Spain.

Chapter 8

Construction methods

D. Peila, C. Todaro, A. Carigi, D. Martinelli, and M. Barbero
Politecnico di Torino

CONTENTS

8.1 INTRODUCTION

Tunnel construction can be described as the set of operations that are carried out to produce a stable underground excavation. These operations are organized in such a way as to produce the greatest length of completed tunnel in the shortest possible time, using the smallest possible number of operations and the minimum number of structural elements for the stabilization of the cavity.

The operations have to guarantee the permanent stability of the tunnel, without creating risks to workers or damage to the objects (both natural and man-made) that exist around the tunnel area, above and below the surface, or to the environment. The construction process can be subdivided into phases as follows:

- excavation, i.e. the detachment of the soil or rock mass from the tunnel core using mechanical tools or explosives and the removal of the material (also known as mucking, muck removal or spoil removal) at the face;
- stabilization of the created cavity. The stabilization process includes the use of the first-phase (also called primary or temporary) supports, generally installed in the newly created span by the excavation process, and of auxiliary measures (i.e. rock or soil reinforcements and improvements usually installed ahead of the tunnel face

DOI: 10.1201/9781003256175-8

and/or around the tunnel). The auxiliary measures have the purpose of stabilizing the free span created by the excavation, helping to correctly control the redistribution of stresses around the tunnel and, finally, managing displacements of the tunnel boundaries and the tunnel face;

• the long-term stabilization of the tunnel with the final lining.

These phases can be carried out in a cyclical scheme, as happens in conventional tunnelling, or following a pseudo-continuous process, as occurs in mechanized full-face excavation, supported by precast segments.

In the following paragraphs, a global overview of the most frequently used construction methods, support technologies and auxiliary measures and their relationships with the tunnel design aspects is provided. A detailed description of the technologies and their application fields is given in the second volume of this book.

Construction methods have a key role in the design process. The excavation system and its sequence (e.g. full-face or partial-face) and the length of the excavation step influence not only the short-term stability and the displacement of the tunnel boundary and of any adjacent structure, but also the rate of the tunnel construction and the safety of the job site.

Therefore, we can conclude that the excavation methods, the supports and the auxiliary methods are the tools that the designer has, and over which he has effective control, to guarantee a safe excavation, to permanently stabilize the tunnel and to mitigate the various hazards. Depending on the type of hazard or constraint of each specific project (as assessed in Chapter 2), the designer should choose the best mitigation measure by comparing the various alternatives in terms of the stability of the excavation itself and of the nearby pre-existing building, the local environment conditions, the health and safety of the workers, the durability of the final product and its lifetime maintenance and, finally, the production and industrialization of the construction process. Once a certain set of mitigation measures are applied, the achieved risk level must be re-evaluated and compared with predefined thresholds, as accepted by all the actors involved in the design and construction process.

This process involves amending and re-implementing the hazards and risks register at each stage to keep the risk level below the acceptable level. As an example, the following case may be considered. A good example of this approach is the case of the stability conditions of the tunnel during full-face excavation. If the design results in a free span and a self-support time of the advancing step that are too small, the designer may choose either to partialize the tunnel face and excavate smaller and more stable drifts, or to reinforce the rock mass/soil around the tunnel and in the core also ahead of the face, or to apply a face-stabilizing pressure using a full-face TBM. This concept is better explained and discussed in Figures 8.1 and 8.2.

8.2 EXCAVATION METHODS

Tunnel excavation methods are usually divided into two main families: conventional methods and full-face mechanized methods (rock or soil TBMs). In urban areas, also the cut and cover method can be applied when the surface is free of interferences and potential sources of underground interference are removed.

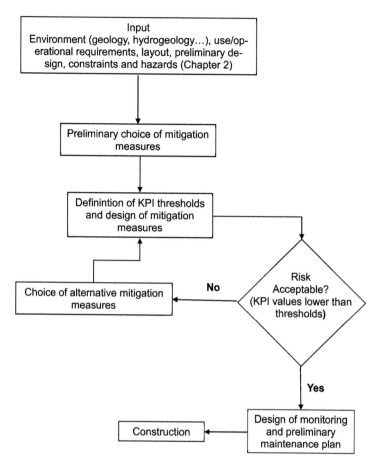

Figure 8.1 Scheme of the design flow chart of the hazard mitigation measures. Where KPI is the Key Performance Indicator, as discussed in Chapter 2.

8.2.1   Conventional method

As defined by ITA WG 19 (2009), excavation with the conventional method implies the use of: explosives (also called drill and blast, the scheme of which is shown in Figure 8.3) or machines (Figure 8.4) such as roadheaders, high-energy impact hammers, rippers and hydraulic breakers (also called drill and split) that can be used in rock masses (Figure 8.4) or excavators with shovels that can be used in soils. A short overview of these techniques will be provided in the following. In a tunnel, explosives and/or machines can both be applied, depending on the variability of the geological formations and their mechanical properties along the alignment, namely their excavatability and strength properties. The choice of the excavation method is mainly driven by both the rock mass mechanical characteristics and the external constraints around the tunnel. As an example, the sensitivity to vibration of buildings or pre-existing unstable slopes (that can be activated by the excavation process) near to the tunnel alignment can influence the choice among drill and blast, roadheader and hydraulic

Identification of the hazard

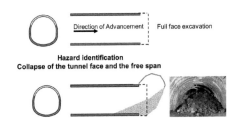

Hazard identification
Collapse of the tunnel face and the free span

Hazard management: to prevent face collapse

Possible options of mitigation measures to face the hazard

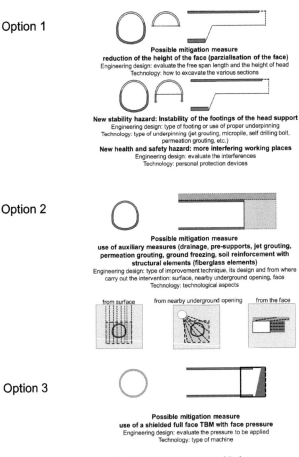

Option 1

Possible mitigation measure
reduction of the height of the face (parzialisation of the face)
Engineering design: evaluate the free span length and the height of head
Technology: how to excavate the various sections

New stability hazard: Instability of the footings of the head support
Engineering design: type of footing or use of proper underpinning
Technology: type of underpinning (jet grouting, micropile, self drilling bolt,
permeation grouting, etc.)
New health and safety hazard: more interfering working places
Engineering design: evaluate the interferences
Technology: personal protection devices

Option 2

Possible mitigation measure
use of auxiliary measures (drainage, pre-supports, jet grouting,
permeation grouting, ground freezing, soil reinforcement with
structural elements (fiberglass elements)
Engineering design: type of improvement technique, its design and from where
carry out the intervention: surface, nearby underground opening, face
Technology: technological aspects

from surface   from nearby underground opening   from the face

Option 3

Possible mitigation measure
use of a shielded full face TBM with face pressure
Engineering design: evaluate the pressure to be applied
Technology: type of machine

New stability hazard: management of the face pressure
Environmental hazard : disposal of the muck (EPB machines)

Figure 8.2 Example of the decision path for the management of some possible choices of mitigation measures to manage the tunnel face instability. (Modified from Anagnostou and Peila (2013).)

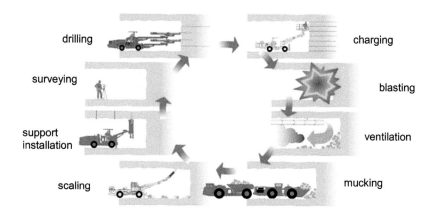

drilling

surveying

support
installation

scaling

charging

blasting

ventilation

mucking

Figure 8.3 Sequence of operations of drill and blast method. (Based on and modified from a scheme proposed by Sandvik.)

breakers for excavation in rock masses. In conventional tunnelling, the excavation process is usually carried out step by step in rounds (cyclic process) and the advancement step depends on the quality of the rock mass, ranging between 1 m (or in a few cases of very poor rock masses even less) and 4–5 m.

The main phases common to each conventional excavation method are the following: (i) excavation of the rock mass in the tunnel core (advancement step); (ii) mucking; and (iii) installation of first-phase supports in the free span and, if needed, ground reinforcement or improvements ahead of the face (depending on the rock mass properties and the related stability conditions before the new advancement step).

The construction can be carried out with full-face (reinforcing the tunnel core, if needed) or with a partial excavation face (ICE, 1996; Galler, 2010; ITA WG 19, 2019) (Figures 8.5–8.8b). Tunnels with variable shapes and geometries can be excavated, depending on the tunnel's final use.

In conventional tunnelling, quite standard and relatively cheap machines are usually used and at almost any time during the tunnelling process, it is possible to access the tunnel face. Generally speaking, this method is a very flexible process from the design point of view, since it can easily adapt to the existing geological and geotechnical conditions and related risk to be handled, by changing the excavation tools or procedure (for example, the geometry of the blast round, or the type of machine), the support technologies and the auxiliary methods.

8.2.1.1   Drill and blast

This excavation method is applied in rock masses, and variable geometries can be easily obtained by simply changing the blast round pattern geometry. This technology has been well known for a long time and has therefore been extensively applied worldwide. It uses simple and relatively cheap devices that can be put into service in a fast way, and the method is flexible to adapt to geological changes since the blast round geometry can be easily changed at almost any time during the tunnelling. The drilling is done

(a)

(b)

(c)

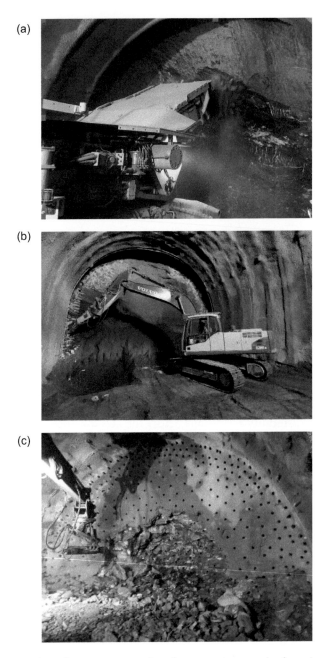

Figure 8.4 Photographs of some conventional excavation methods using machines. (a) Roadheader working in an argilloschist formation; (b) high-energy impact hammer working in a tuff tunnel face; (c) drill and split method in a hard gneiss tunnel face. (Courtesy of Italferr S.p.A. and AK.)

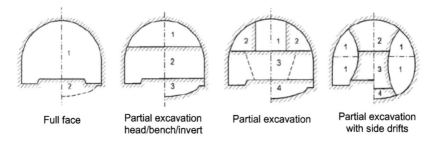

Full face   Partial excavation   Partial excavation   Partial excavation
              head/bench/invert   with side drifts

Figure 8.5 Example of full-face and different choices of partial excavations. The numbers indicate the sequence of the excavation stages.

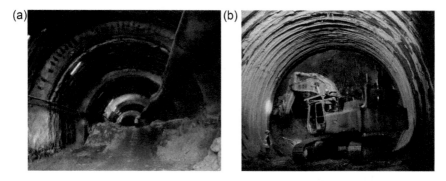

Figure 8.6 Photographs of an excavation with head and bench excavation and use of steel pipe umbrella arch as pre-support in a moraine and a full-face excavation with face reinforcing.

Figure 8.7 Example of Chakara Klabin Station in Sao Paulo (over 350 m^2) constructed with partial excavation with side drifts and systematic pre-support in a poor rock mass and soil, below water table at low overburden. (Courtesy of Geodata S.p.A.)

with drilling hammers installed on jumbos, and modern computerized machines allow very precise patterns of drill holes to be achieved.

The logistics of the job site is usually simpler than that required for a full-face TBM. It is suitable to be used for short tunnels, when the use of a rock TBM requires too long a time for the machine to be available and assembled on the job site. For other conditions, a comparison of the possible production should be carried out for an optimal choice (Grandori, 2016). The drill and blast method is suitable to deal with the technical hazard of hard and abrasive rock masses that could be difficult to excavate with full-face TBMs or roadheaders. The use of drill and blast is negatively affected by the presence of a large number of discontinuities in the rock mass and by large water inflow at the face, which can make the explosive charging problematic. Furthermore, the blast can resituate unstable blocks at the face and around the tunnel boundary that then have to be removed through a careful scaling after the blast. In the case of water inflow, a drainage system has to be installed.

As an inherent hazard of the drill and blast method, the blast causes vibrations. This can be a hazard if there are sensitive structures nearby, and if it produces dust and noise that can be critical at the portal in urbanized areas. These environmental hazards towards the external environment should be managed with a proper design (Grasso et al., 1995). Drill and blast can be combined with the rock TBM excavation to enlarge the TBM bore (Lunardi, 2008). With this method, it is possible to minimize the vibrations since the cut has already been done with a rock TBM and the production blast can be fractioned to a large extent to reduce the amount of charge per delay (i.e. the amount of charge that explodes in the same moment).

This technology is problematic to apply when polluting minerals such as asbestos are present in the rock mass since it can be difficult to manage polluted dust in the tunnel, and special care should be taken when methane is present.

This excavation method can be combined with different types of support and auxiliary methods to deal with the stability hazards.

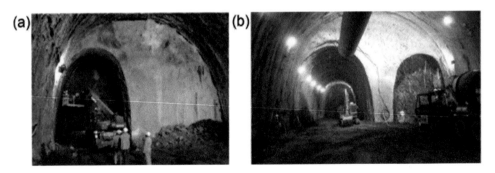

Figure 8.8 (a) Example of the excavation of the lateral drifts of the large cavern 'Camerone C GN13' that connects a double-track tunnel with two single-track tunnels in the high-speed train connection between Genoa and Milan. The cavern has been excavated with a partialized section mainly in an argilloschist formation (argille a palombini) with an overburden ranging from 60 to 100 m. The cross section shown in the photograph is of 395 m^2, while the lateral drifts have a cross section of 105 m^2. (Courtesy of Italferr S.p.A.)

(Continued)

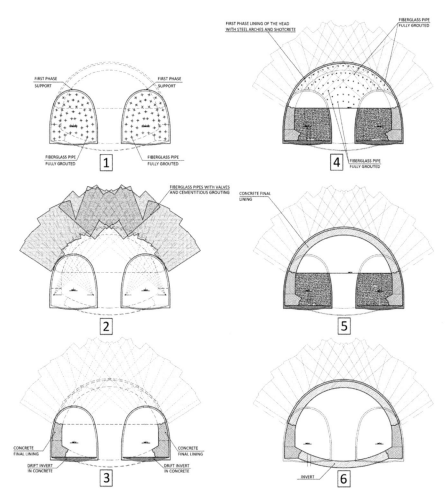

Figure 8.8 (Continued) (b) Example of the excavation of the large cavern 'Camerone C GN13' that connects a double-track tunnel with two single-track tunnels in the high-speed train connection between Genoa and Milan. In the following the construction steps: (i) face reinforcement of the side drifts works and their excavation; (ii) consolidation works of an arch at the cavern extrados, performed from side drifts; (iii) concrete casting of tunnel side walls; (iv) side drifts backfilling, consolidation works of cavern head and its excavation; (v) concrete casting of tunnel crown; and (vi) removal of the lateral drift backfilling, excavation of the invert and closure of the final lining with the casting of the invert.

8.2.1.2 Drill and split

This method can be used in hard rocks, but is rarely applied due to the low production that may be obtained. As a main advantage, it allows the vibrations to be nullified; hence, it is helpful when the nearby structures are highly sensitive to vibration and a rock TBM cannot be applied, for example for short-length tunnels. It uses simple, easy-to-find and relatively cheap devices that can be put into service in a fast way since the required devices can be found easily. The logistics and labour work constraints involved are also limited. Variable geometries can be obtained and they require a high precision in the hole pattern, so the use of completely automated jumbos is strongly recommended. The local stability at the face and the presence of high water inflow can negatively affect the use of the method, if these hazards are not properly managed. A good and recent example is the excavation of the under-passing of some gasoline deposits by the tunnels of the Follo Line in Oslo (Santarelli & Ricci, 2019).

8.2.1.3   Roadheader

This excavation method uses a mechanical technology that was originally developed in the mining industry and has since been applied in the civil tunnelling industry. The production achievable per day is reasonable, but it is strongly dependent on the quality of the rock mass and weight and the power installed on the cutterhead, as well as on the tunnel cross section. Variable geometries may be more difficult to obtain than by using the drill and blast method. The roadheader machine is usually more expensive than the jumbos, and it requires expert workers to be properly managed. Excavation with a roadheader minimizes the induced vibrations in the rock mass and hence may be a valuable option when there is any vibration-sensitive structure nearby.

The roadheader is not suitable in very hard and abrasive rock masses due to very low productivity and the high wear of the tools.

This excavation machine can be easily combined with different types of support and auxiliary measures to manage the face and cavity stability hazards, local instability and high water inflow that can negatively affect the excavation.

Generally speaking, the performance is influenced by the following key parameters (Bilgin et al., 2014):

- Machine parameters: machine type, weight and dimensions, boom force capacities (shearing, lifting and lowering), cutterhead type (transversal and axial), cutterhead power, bit type and its metallurgical properties.
- Geological–geotechnical parameters of the intact rock and of the rock mass: strength (uniaxial compressive strength, tensile strength, elasticity modulus and cohesion); texture and abrasivity (mineral/quartz content and grain size, microfractures, etc.); cuttability; brittleness; rock mass properties (RQD, bedding, foliation and fault zones, joint sets' orientation, spacing, filling, etc.); hydrogeology (water pressure and water inflow); and adverse geological conditions (squeezing, swelling and blocky ground).
- Operational parameters: tunnel shape and dimensions, tunnel dip, muck transport, utility lines (power, water and air supply), workers' availability and their formation.

8.2.1.4   High-energy impact hammer

This excavation method was originally developed for scaling or for local demolition. It has since been applied for tunnel production excavation when the rock mass conditions are suitable (i.e. when they can be fragmented by the action of the impact tool) and vibration control is important. The largest number of Italian applications has been for the excavation of limestone and stratified gneiss and mica schists. In these rock masses, an average production ranging from 3 up to 8 m/day with a cross section of 80–100 m^2 is achievable. This technology is not suitable in hard and compact rock masses.

Variable geometries can be easily obtained and can be used to excavate special sections such as niches or cross-cuts.

8.2.1.5 Excavator with shovels or with ripper tooth

This excavation method can be used only in soils. To ensure the stability of the excavation, it is usually combined with the systematic use of auxiliary methods and supports. The ripper tooth can be installed on the excavator boom to detach cohesive and compact soils.

8.2.2   Full-face mechanized tunnelling method

The other major excavation method is the full-face mechanized tunnelling, where the excavation is carried out using full-face tunnel boring machines: rock TBMs or soil mechanized shields (also called soil TBMs with different technologies for face support). These machines are able to excavate a circular tunnel, and the phases of excavation and mucking are carried out by the machine itself. In rock TBMs, the cutterhead tools (usually rolling cutters) are pushed against the rock mass and they detach the rock chips that are then collected by the buckets with which the cutterhead is provided and discharged onto the conveyor belt inside the machine. This belt then discharges the muck onto the transport conveyor belt installed along the tunnel or onto other types of transport devices such as trains or trucks. When tunnelling in soil, the cutterhead tools (such as scrapers, teeth or rippers) detach the soil, which, through the cutterhead openings, enters the pressure chamber from where it is removed (by a fluid transport or by a screw conveyor) and is transported outside the tunnel. A complete description of the machines can be found in Maidl et al. (2012) and in the second volume of this book.

In rock TBMs, the key issue is the pressure needed on cutters to achieve rock demolition, while in soil TBMs, the most important factor is the stability of the tunnel boundary (both cavity and face); based on the stabilizing action, modern machines can be subdivided as shown in Table 8.1 (Maidl et al., 2008; Bilgin et al., 2014; Maidl et al., 2012, 2013a, b; DAUB, 2020; AFTES-GT4, 2019) and in Figures 8.9–8.12. A comparison of the range of applications of the machines following the scheme proposed by DAUB (2020) is presented in Tables 8.2 and 8.3.

Face stabilization, if needed, is directly provided by the machine in the pressure chamber (also called the bulk chamber or plenum), thanks to a face pressure applied with a fluid (usually a slurry) or with the excavated soil itself properly conditioned, that is to say, mixed with chemical additives that are able to get a pulpy and plastic behaviour of the muck (Peila et al., 2016a, b; Thewes & Budach, 2010; Martinelli et al., 2019). The pressure at the face should be properly designed to minimize the face volume loss (i.e. reducing the extrusion of the face) and therefore to minimize the possible movements around the tunnel (Guglielmetti et al., 2008; Maidl et al., 2012). The control of the

Table 8.1 Most frequently used full-face machines and their support action

Machine support action			Water control	Excavation	Reaction force	Machine type
Place	System			Tools		
	Cavity	Face				
None			None	Full-face cutterhead with disc cutters	Gripper (single or double)	Open rock TBM
Cavity	Shield	None	None	Full-face cutterhead with disc cutters and/or soil tools	Thrust jacks	Single-shield rock TBM
	Shield	None	None	Full-face cutterhead with disc cutters and/or soil tools	Thrust jacks and grippers	Double-shield rock TBM
Face & cavity	Shield	Bentonite slurry	Yes	Full-face cutterhead with soil tools and disc cutters	Thrust jacks	Slurry and Mixshield TBM
	Shield	Conditioned soil	Yes (closed mode)	Full-face cutterhead with soil tools and disc cutters	Thrust jacks	EPB shield TBM (open and closed mode)
	Shield	Slurry or bentonite slurry or conditioned soil	Yes	Full-face cutterhead with soil tools and disc cutters	Thrust jacks	Variable density shield TBM (open, EPB, high-density slurry and slurry mode)

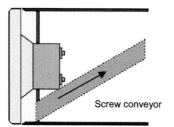 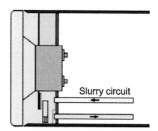

Figure 8.9 Scheme of the EPB and slurry shield (also called hydro-shield) machines. In EPB machines working in closed mode, the screw conveyor is used for primary muck discharge. The advance speed and discharge volume regulation is used to control pressure. In slurry machines working in closed mode, the slurry circuit is used for primary muck discharge and the air bubble is used for face pressure control.

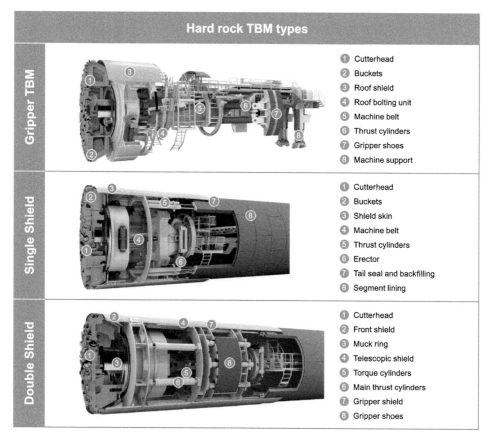

Figure 8.10 Pictures of the rock TBM indicating the key parameters. (Courtesy of Herrenknecht.)

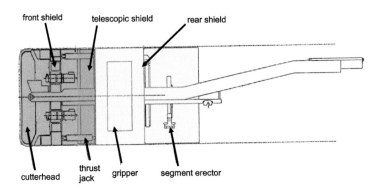

Figure 8.11 Scheme of the double-shield TBM.

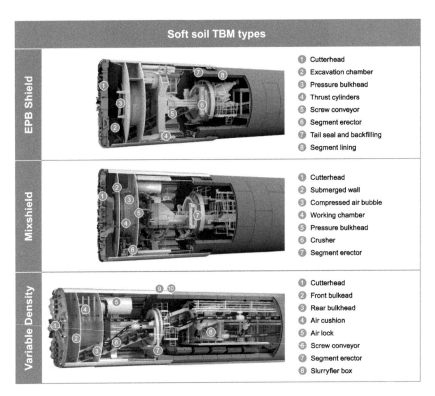

Figure 8.12 Schemes and indication of the main components of the soil TBM. (Courtesy of Herrenknecht.)

settlements is managed with a suitable backfill injection (Todaro et al., 2020; Thewes & Budach, 2009) and, sometimes, with injections of bentonite slurry along the shield. Since the face stabilization is directly provided by the material inside the pressure chamber, appropriate design of the face pressure allows both the face stability to be controlled and the volume loss to be minimized (Maidl et al., 2012; Anagnostou & Kovari, 1994, 1996).

Table 8.2 Application fields of the rock TBM machine (adapted and modified from DAUB (2020))

Parameter	Parameter range					Machine type Rock TBM
Soil						
Fines content (<0.06 mm)	5%	5%-15%	15%-40%	40%		Gripper
	-	-		NR		Single-shield
			o	o		Double-shield
Permeability (m/s)	Very high $>10^{-2}$	High 10^{-2} to 10^{-4}	Medium 10^{-4} to 10^{-6}	Low $<10^{-6}$		Gripper
	-	-		NR		Single-shield
			o	o		Double-shield
Consistency index (Ic)	Very soft 0–0.5	Soft 0.5–0.75	Stiff 0.75–1.0	Very stiff 1.0–1.25	Hard 1.25–1.5	Gripper
	NR	NR	NR	NR	NR	Single-shield
				o		Double-shield
Relative density	Dense	Medium dense	Loose			Gripper
	+	NR	NR			Single-shield
	-	-				Double-shield
Face pressure (bar)	0	1–4	4–7	7–15		Gripper
	o	NR	NR	NR		Single-shield
	-	-		-		Double-shield
Swelling potential	None	Little	Medium	High		Gripper
	+	+	o	NR		Single-shield
	+	+	o	-		Double-shield
Abrasivity (equivalent quartz content) (%)	0–5	5–15	15–35	35–75	75–100	Gripper
	+	+	NR	NR	NR	Single-shield
	+	+	+	o	o	Double-shield

(*Continued*)

Table 8.2 (Continued) Application fields of the rock TBM machine (adapted and modified from DAUB (2020))

Rock

Uniaxial compressive strength (MPa)

Machine type Rock TBM	0–5	5–25	25–50	50–100	100–250	>250
Gripper	-	o	+	+	+	o/-
Single-shield		o	+	+	+/o	-
Double-shield		o	+	+	+	o/-

RQD

Machine type Rock TBM	Very poor 0–25	Poor 25–50	Fair 51–75	Good 75–90	Very good 91–100
Gripper	o	+	+	+	o
Single-shield	o	+	+	+	o/-
Double-shield	o	+	+	+	o

RMR

Machine type Rock TBM	Very poor <20	Poor 21–40	Fair 41–60	Good 61–80	Very good 81–100
Gripper	o	+	+	+	o
Single-shield	o	+	+	+	o
Double-shield	o	+	+	+	o

Water inflow per 10 m of tunnel (L/min)

Machine type Rock TBM	Null	0–10	11–25	25–125	125
Gripper	+	+	+	o	-
Single-shield	+	+	+	o	-
Double-shield	+	+	+	o	-

Abrasivity (CAI)

Machine type Rock TBM	Little abrasive 0.3–0.5	Slightly abrasive 0.5–1	Abrasive 1–2	Very abrasive 2–4	Extremely abrasive 4–6
Gripper	+	+	+	o	
Single-shield	+	+	+	o	
Double-shield	+	+	+	o	

Swelling potential

Machine type Rock TBM	None	Poor	Fair	High
Gripper	+	+	o	o
Single-shield	+	+	o	o
Double-shield	+	+	o	o

Face pressure (bar)

Machine type Rock TBM	0	1–4	4–7	7–15
Gripper	+	-	-	-
Single-shield	+	-	-	-
Double-shield	+	-	-	-

The field of application can be enlarged with the use of auxiliary methods mainly to cross poor rock masses.
Key: +, main field of application; o, extended field of application; -, limited application; NR, not recommended.

Table 8.3 Application fields of the soil EPB-TBM (Adapted and modified from DAUB (2020))

Parameter	Parameter range					Machine type Soil TBM
Soil						
Fines content (<0.06 mm)	<5%	5%–15%	15%–40%	>40%		Slurry shield
	+	+	+	o		EPB
	−	o	o/+	+		Variable density
		o	+	o		
Permeability (m/s)	Very high >10⁻²	High 10⁻² to 10⁻⁴	Medium 10⁻⁴ to 10⁻⁶	Low <10⁻⁶		Slurry shield
	−	+	+	o		EPB
	−	−	o	+		Variable density
	−	o	+	o		
Consistency index (Ic)	Very soft 0–0.5	Soft 0.5–0.75	Stiff 0.75–1.0	Very stiff 1.0–1.25	Hard 1.25–1.5	Slurry shield
	−/o	o	o	o	o	EPB
	o	o	o	o	o	Variable density
	−/o	o	o	o	o	
Relative density	Dense	Medium dense	Loose			Slurry shield
	+	+	o			EPB
	+	+	+			Variable density
	+	+	+			
Face pressure (bar)	0	1–4	4–7	7–15		Slurry shield
	o	o	o			EPB
	o	+	+	−		Variable density
	o	+	+	−		
Swelling potential	None	Little	Medium	High		Slurry shield
	+	+	o	−		EPB
	+	+	o	−		Variable density
	+	+	o	−		
Abrasivity (equivalent quartz content) (%)	0–5	5–15	15–35	35–75	75–100	Slurry shield
	+	+	+	o	o	EPB
	+	+	o	o	−	Variable density
	+	+	+	o	o	

(Continued)

Table 8.3 (Continued) Application fields of the soil EPB-TBM (Adapted and modified from DAUB (2020))

Rock

Uniaxial compressive strength (MPa)

Machine type (Soil TBM)	0–5	5–25	25–50	50–100	100–250	>250
Slurry shield	o	o	o	o	o	-
EPB	o	o	o	-	-	-
Variable density	o	o	o	o	o	-

RQD

Machine type (Soil TBM)	Very poor 0–25	Poor 25–50	Fair 51–75	Good 76–90	Very good 91–100
Slurry shield	o	o	o	o	o
EPB	+	o	o	o	o
Variable density	o	o	o	o	o

RMR

Machine type (Soil TBM)	Very poor <20	Poor 21–40	Fair 41–60	Good 61–80	Very good 81–100
Slurry shield	o	o	o	o	o
EPB	+	o	o	-	o
Variable density	o	o	o	o	o

Water inflow per 10 m of tunnel (L/min)

Machine type (Soil TBM)	Null	0–10	11–25	25–125	>125
Slurry shield	o	o	o	o	o
EPB	o	o	o	o	o
Variable density	o	o	o	o	o

Abrasivity (CAI)

Machine type (Soil TBM)	Little abrasive 0.3–0.5	Slightly abrasive 0.5–1	Abrasive 1–2	Very abrasive 2–4	Extremely abrasive 4–6
Slurry shield	+	+	o	o	o
EPB	+	+	o	o	-
Variable density	+	+	o	o	o

Swelling potential

Machine type (Soil TBM)	None	Poor	Fair	High
Slurry shield	+	+	o	-
EPB	+	+	o	-
Variable density	+	+	+	o

Face pressure (bar)

Machine type (Soil TBM)	0	1–4	4–7	7–15
Slurry shield	o	+	+	+
EPB	o	+	o	-
Variable density	o	+	+	o

The field of application can be enlarged with the use of auxiliary methods mainly to cross poor rock masses.
Key: +, main field of application; o, extended field of application; -, limited application; NR, not recommended.

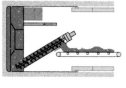
Closed Mode – EPB operation
- regular mode of operation
- face support through conditioned soil pressure in the chamber
- max operation pressure 6-8 bar, depending on soil conditioning

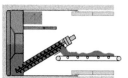
Closed Mode – EPB & compressed air operation
- exceptional mode of operation
- no significant face support through conditioned soil pressure
- control of water inflow
- max operational pressure 2.5 bar, depending on soil conditioning

Open Mode – requires stable face conditions
- atmospheric excavation chamber
- rapid chamber isolation possible (discharge gate)
- muck pile in the chamber for screw conveyor management
- No control of water inflow

Figure 8.13 Scheme of the EPB working in open and closed modes. (Courtesy of Herrenknecht, Cresto 2020.)

The EPB-TBMs can operate with the pressure chamber either full of material (closed mode) or empty (open mode) (Figure 8.13). The machine can work in open mode only if the excavation face is stable and there is no need to control the underground water, while it has to be kept full if dealing with poor rock masses, soils or water inflow hazards. Sometimes, auxiliary methods ahead of the tunnel face can be used to properly manage local tunnel stability and/or the water inflow, but it is important that the machine is equipped with drilling equipment with correctly designed mechanical properties to achieve a reasonable drilling speed.

The classification reported in Table 8.1 is a simplification, since many hybrid conditions can be found: soil machine working principles (both slurry and EPB mode) can be used when tunnelling in stable rock masses to manage the risk of local instabilities (such as local faults or very fractured zones) and/or to control the water inflow or polluting/explosive gases from the rock mass (Bandini et al., 2017). For example, the multimode machines are able to change their way of operation from rock TBM to EPB and slurry modes by changing inside the machine the devices used that are already available, while the variable density machines are able to change their operational mode in a continuous way (Figure 8.14).

The environmental hazards related to the muck management and its final disposal when working with rock TBMs are similar to the drill and blast method. When the soil TBMs are used, the possible presence of chemical products in the muck can be carefully studied at the design stage (a complete discussion of this topic is provided in Chapter 5). Regarding the interference between the excavation and the build environment, since these machines are able to minimize the vibration during the excavation process, they are a suitable tool when this constraint is present.

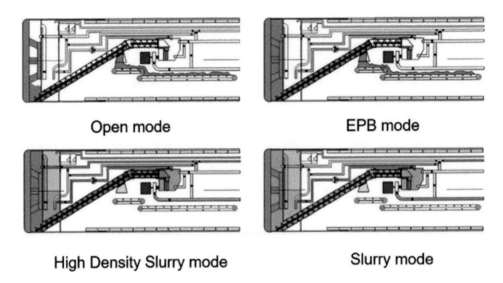

Open mode   EPB mode

High Density Slurry mode   Slurry mode

Figure 8.14 Schemes of the variable density machine operation modes. (Courtesy of Herrenknecht.)

The support installation in gripper rock TBMs follows a sequence similar to that of conventional tunnelling.

The first-phase support is installed (if needed) immediately behind the cutterhead, and it may be made of rock bolts, wire mesh, steel arches and shotcrete, while the final lining (if used) is cast in place far from the excavation face.

In shielded TBMs, the final lining is directly installed under the protection of the shield using precast segments. In single-shield TBMs, the segment lining is installed in alternate phases with the excavation stage, since the TBM pushes itself forward through jacks that react against the lining. In double-shield TBMs, the installation of the segment lining can be done during the excavation since the machine can be gripped to the rock mass, thanks to the telescopic shield assembly.

The choice of a full-face TBM versus the use of a conventional method is a complex task that is influenced by many parameters, and a complete discussion of this topic can be found in Grandori (2017), which highlights the following key parameters: the technical feasibility of the TBM excavation, the construction cost and time, the investment for the TBM itself and for the required plant, the quality of the tunnel, the environmental aspects and the workers' health and safety.

In addition to the parameters listed in Tables 8.2 and 8.3, the following technical conditions should be considered in depth: in rock masses – extremely hard rock that is difficult to bore or extremely abrasive (particularly for very large machine diameters, i.e. bigger than 12 m), a very large tunnel diameter relative to the machine technology and mechanical aspects, the occurrence of faults with fractured zones and with water under pressure, squeezing the rock mass at great depth and thermal waters; and in soil – high water pressure at great depth, the occurrence of large boulders embedded in weak soil and very low overburden.

8.3 SUPPORT TECHNOLOGIES

Tunnel supports can be divided into two main classes: first-phase (or primary) supports and final supports (or final lining). The first-phase support is installed in contact with the rock mass immediately after the excavation to guarantee the safety of the workers, and the final lining is installed to guarantee the stability during the lifetime of the structure. Between these two supports, a waterproofing layer is installed to prevent water leakage in the tunnel. The waterproof layer may be installed by anchoring and welding PVC membranes or (sometimes) spraying a waterproofing material on the boundaries of the cavity between the first and the final linings. In the case of segment lining, the waterproofing of the tunnel is achieved through the use of gaskets.

A scheme of the supports system is given in Figure 8.15. Sometimes, in rock masses of very good quality it is possible (even if infrequently) not to install the final lining and/or the first-phase supports and to leave the rock mass unsupported. The stabilization function is achieved in a way that depends on the type of support chosen, the geometry and the position of installation.

The supports are designed to bear the loads applied by the ground and to control and manage the displacements which the rock mass is naturally subject to after the creation of the cavity. As already stated, in shielded machines the first-phase support is not installed because the stability of the void is granted by the presence of the shield until the final lining, i.e. the segment lining, is installed and locked in place by the backfill injection.

The primary supports must be considered as measures for the mitigation of stability hazards since they are able to apply a support stabilization action into the underground excavated free span. Their action depends on the type of support chosen, their geometry and how and when they are applied in the excavation process. Such supports (today mainly steel arch, rock bolts and shotcrete) can be installed only inside the free span, and therefore, they cannot manage the pre-convergence of the cavity, i.e. the displacements that occur ahead of the tunnel face.

These supports are also designed to guarantee the safety of the workers during the construction process and to obtain the stability of the tunnel until the final lining is installed.

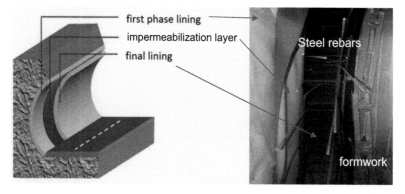

Figure 8.15 General scheme of the linings in conventional tunnelling before casting the concrete.

8.3.1 First-phase lining

The most frequently applied technologies are steel ribs (or steel arches or steel sets), rock bolts and shotcrete (Franzén, 1992; Thomas, 2020) and wire mesh, depending on the properties of the ground conditions (Figures 8.16–8.18). The shotcrete may be reinforced with steel and/or plastic fibres.

In each excavation step, these elements are installed for workers' safety and according to the structural analysis. Usually, the baseline construction plan indicates the support types needed for each homogeneous zone in the geotechnical model, together with criteria for possible variations on site and the warning criteria and remedial measures for cases in which the acceptable limits of behaviour are exceeded.

In the last decades, the use of sprayed concrete lining systems has become very common in soft ground because of the flexibility it can offer in terms of the shape of the tunnel and the combination of support measures (ICE, 1996; ITA WG 12, 2010; Kovari, 2003a, b). This technology consists of spraying a cement–water mixture at high pressure onto the excavation wall. Different additives are added to the mixture to increase the adhesion, setting time and early strength of the shotcrete. Fibres, wire meshes and bars can be added to improve the shotcrete behaviour.

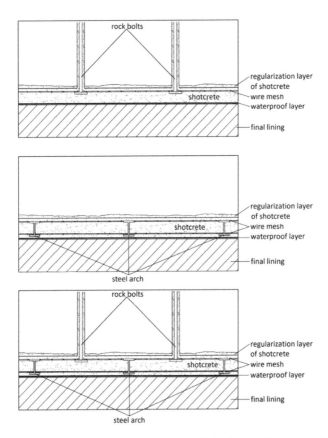

Figure 8.16 Scheme of the possible combinations of the basic elements of the first-phase lining.

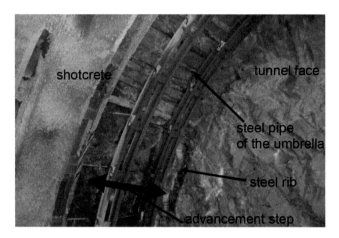

Figure 8.17 Photograph of a first-phase lining of a road tunnel, in a highly fractured rock mass, with steel arches and shotcrete under the protection of a steel pipe umbrella arch.

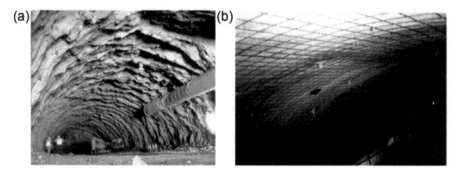

Figure 8.18 Photographs of the support obtained with shotcrete and rock bolts (a) and using rock bolts and wire mesh in an open rock TBM excavation (b).

Steel ribs are installed in the free span at a certain distance from the face. These can be classified as passive interventions since they only respond to loads imposed by the rock due to its inward movement.

Steel ribs can have different cross sections, and the choice is usually made considering the required structural resistance, the availability of the product on the local market and the possibility of bending the profile to the required curvature. They are designed to fit the geometry of the excavation, but in order to be installed, they have to be a little smaller than the excavation. Then the contact between the steel ribs and the rock mass is usually guaranteed using wooden contact and/or shotcrete.

The steel ribs profiles for the tunnel support should fulfil the requirements of a large moment of inertia with small cross-sectional areas and weights, to ensure high strength and easy handling. The steel arches can be closed into a full round or into a horseshoe shape to grant superior performance. Usually, each steel arch is composed of two or three parts for easier installation and then connected appropriately, and sideways buckling has to be prevented by longitudinal stiffening. When the steel arches do not form a closed ring, their feet can be fixed to the rock mass by rock bolts, micropiles or jet grouting columns.

The most frequently used profiles are IPN, HEB (single or coupled), UPN, lattice girders (Hoek & Brown, 1982; Hoek & Wood, 1989; Hoek, 1998, 2001) or a pipe cross section filled with shotcrete mix after erection. When a deformable structure has to be used, the TH profile is the most frequently used and a good discussion of the behaviour of the deformable supports can be found in Anagnostou and Cantileni (2007) and in Kovari (2012). Timber is nowadays used only in special cases, such as partial collapses, and as an emergency support after collapses and, sometimes, in small-diameter tunnels.

Many different types of rock bolts are available. They can be classified into two main groups depending on the way they are connected to the rock mass: fully connected along the whole bolt or with punctual anchorage at the bolt end. The first ones realize the connection through friction or grouting, and the second have the main advantage that, being anchored inside the rock mass and at the tunnel boundary, they may be pre-stressed, if required in the design (Hoek & Brown, 1982; Hoek & Wood, 1989; Kuesel et al., 1996). The tensioned rock bolts apply an additional compressive stabilizing force to the surrounding ground or along discontinuities, providing an immediate confinement of the tunnel boundary, while non-tensioned rock bolts provide additional shear strength that is mobilized for a certain level of displacements into the ground or along the rock discontinuities.

Punctual anchored bolts are inserted into the drilled hole and fixed to the ground by the expansion mechanical element (or, sometimes, by local grouting). Mechanically connected bolts can be used in hard to moderately hard rocks. Rock bolts with corrosion protection are available and, after they are connected with the rock mass using a mechanical anchorage and then after they are completely grouted, they can be considered as permanent elements, i.e. to be considered as part of the final lining. When rock bolts are used for permanent supports, systematic quality controls should be carried out after they are installed to test the integrity of the cementation.

Grouted bolts can be made of both steel and fibreglass, and they are grouted in the hole along the entire length by cement mortar or chemical grouting. They can be used also in soft or poor rocks, where they create an improved rock mass layer around the tunnel. In any case, the grout has to be easily workable, pumped and injected.

Self-drilling bolts are available, constituted by steel pipes equipped with a drill bit that will not be recovered, and are directly drilled into the ground and then filled with grout injection through the steel pipe. They are used in alluvial, moraine and non-cohesive loose soils and in rock masses that can be drilled with the type of available drill bits.

A friction bolt is a version of a fully connected bolt that is connected with the rock mass, thanks to a mechanical action of the whole bolt in the drill hole, such as the Swellex bolt. It is a double-folded steel tube which is expanded into the hole by high-pressure water. During this process, the bolt changes its shape and expands, providing a contact with the hole surface and ensuring a friction action that guarantees the connection.

The bolts can be installed radially to the tunnel axis following a systematic pattern and create a reinforced ring around the cavity, or individually in a specific position to inhibit the movement of rock blocks, wedges or slabs. The thickness of the ground improved ring is linked with the length of the bolts and should be properly designed.

8.3.2   Final lining

The final lining in conventional tunnelling is usually cast in place using mobile form-works below the first-phase lining (Figure 8.19) that has already stabilized the tunnel displacement. Depending on the stability of the rock mass, without considering the contribution of the first-phase lining, the final lining may be reinforced or not, and the type, geometry and amount of reinforcement is chosen on the basis of the struc-tural design. If reinforcement is required, steel rebars, wire mesh and steel fibres are typically used (Hoek et al., 2008).

The final lining can be installed with or without the invert, depending on the ground behaviour and the design approach, and may be installed far away or very close to the excavation face (Lunardi, 2008) (Figure 8.20).

During the concrete casting, special attention has to be paid to avoid the forma-tion of cavities at the tunnel crown, i.e. at the extrados of the lining. This problem can be solved and eliminated through grouting the gap by injecting cement mortar.

In some cases, the final lining can be obtained with a permanent layer of shot-crete. This type of intervention is well suited to short or irregular-shaped tunnels with many junctions. A good overview of this technology can be found in ITA WG 12 & ITAtech (2020).

Segmental linings are used with full-face mechanized TBM tunnelling. They consist of the assembly of precast elements of suitable shape assembled behind the protection of the shield of the TBM for both rock and soil (ITA WG 2, 2019) (Figure 8.21). The precast segments may have different geometries, but the most com-monly used is the universal ring. They are bolted or connected by pins. The universal ring allows the same element types to be used for the entire length of the alignment, achieving different curvatures by appropriately changing the relative orientation of the rings.

Figure 8.19 Example of the lining cast in place and of the waterproofing layer.

(a)

(b)

Figure 8.20 Example of a conventional tunnel where there is a large (a) distance between the final lining and the tunnel face (courtesy of Mapei S.p.A.) and (b) where the invert is cast very close to the tunnel face. (Courtesy of Italferr S.p.A.)

The precast segments may have one or two waterproofing gaskets, depending on the requirements of watertightness and the ground water pressure. The gap between the segment linings, left by the shield moving forward during the excavation, is continually filled with an injection of inert materials or with a special mortar. The gap-filling process is necessary in order to connect the lining with the ground and prevent movements of the segments during advancement of the machine and settlements on the surface.

In addition to a structural analysis for the ground loads and the TBM jacking system loads applied to the segments, the design of segment ring also requires to consider the whole process of manufacture, storage, delivery, handling and erection, as well as the stresses generated by sealing systems and bolts or other erection aids (a detailed description of this topic is provided in Chapter 9).

Figure 8.21 View of a tunnel made of segments and detail of the machine jacks acting against the already installed ring.

8.4   MOST FREQUENTLY USED AUXILIARY METHODS

Ground reinforcements and improvements, drainage, pre-confinements and pre-supports are interventions that are carried out both inside the geological materials that must be excavated (into the tunnel core) and around the future tunnel cavity, to ensure the stability of the tunnel, to manage the stresses around the tunnel and ahead of the tunnel face and to limit the ground deformations and finally to guarantee the health and safety of the workers. They are usually used as mitigation measures to manage the displacement of the boundary and the stress release induced by the excavation, to stabilize the cavity in soft ground and to reduce the permeability of soils and fractured rock masses (Van Impe, 1989; Pelizza & Peila, 1993; Anagnostou & Ehrbar, 2013).

The main problem when tunnelling through difficult geotechnical conditions with conventional methods is the control of deformation: without support or reinforcement, the ground plasticizes and tends to sink into the opening (falling of material from the upper part of the tunnel face, tunnel face extrusion and tunnel face failure). To stop this de-stress (also called decompression) phenomenon, it could be useful to use a pre-confinement technique (that is to say a measure that acts to facilitate the formation of an arch effect in the ground already ahead of the tunnel face). When tunnelling below the water table also with rock TBMs, it could be necessary to make proper injections to control the water ingress towards the tunnel or to drain it with a pattern of drainage.

The main actions of the various methods are: to modify the entity of the displacements at the face (also called face extrusion), and the radial displacement ahead of the face (also called pre-convergence) and behind the face (also called convergence) before applying supports inside the cavity; to guarantee the stability of the excavation face and of the free span; to guarantee the stability of local portions of the rock mass in particular geological and geotechnical conditions; and, finally, to control the water flow by preventing it or managing it. Each technological solution plays a different role (or several roles at the same time) in the stress–strain control at the boundary of the tunnel or ahead of the face during or after the excavation stage. An updated overview of the various technologies can be found in ITACET (2013).

The improvement interventions act by inducing a better mechanical or hydraulic performance of the soils and rock masses around the excavation. The most frequently used technologies are as follows:

- Permeation grouting (also called conventional grouting or low-pressure grouting or injection): a lasting intervention that produces its action with the injection of grout mixes into the soil at a pressure usually ranging between 10 and 40 bar. Special care should be given to the choice of the injection pressure to allow the correct permeation of the soil, but also not to induce hazard to nearby structures such as uplifting of the surface buildings' foundations. The mixes are able to fill the voids of the soil, thereby reducing the permeability and improving the soil's geotechnical and mechanical parameters (Chieregato et al., 2014; Todaro, 2021). The grout mixes (suspensions, solutions or emulsions) can be either cement based or chemical based, and they are usually injected using manchette pipes (Cambefort, 1967, 1977; Tornaghi, 1978, 1981; Nonvellier, 1988; Bell, 1993; Littlejohn, 2003a; AFTES, 2006; Houlsby, 1992). Permeation grouting is also used to inject rock masses to fill the joints, thus preventing water circulation and water inrush, and sometimes, to improve the quality of rock masses (ISRM, 1996; Lombardi & Deere, 1993; Lombardi, 2003; Stille, 2015).
- Jet grouting: a lasting intervention that is carried out by injecting a cement-based grout mix into the soil at a high pressure (over 100 bars). Different injection layouts have been developed, and the available techniques can be divided according to how the various fluids are injected from the monitor: single fluid, double fluids or triple fluids that use, respectively, grout mix (water + cement) only, air and grout mix, or grout mix plus air and water. The choice of each specific technique depends on the performance to be obtained, but more importantly, the geometry of the intervention and the available position from which to carry out the injection. It is important to remark that to create a sub-horizontal column, only the single-fluid technique can be used and for this reason the most frequently used technology in tunnelling is the single fluid. The final result of the injection is a column of treated soil with better geotechnical characteristics than the natural ground itself (Croce et al., 2014).
- Compensation grouting: an intervention that is carried out by injecting a cement-based grout mix into the soil with the goal of compacting the soil and displacing it in a way to compensate for the surface settlements induced by the excavation. The injection pipes are usually installed before the tunnel excavation and used if and when necessary (Littlejohn, 2003b).
- Freezing: a temporary intervention which can be applied when the soil is wet. The aim of this technique is to create an area of frozen ground that is more resistant than the original ground and that has no water flow inside (Johansen & Frivik, 1980; Jones, 1980; Harris, 1988, 1995; ISFG WG 2, 1991; Pimentel et al., 2012).

Ground reinforcements are interventions that are carried out by introducing into the soil, usually by drilling holes, structural elements that are more resistant and rigid than the soil (i.e. bolts, steel cables, or steel or fibreglass pipes or bars) and that are connected to the ground with proper technologies (frequently using grout mixes), with the purpose of obtaining a better global behaviour of the reinforced ground with reference to the specific hazard that has to be dealt with. By using a resistant element of notable geometry and well-known characteristics, the various applications offer the maximum adaptability to the soil variations (i.e. changing the dimensions and the number of the structural elements). The use of structural reinforcing elements guarantees that the design intervention is carried out correctly, while the durability of the reinforcing elements is one of the problems if the intervention must be long-lasting. There are several ground reinforcing schemes that can be used according to the structural requirements of the design (Peila & Pelizza, 2013): elements (made of steel or fibreglass) transversal to the tunnel installed from the surface (with a comb and fan vault layout) or from an existing underground void (an existing tunnel or from a pilot tunnel) with elements installed transversal to the tunnel. When the reinforcing elements (bolts fully grouted or pre-tensioned if they are installed in the free span, which can be classified as support interventions, steel cables, steel pipes or bars, or glass-fibre or carbon-fibre pipes or bars) are installed from the tunnel itself, they usually have a radial geometry. If the reinforcing elements are installed ahead of the tunnel face, they act as pre-confinement.

The pre-confinement and pre-support techniques are interventions developed and installed ahead of the tunnel face to improve the stability in the tunnel advancement span and to manage the stress release and displacements (these can also be obtained by using a layout of ground improving/reinforcement techniques or structural elements such as shotcrete/concrete vaults) (Figures 8.22–8.24 and Table 8.4):

- Steel pipe umbrella (also called pipe roof umbrella, pipe roof support, umbrella arch method, forepoling, spiling, steel pipe canopy and lances): an umbrella of steel pipes or bars with a truncated conic shape ahead of the tunnel face. This intervention can be created with long pipes (Figure 8.22) or with short bars, the

Figure 8.22 Frontal view of the steel pipe umbrella and reinforcements installation. 'Tunnel of Cavallo' on high speedway A 14. (Courtesy of Trevi S.p.A.)

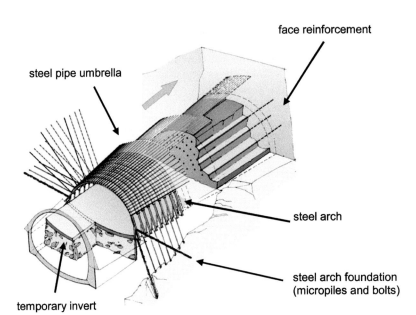

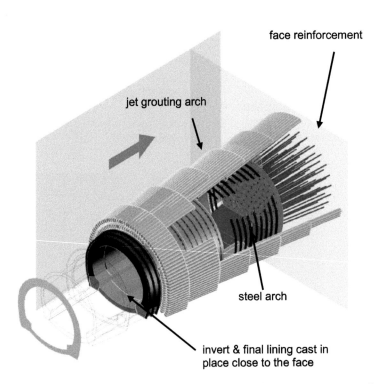

Figure 8.23 Example of the three-dimensional view of the auxiliary measures installed in an excavation carried out with head and bench (top – courtesy of Geodata S.p.A. – modified) and in a full-face excavation (bottom – courtesy of Rocksoil S.p.A. – modified).

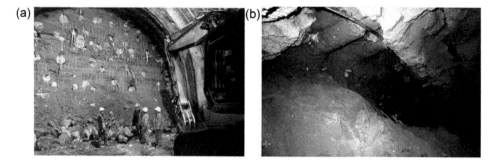

Figure 8.24 Example of the use of jet grouting ahead of the tunnel face. (a) Sub-horizontal face reinforcement in Firenzuola tunnel (silt and silty sand, diameter 13.90 m) and (b) details of the columns of the jet grouting arch in the double-track railway tunnel of Monte Olimpino 2 (alluvial soil, marl and sandstones). (Courtesy of Rocksoil S.p.A.)

Table 8.4 Application fields of the auxiliary methods that can be installed ahead of the tunnel face

Auxiliary method	Cohesive soil	Sand & gravel	Soil with boulders	Fractured rock mass
Grouting	⊙○	●	○	●
Jet grouting	⊙	●	⊙	
Freezing	●	●	●	●
Drainage	⊙	●	●	●
Fibreglass elements (face and tunnel boundary)	●	●⊙	○	⊙
Pre-cut	●	●		
Steel pipe umbrella	●	⊙	⊙	●

The listed interventions can be combined in order to fulfil the required needs.
  Grouting, jet grouting, freezing and dewatering can normally be applied also when tunnelling under the water table, but the drilling operation must be carried out with special care, for example with the use of a preventer, and should manage the water and soil ingress.
Key: ●, applicable; ⊙, applicable with special interventions; ○, difficult but possible.

latter frequently being installed using self-drilling bolts (Carrieri et al., 2004; Peila & Pelizza, 2013; Pelizza et al., 2015). This is a typical pre-support technique since each pipe acts independently as a single beam and there is no significant arch effect in the structure.

• Ground reinforcement using fibreglass fully grouted around the tunnel (also called 'coronella'): the ground is reinforced around the tunnel with a cylindrical and/or truncated conic shape, using grouted fibreglass pipes or bars. These elements are frequently provided with manchette pipes that can be used for both grouting the structural element and injection of the soil (Lunardi, 2015; Cassani, 2013).

• Longitudinal reinforcing elements installed ahead of the tunnel face in the core to be excavated (also called face reinforcement or face improving): fibreglass pipes or bars and, sometimes, steel pipes or bars. With reference to this aspect, it is useful to report what is reported by ITA WG 19 (2009):

> Face bolts are often necessary to stabilize or reinforce the face. Depending on the relevant hazard scenario, the bolt type and length have to be determined in the design. Practically any bolt type or length is possible. As protection against rock fall, spot bolts may be sufficient whereas in difficult ground conditions (e.g. squeezing rock and soils) systematic anchoring with a high number of long, overlapping steel or fiberglass bolts may be necessary. Face bolts are placed during the excavation sequence, if necessary in each round or in predefined steps.
>
> (Lunardi, 1995, 2015; Peila, 2014; Pelizza et al., 2011;
> Cassani, 2013; Perazzelli & Anagnostou, 2017)

- An improved arch or body around the tunnel obtained with the use of a designed geometry of manchette pipes, to properly grout the soil.
- Jet grouting arch (or canopy): a structure of sub-horizontal jet grouting columns at the crown. The arch supports the soil during the excavation, stabilizing the free span and the face, and acts to manage the pre-convergence. In shallow tunnels, the arch can be obtained by a pattern of sub-vertical columns. Frequently, the jet grouting arch is combined with sub-horizontal columns or with fibreglass pipes or bars installed on the tunnel face to stabilize it. The jet grouting columns can be reinforced with structural elements such as steel pipes or bars (Casale, 1986; Lunardi et al., 1986; Croce et al., 2014).
- Cellular arch or arch of microtunnels: used for short and large underground excavations: before the underground excavation is carried out, a supporting structure is made with many microtunnels connected together. This technique can be applied if the overburden is so thin that it will not allow the use of other supporting techniques or it is necessary to control to a high extent the possible subsidence of the surface, for example when under-passing very sensitive buildings or infrastructures (Lunardi, 2008; Miliziano et al., 2019).
- Pre-cut: a lasting intervention that is carried out by excavating a tile cut around the tunnel ahead of the tunnel face using a chainsaw blade machine. The cut is filled with a concrete of high mechanical characteristics and reinforced with steel fibre. The method was developed originally to be used in clay and then applied also to weak rocks. It is frequently applied combined with face reinforcement (Arsena et al., 1991; van Walsum, 1991).
- Pre-tunnel: a concrete lining (up to 1.5 m in thickness) is installed along the tunnel perimeter. The structure of pre-lining may be used as the final lining or is integrated with the final lining. The cut is made with a special cutting boom. This technique has also been applied to widen existing tunnels while keeping them in operation (Peila et al., 1995; Lunardi, 1997).

Finally, drainage can be used to manage and control the water inflow and to contribute to improving the tunnel stability, thanks to the reduction in the water pressure. Drainage is a feasible and cheap technology and is usually done using small-diameter drains (Vielmo, 1986; Zingg & Anagnostou, 2016).

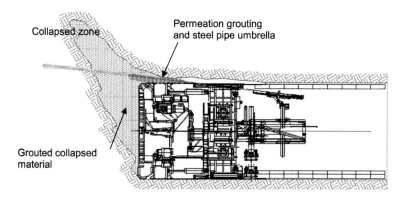

Figure 8.25 Example of the grouting ahead of a shielded rock TBM, through the shield to grout and improve a collapsed area.

Some of the described technologies (mainly permeation grouting, steel pipe umbrella and fibreglass reinforcements ahead of the face) can be applied also in full-face mechanized tunnelling depending on the conditions being faced (Barla & Pelizza, 2000; Peila & Pelizza, 2009; McFeat-Smith & Concilia, 2000; Grandori & Romualdi, 2004; Bilgin et al., 2016). In full-face mechanized tunnelling, the interventions are done ahead of the tunnel face. In this last case, it is necessary for the shield to be equipped with holes usually provided by a preventer if working below the water table, and this intervention is done when the machine has to go through poor rock masses, to cross fault zones or to face a collapse ahead of the tunnel face (Figure 8.25). Permeation grouting or compensation grouting can also be used to create a protection of existing buildings to mitigate the settlement hazard.

The drilling tool installed on the machines can also be used for investigation ahead on the TBMs, which is a key issue particularly when tunnelling at great depths or where the investigations from the surface could be complex (AFTES, 2006, 2019).

Frequently, with reference to the use of auxiliary methods combined with mechanized tunnelling, permeation grouting or jet grouting is used in urban areas to create an injected dice behind a diaphragm wall used to construct the station to guarantee a safe entry and exit of the full face of the underground station, preventing face instability and/or water ingress. The same permeation grouting can be used to create a reinforced soil volume where the machine can stop for cutterhead maintenance also below the water table to reduce the need for compressed air operations. Other examples of permeation grouting include creating additional protection for surface buildings when under-passed by the full-face TBM. Sometimes, compensation grouting can be used to properly manage the settlements, particularly when the tunnel is close to very sensitive buildings.

ACKNOWLEDGEMENTS

Herrenknecht AG and A. Cresto are gratefully acknowledged for the help, the valuable comments and the suggestions.

AUTHORSHIP CONTRIBUTION STATEMENT

The chapter was developed as follows. Peila: chapter coordination and §8.1,8.2,8.3,8.4; Todaro and Carigi: §8.1,8.2,8.3,8.4; Martinelli: §8.3; Barbero §8.3. All the authors contributed to chapter review. The editing was managed by Peila, Todaro and Carigi.

REFERENCES

AFTES (2006) La conception et la réalisation des travaux d'injection des sols et des roches, Recommandation de l'AFTES, GT8R2F1. [Online] Available from www.aftes.asso.fr [Accessed 19 November 2020].

AFTES – GT 4 (2019) Etat de l'art concernant les évolutions des tunneliers et de leurs capacités de 2000 à 2019. Recommandation de l'AFTES, GT4R6F1. [Online] Available from www.aftes. asso.fr [Accessed 20 January 2021].

Anagnostou, G. & Cantileni, L. (2007) Design and analysis of yielding support in squeezing ground. In: Sousa, L.R., Olalla, C. & Grossmann, N.F. (eds.) *ISRM 2007: 11th ISRM Congress, The Second Half Century of Rock Mechanics: Proceedings of ISRM 2007*, July 2007, Lisbon, Portugal: CRC Press.

Anagnostou, G. & Ehrbar, H. (2013) Auxiliary measures in tunneling. *Geomechanics and Tunnelling*, 6, 186.

Anagnostou, G. & Kovari, K. (1994) The face stability of Slurry-shiled-driven Tunnels. *Tunnelling and Underground Space Technology*, 9(2), 165-174.

Anagnostou, G. & Kovari, K. (1996) Face stability conditions with earth-pressure-balanced shields. *Tunnelling and Underground Space Technology*, 11(2), 165-173.

Anagnostou, G. & Peila, D. (2013) Basic considerations on auxiliary measures [Lecture]. Seminar for Continuous Professional Education. Ground Improvement, pre-support and reinforcement, WTC 2013, Geneva, 31 May.

Arsena, F.P., Focaracci, A., Lunardi, P. & Volpe, A. (1991) La prima applicazione in Italia del pretaglio meccanico. In: Arsena, F.P., Focaracci, A., Lunardi, P. & Volpe, A. (eds.) *International Congress on Soil and Rock Improvement in Underground Works: Proceedings of International Congress on Soil and Rock Improvement in Underground Works*, 18–20 March 1991, Milan, Italy. Vol. 2, pp. 549–556.

Bandini, A., Berry, P., Cormio, C., Colaiori, M. & Lisardi, A. (2017) Safe excavation of large section tunnels with Earth Pressure Balance Tunnel Boring Machine in gassy rock masses: The Sparvo tunnel case study. *Tunnelling and Underground Space Technology*, 64, 85–97.

Barla, G. & Pelizza, S. (2000) TBM tunnelling in difficult ground conditions. *GeoEng2000: An International Conference on Geotechnical & Geological Engineering: Proceedings of GeoEng2000*, 19–24 November 2000, Melbourne, Australia: Technomic.

Bell, A.L. (1993) *Grouting in the Ground*. London: Thomas Telford.

Bilgin, N., Copur, H. & Balci, C. (2014) *Mechanical Excavation and Civil Industries*. Boca Raton: CRC Press.

Bilgin, N., Copur, H. & Balci, C. (2016) TBM excavation in difficult ground conditions. Case studies from Turkey. Ernst & Sohn.

Cambefort, H. (1967) *Injection des sols*. Vol. I and II. Paris: Editions Eyrolles.

Cambefort, H. (1977) The principles and applications of grouting. *Quarterly Journal of Engineering Geology and Hydrogeology*, 10(2), 57–95.

Carrieri, G., Grasso, P.P., Fiorotto, R. & Pelizza, S. (2004) Venti anni di esperienza nell'uso del metodo dell'ombrello di infilaggi come sostegno per lo scavo di gallerie. *Gallerie e Grandi Opere Sotterranee*, 72, 41–61.

Casale, R. (1986) Il consolidamento mediante jet grouting nelle realizzazioni in sotterraneo. *Proceedings of International Congress on Soil and Rock Improvement in Underground Works*, 1986, Milan, Italy. Vol. 2.

Cassani, G. (2013) "Val Fortore State Road 2012" – Excavation of tunnels in structurally complex formations. *Geomechanics and Tunnelling*, 6, 197–214.

Chieregato, A., Oñate Salazar, C.G., Todaro, C., Martinelli, D. & Peila, D. (2014) Test di laboratorio di iniezione per l'impermeabilizzazione e consolidamento di terreni granulari per mezzo di materiali innovativi. *Geoingegneria Ambientale e MinerariaVolume*, 141(1), 63–68.

Cresto, A. (2020) Multimode TBMs [Lecture]. Master Course Tunnelling and Tunnel Boring Machines, Politecnico di Torino, 24 April.

Croce, P., Flora, A. & Modoni, G. (2014) *Jet Grouting*. Boca Raton: CMC Press.

DAUB (2020) Empfehlungen zur Auswahl von Tunnelbohrmaschinen. Koln.

Franzén, T. (1992) Shotcrete for underground support: a state-of-the-art report with focus on steel-fibre reinforcement. *Tunnelling and Underground Space Technology*, 7(4), 383–391.

Galler, R. (2010) *NATM. The Austrian Practice of Conventional Tunnelling*. Salzburg: OEGG (Austrian Society for Geomechanics) Publishing.

Grandori, R. (2016) Frese scudate di grande diametro per lo scavo in roccia, confronto produzioni e criteri di scelta al variare del diametro e della geologia. *Gallerie e Grandi Opere Sotterranee*, 120, 41–61.

Grandori, R. (2017) Vantaggi e limiti dello scavo meccanizzato: quando con la macchina e quando senza. *Gallerie e Grandi Opere Sotterranee*, 124, 39–44.

Grandori, R. & Romualdi, P. (2004) The Abdalajis tunnel (Malaga-Spain). The new Double Shield Universal TBM challenge. *Proceedings of Congress on Mechanized Tunnelling: Challenging case Histories, GEAM/SIG*, Turin, pp. 35–42.

Grasso, P., Brino, L., Xu, S., Pelizza, P. & Lanciani, M. (1995) Excavation of a shallow tunnel in a complex formation under severe environmental constraints. *ITA '95: Proceedings of the ITA 1995*, 6–11 May 1995, Stuttgart, Germany.

Guglielmetti, V., Grasso, P.G., Mahtab, A. & Xu, S. (2008) *Mechanized Tunnelling in Urban Areas*. London: Taylor & Francis.

Harris, J.S. (1988) State of the Art Report: Tunnelling using artificially frozen ground. *Ground Freezing 88: Proceedings of the 5th International Symposium on Ground Freezing*, 6–27 July 1988, Nottingham: Balkema.

Harris, J.S. (1995) *Ground Freezing in Practice*. London: Thomas Telford.

Hoek, E. (1998) Tunnel support in weak rock. [Keynote lecture]. *Symposium of Sedimentary Rock Engineering*, 20–22 November, Taipei, Taiwan.

Hoek, E. (2001) Big tunnels in bad rock. *Journal of Geotechnical and Geoenvironmental Engineering*, 129(9), 726–740.

Hoek, E. & Brown, E. (1982) *Underground Excavation in Rock*. London: IMM; E &FN Spon.

Hoek, E., Carranza-Torres, C., Diederichs, M. & Corkum, B. (2008) Integration of geotechnical and structural design in tunneling. *The 2008 Kersten Lecture, Keynote address of the 56th Annual Geotechnical Engineering Conference*, Minneapolis, 29 February 2008.

Hoek, E. & Wood, D. (1989) Rock support, World Tunnelling and Subsurface Excavation. Republished and translated in Italian by Pelizza, S. (1990) Armatura degli scavi in roccia. *Gallerie e Grandi Opere Sotterranee*, 31, 37–51.

Houlsby, A.C. (1992) *Construction and Design of Cement Grouting*. New York: John Wiley.

ICE (1996) *Sprayed Concrete Linings for Tunnels in Soft Ground*. London: Thomas Telford.

ISFG Working Group 2 (1991) Frozen ground structures – Basic principles of design. *Proceedings of the 6th International Symposium on Ground Freezing*, Beijing, 24–26 June. Balkema.

ISRM (1996) Final Report. Commission on Rock Grouting. [Online] Available from https://www.isrm.net/gca/?id=1020 [Accessed 21 November 2020].

ITA-AITES WG 14 & WG 19 (2016) Recommendations on the development process for mined tunnel. [Online] Available from https://about.ita-aites.org/publications/wg-publications/content/19-working-group-14-mechanized-tunnelling [Accessed 10 October 2020].

ITACET (2013) Ground improvement for support & reinforcement. *Seminar for Continuous and Professional Education. World Tunnel Congress* 2013, 31 May–1 June 2013. Geneva, Switzerland.

ITA WG 2 (2019) Guidelines for the design of segmental tunnel linings. [Online] Available from https://about.ita-aites.org/publications/wg-publications/content/7-working-group-2-research [Accessed 12 September 2020].

ITA WG 12 (2010) Shotcrete for rock support. A summary report on State-of-the-art, ITA report N° 5. [Online] Available from https://about.ita-aites.org/publications/wg-publications/content/17-working-group-12-sprayed-concrete-use [Accessed 20 November 2020].

ITA WG 12 & ITAtech (2020) Permanent sprayed concrete linings, ITA REPORT N° 24. [Online] Available from https://about.ita-aites.org/wg-committees/itatech/publications [Accessed 20 November 2020].

ITA WG 19 (2009) General report on conventional tunneling. [Online] Available from https://about.ita-aites.org/publications/wg-publications/content/24/working-group-19-conventional-tunneling [Accessed 12 September 2020].

Johansen, O. & Frivik, E.P.E. (1980) Thermal properties of soil and rock materials. *Proceedings of the 2nd International Symposium on Ground Freezing.* Trondheim, 24–26 June. Published by Norwegian Institute of Technology.

Jones, J.S. Jr. (1980) State of the Art Report – Engineering practice in artificial ground freezing. *Proceedings of the 2nd International Symposium on Ground Freezing.* Trondheim, 24–26 June. Published by Norwegian Institute of Technology.

Kovari, K. (2003a) History of the sprayed concrete lining method-part I: Milestones up to the 1960s. *Tunnelling and Underground Space Technology,* 18(1), 57–69.

Kovari, K. (2003b) History of the sprayed concrete lining method-part II: Milestones up to the 1960s. *Tunnelling and Underground Space Technology,* 18(1), 71–83.

Kovari, K. (2012) Supporti cedevoli nello scavo di gallerie. *Proceedings of the "Seminario dei 30 anni di rocksoil".* EdiCem, Milano.

Kuesel, T.R., King, E.H. & Bickel, J.O. (1996) *Tunnel Engineering Handbook.* Berlin: Springer.

Littlejohn, S. (2003a) The development of practice in permeation and compensation grouting: A historical review (1802–2002): Part 1 Permeation grouting. In: Reston, V.A. & American Society of Civil Engineers (eds.) *Proceedings of Third International Conference on Grouting and Ground Treatment,* 10–12 February 2003, New Orleans, LO.

Littlejohn, S. (2003b) The development of practice in permeation and compensation grouting: A historical review (1802–2002): Part 2 Compensation grouting. In: Reston, V.A. & American Society of Civil Engineers (eds.) *Third International Conference on Grouting and Ground Treatment: Proceedings of Third International Conference on Grouting and Ground Treatment,* 10–12 February 2003, New Orleans, LO.

Lombardi, G. (2003) Grouting of rock masses. [Keynote lecture]. *3rd International Conference on Grouting and Grout Treatment,* New Orleans.

Lombardi, G. & Deere, D. (1993) Grouting design and control using the GIN principle. *International Water Power & Dam Construction,* 45(6), 15–22.

Lunardi, P. (1995) Fiber glass tubes to stabilize the face of tunnels in difficult cohesive soils. *Materials and Engineering,* 6(1–2), 107–165.

Lunardi, P. (1997) Pretunnel advance system. *T&T International,* October 1997, 35–38.

Lunardi, P. (2008) *Design and Construction of Tunnels.* Berlin: Springer-Verlag.

Lunardi, P. (2015) Extrusion control of the ground core at the tunnel excavation face as a stabilisation instrument for the cavity. Muirwood Lecture 2015. [Online] Available from https://about.ita-aites.org/publications/muir-wood-lecture [Accessed 10 September 2020].

Lunardi, P., Mongilardi, E. & Tornaghi, R. (1986) Il preconsolidamento mediante jet-grouting nella realizzazione di opere in sotterraneo. *Proceedings of International Congress on Soil and Rock Improvement in Underground Works*, 1986, Milan, Italy. Vol. 2, pp. 601–612.

Maidl, B., Herrenknecht, M., Maidl, U. & Wehrmeyer, G. (2012) *Mechanized Shield Tunneling*. Berlin: Ernst & Sohn.

Maidl, B., Schmid, L., Ritz, W. & Herrenknecht, M. (2008) *Hardrock Tunnel Boring Machines*. Berlin: Ernst & Sohn.

Maidl, B., Thewes, M. & Maild, U. (2013a) *Handbook of Tunnel Engineering. Volume I: Structures and Methods*. Berlin: Ernst & Sohn.

Maidl, B., Thewes, M. & Maild, U. (2013b) *Handbook of Tunnel Engineering. Volume II: Basics and Additional Services for Design and Construction*. Berlin: Ernst & Sohn.

Martinelli, D., Todaro, C., Luciani, L., Peila, L., Carigi, A. & Peila, D. (2019) Moderno approccio per lo studio del condizionamento dei terreni granulari non coesivi per lo scavo con macchine EPB. *Gallerie e Grandi Opere Sotterranee*, 128, 19–28.

McFeat-Smith, J. & Concilia, M. (2000) Investigation, prediction and management of TBM performance in adverse geological conditions. *Gallerie e Grandi Opere Sotterranee*, 62, 21–28.

Miliziano, S., Caponi, S., Carlaccini, D. & De Lillis, A. (2019) Design of an underground railway station beneath a historic building in Rome and class A predictions of the induced effects. *Gallerie e Grandi Opere Sotterranee*, 132, 9–26.

Nonvellier, E. (1988). *Grouting Theory and Practice*. New York: Elsevier.

Peila, D. (2014) Face reinforcement – Execution aspects and experiences from Italy. *Proceedings of Swiss Colloquium*. Lucerne, Switzerland, pp. 214–223.

Peila, D., Oreste, P.P., Rabajoli, G. & Trabucco, E. (1995) The Pretunnel method, a new Italian technology for full-face tunnel excavation: A numerical approach to design. *Tunnelling and Underground Space Technology*, 10(3), 367–374.

Peila, D. & Pelizza, S. (2009) Ground probing and treatments in rock TBM tunneling. *Journal of Mining Science*, 45, 602–619.

Peila, D. & Pelizza, S. (2013) Ground reinforcement and steel pipe umbrella system in tunneling. *Advances in Geotechnical Engineering and Tunnelling*, 8, 93–132.

Peila, D., Picchio, A. & Chieregato, A. (2016a) Earth pressure balance tunnelling in rock masses: Laboratory feasibility study of the conditioning process. *Tunnelling and Underground Space Technology*, 35, 55–66.

Peila, D., Picchio, A., Martinelli, D. & Negro, E.D. (2016b) Laboratory tests on soil conditioning of clayey soil. *Acta Geotechnica*, 11(5), 1061–1074.

Pelizza, S., Alessio, C. & Kalamaras, G. (2015) Recent advances in the umbrella arch-method: case-histories in adverse tunnelling conditions. *Gallerie e Grandi Opere Sotterranee*, 115, 32–41.

Pelizza, S. & Peila, D. (1993) Soil and rock reinforcements in tunneling. *Tunnelling and Underground Space Technology*, 8(3), 357–372.

Pelizza, S., Sodero, G., Belcastro, D., Silletta, L., Sanna, G., Gemelli, A., Caruso, F. & Mignelli, C. (2011) Metodo innovativo per la prearmatura in avanzamento ed il rinforzo del fronte con tubi in acciaio autoperforanti per la costruzione di una galleria in sabbie instabili sull'autostrada SA-RC. *Gallerie e Grandi Opere Sotterranee*, 100, 91–98.

Perazzelli, P. & Anagnostou, G. (2017) Analysis method and design charts for bolt reinforcement of the tunnel face in purely cohesive soils. *Journal of Geotechnical and Geoenvironmental Engineering*, 143(9).

Pimentel, E., Papakostantinou, S. & Anagnostou, G. (2012) Numerical interpretation of pressure distributions from three ground freezing applications in urban tunneling. *Tunnelling and Underground Space Technology*, 28, 57–69.

Santarelli, S. & Ricci, P. (2019) Innovative methods for excavation and ground improvement in Oslo. In: In: Peila, D., Viggiani, G. and Celestino, T. (eds.) *ITA-AITES 2019 World Tunnel Congress: Tunnels and Underground Cities: Engineering and Innovation meet Archaeology, Architecture and Art: Proceeding of WTC 2019*, 3–9 May 2019, Naples, Italy. London, Taylor & Francis.

Stille, H. (2015) *Rock Grouting – Theories and Applications*. Stockholm: BEFO.

Thewes, M. & Budach, C. (2009) Grouting of the annular gap in shield tunneling – An important factor for minimisation of settlements and production performance. In: Kocsonya, P. (ed.) *WTC 2009: Safe Tunnelling for the City and for the Environment: Proceedings of ITA-AITES World Tunnel Congress 2009, WTC 2019*, 22–28 May 2009, Budapest, Hungary.

Thewes, M. & Budach, C. (2010) Soil conditioning with foam during EPB tunneling. *Geomechanik und Tunnelbau*, 3(3) 256–267.

Thomas, A. (2020) *Sprayed Concrete Lined Tunnels*. Boca Raton: CRC Press, Taylor & Francis.

Todaro, C. (2021) Grouting of cohesionless soils by means of colloidal nanosilica. *Case Studies in Construction Materials*, 15, e00577.

Todaro, C., Godio, A., Martinelli, D. & Peila, D. (2020) Ultrasonic measurements for assessing the elastic parameters of two-component grout used in full-face mechanized tunneling. *Tunnelling and Underground Space Technology*, 106, 103–130.

Tornaghi, R. (1978) Iniezioni. Proceeding of "Seminario su consolidamento di terreni e rocce in posto nell'ingegneria civile", 26–27 May 1978. Stresa, Italy.

Tornaghi, R. (1981) Criteri generali di studio e controllo dei trattamenti mediante iniezioni. *Proceedings of: X Ciclo annuale di conferenze dedicate ai problemi di ingegneria delle fondazioni*, 17–19 November, Turin.

US Army Corps of Engineers (1997) Tunnels and shafts in rock, Engineer Manual 1110-2-2901. [Online] Available from https://www.publications.usace.army.mil/Portals/76/Publications/EngineerManuals [Accessed 14 January 2021].

Van Impe, W.F. (1989) *Soil Improvement Techniques and Their Evolution*. Balkema.

Van Walsum, E. (1991) Mechanical pre-cutting, a rediscovered tunneling technique. *Rock Mechanics and Rock Engineering*, 24, 65–79.

Vielmo, I. (1986) I drenaggi nella realizzazione di opere in sotterraneo. Sviluppi e tendenze in Italia. *Proceedings of International Congress on Soil and Rock Improvement in Underground Works*, 1986, Milan, Vol. 2.

Zingg, S. & Anagnostou, G. (2016) Static effects, feasibility and execution of drainages in tunneling. *Eidgenössisches Departement für Umwelt, Verkehr, Energie und Kommunikation UVEK, Bundesamt für Strassen*, 1587, Bern, DOI: 10.3929/ethz-a-010814194.

Chapter 9

Computational methods

C. di Prisco
Politecnico di Milano

D. Boldini and A. Desideri
Sapienza Università di Roma

E. Bilotta and G. Russo
Università di Napoli Federico II

C. Callari
Università degli Studi del Molise

L. Flessati
Politecnico di Milano

A. Graziani
Università degli Studi Roma Tre

A. Meda
Università degli Studi di Roma Tor Vergata

CONTENTS

DOI: 10.1201/9781003256175-9

9.1   MODELLING APPROACHES AND CALCULATION METHODS

9.1.1   Introduction

The aim of any computational analysis involving geotechnical problems consists in describing and simulating the hydro-thermo-chemo-mechanical (HTCM) processes occurring in a determined volume in an assigned period of time. This requires the solution of a system of partial derivative differential equations. To this aim, the analyst has to consider the response of the previously defined limited spatial domain over the specified time period, typically starting before the construction of the tunnel and ending many years later. Each numerical analysis simulates only a specific HTCM process. This allows us to reduce abruptly the number of unknowns, the time period and, therefore, the associated computational costs. All the simplifications deriving from this choice have to be justified by the analyst. Analogously, the system evolution preceding the considered time period has to be summarized by imposing appropriate initial conditions. Very often, thermo- and chemo-mechanical couplings are not accounted for, if not in some special cases, as for instance in energy and nuclear waste storing tunnels.

As previously mentioned, equations need to be integrated over a finite spatial domain. In principle, the dimensions of the spatial domain should be as larger as possible, but to reduce computational costs, a limited domain is analysed, whose dimensions have to depend on the peculiarities of the investigated problem concerning, for instance, the physical/mechanical processes, the constitutive relationships (as is well known when irreversible strains develop, the spatial diffusion of the local perturbation reduces) and the chosen computational approach/strategy. In tunnelling problems, the size of the spatial domain is tailored on (i) the tunnel diameter (D) (Figure 9.1), interpretable as a sort of 'engineering internal length', and (ii) the 'geological internal length' in case of complex stratigraphic/topographic conditions. The reliability of any numerical analysis should be assessed by performing an ad hoc parametric study aimed at demonstrating the negligible influence of boundary conditions on the numerical results. This would imply the necessity of declaring the variables to be considered as representative and to provide a threshold for the resulting numerical error. The selection of the spatial domain also implies a substructuring approach, inevitably associated with a simplification of the interaction of the spatial subdomain taken into account with the 'external world'. For instance, when a tunnel is excavated in an urban area, the presence of pre-existing buildings may be simulated by substituting them with a vertical stress distribution at the ground

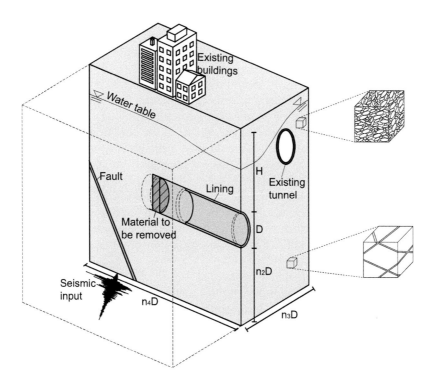

Figure 9.1 Schematization of the problem.

level. By simplifying abruptly the problem, the settlements, calculated where vertical stresses are initially applied, can be later employed to evaluate the building structural response.

Once the spatial domain is chosen and boundary conditions are imposed, the dimension (l) of REV (representative elementary volume) needs to be defined, by introducing a dimensional limit between 'macro'- and 'micro'-scales. l has to be correlated with L (a representative dimension of the spatial domain), this latter providing a characteristic length for the so-called 'mega'-scale. Soils are typically schematized as continuum media, given the high ratio between tunnel diameter and solid grain dimensions. Analogously, the same assumption is carried out for rock masses characterized by an average joint spacing small enough in comparison with the characteristic length of the construction. In this case, homogenized mechanical parameters have to be properly defined to take into account the properties of both rock material and joints (Chapter 4), thus allowing to simulate the global response to tunnelling, such as the average cross-section convergence or the face overall stability and the lining system deformation. According to this 'scale hierarchy', at the macrolevel the spatial domain (for instance the soil/rock mass to be excavated) is assumed to be filled by a generally heterogeneous smeared multi-phase continuum. In contrast, discontinuities characterized by a spacing larger than l are to be considered as spatial heterogeneities possibly filled by fluids (either gas or water). In many cases, the precise description of both stratigraphy and heterogeneities is missing. An alternative strategy to describe discontinuities consists in introducing interfaces. When joint/discontinuity spacing and tunnel diameter have comparable values, convenient is the use of a discontinuum approach (Section 9.1.5.3), according to which, in general, a finite number of rigid blocks interact along a discrete number of interfaces.

The multi-phase nature of the medium implies the definition of a large number of variables, all depending on space and time. The goal of the numerical analysis consists in the definition of a set of transforming functions relating the unknown variables to the controlled ones. In HTCM processes, to the common geotechnical static/kinematic variables σ_{ij} (stress tensor), fluid pore pressures, ε_{ij} (strain tensor) and u_i (displacement vectors), state variables (for instance saturation), temperature T and chemical species concentration have to be added. The solution of the problem is obtained by writing a system of partial differential equations, depending on the HTCM problem to be analysed. For instance, when a 'simple' one-phase mechanical problem is approached, these equations express the balance of momentum, the compatibility and the constitutive relationship, but for a bi-phasic continuum, the fluid mass balance needs to be added (Section 9.2.1).

As far as the period of time is concerned, crucial is the role of the strategy employed to define the initial conditions, summarizing, as better as possible, the geological/anthropic history of the domain. In some cases, the geological history is abruptly simplified; in other cases, the initial values of variables are derived from in situ experimental data. In many cases, initial conditions are unknown, thereby requiring the analyst to consider them as parameters and to assess their influence on numerical results. The duration of the numerical analyses, in terms of physical time to be considered, is strictly related to the characteristic time of the processes accounted for. For instance, in case the construction of the tunnel is concerned, the characteristic time coincides with construction time, but if a coupled hydro-mechanical problem (Section 9.2.1) is investigated, the characteristic time is likely to be dominated by the consolidation process.

Any HTCM process is associated with an evolution of either boundary conditions, or geometry of the heterogeneities within the spatial domain. When the problem is numerically simulated, the analyst has to transform such an evolution into functions defining the control variables' time history. In case of tunnels, perturbations applied to the boundaries of the spatial domain are, for example, (i) the change in the stress distribution at the ground level in case of new buildings, (ii) the evolution of displacements due to active landslide and seismic events, (iii) the water infiltration at the ground level and (iv) the modification in the pore water pressure distribution, for instance, due to seasonal phreatic level oscillations. The evolution of heterogeneities (geometry) is particularly crucial for tunnelling, since excavation, soil improvement and lining installation are the most relevant processes to be typically simulated.

As previously mentioned, each numerical analysis is performed to simulate a precise HTCM process. Therefore, for each numerical analysis, the introduction of a series of simplifying hypotheses taking into consideration the following items is convenient:

• Number of dimensions: in principle, tunnel excavation is a time-dependent three-dimensional problem, but, in particular cases, the time dependency can be disregarded and the number of spatial dimensions reduced. For instance, the employment of simplified two-dimensional models is particularly common during the preliminary lining design (Sections 9.2.2.3 and 9.2.2.4). In fact, the typical 3D schematization required to reproduce the tunnelling process is frequently scaled down to 2D plane strain models allowing, although characterized by more uncertainties in relation to the proper representation of the construction phases, the reduction in the computational time and the potential execution of a larger number of parametric analyses.
• Number of phases: in the simplest case, only one phase can be considered (e.g. intact rocks, dry materials and fine-grained soils under undrained conditions) and the unknown variables (stress tensor, strain tensor and displacement vector) can be determined by means of the standard solid continuum mechanics equations. More common are the bi-phasic approaches valid for saturated media, for which the effective stress principle is valid. In this case, also the pore water pressure distribution has to be determined. In case more than one fluid is present, also equations, describing the evolution of state variables (e.g. saturation) with static variables, have to be introduced.
• Small or large strains: the definition of both kinematic and static variables is dependent on the choice of taking into consideration or not the large strains developing at the REV scale. In practical applications, large strain approaches are not employed in geotechnical numerical analyses since the evolution of the material state is described, if necessary, by updating state variables (e.g. void ratio).
• Constitutive relationships for continua: for the solid phase, in most of the applications, incremental elastic–plastic constitutive relationships are adopted, but in some cases, also elastic or elastic-viscoplastic (Section 9.4.1) constitutive models can be employed. For instance, this last approach is usually implemented to simulate creep phenomena and long-term response of tunnel linings. As will be clarified in the following, in some very specific applications, rigid constitutive relationships

are used (limit equilibrium method, Section 9.1.4.1, and discrete element method, Section 9.1.5.3). As far as the fluid phase is concerned, the employment of the Darcy law is the most common, but in case of rock masses, the presence of discontinuities at the REV scale may suggest the use of more complex theories (e.g. Louis, 1969).

- Constitutive relationships for discontinuities: the most common interface constitutive relationship for the solid phase is frictional, which in the limit cases of nil or infinite friction angle values can become either smooth or rough. For the liquid phase, impervious and drained (nil pore pressure) conditions can be introduced. As in the case of continuum, the characteristic length of discontinuities, depending on both the engineering problem taken into account and the discontinuity microstructure, governs the choice of the constitutive relationship and the calibration of the constitutive parameters.

- Inertial effects: in most of the applications, inertia forces can be disregarded. This is not the case when a seismic perturbation (Section 9.4.3) or drill and blast excavations (Section 9.3.2.3) are simulated. The simulation of wave propagation within the continuum makes significantly more complex the numerical analysis, which is severely affected by the imposed boundary conditions and time/spatial discretization (Section 9.4.3).

- Small or large displacements: in many practical applications, the balance of momentum equations need to be written with respect to the updated geometrical configuration of the system. For instance, this is particularly relevant when either the excavation phases are simulated, or the tunnel face collapse is of interest. Nowadays, numerous are the numerical methods capable of dealing with large displacements, as the distinct element method or the particle finite element method.

Depending on all the ingredients listed above, by assuming a continuum-based approach, and in particular if a large number of simplifying hypotheses are introduced, in some special cases, the analytical solution (Figure 9.2 and Section 9.1.2) of the system of differential equations mentioned above may be obtained. As an example of these analytical solutions are the well-known characteristic curves for either cylindrical or spherical tunnel cavities (Sections 9.2.4.2.2 and 9.2.5.2.2).

The numerical problem can be solved by employing different strategies (Figure 9.2), as is detailed in the following.

The integrated/derived solutions are closed-form expressions derived from the interpretation of numerical analyses (finite element or limit equilibrium method). Examples are the evolution of tunnel convergence with the distance from the tunnel face or face extrusion as a function of the mean applied stress (Section 9.1.3).

Solutions for stability analyses are based on either the limit equilibrium method (LEM), or the limit analysis theory. Their use requires (i) the definition of either different possible failure mechanisms (LEM or kinematic approach of limit analysis), or admissible stress fields (static approach of limit analysis) and (ii) the minimization/maximization algorithm to individuate the most probable failure condition for the system (Section 9.1.4).

The HTCM problem can be solved numerically (Section 9.1.5) by using different approaches (numerical analyses). From a mechanical point of view, the first distinction to be mentioned is the introduction in the numerical model of interfaces modelling

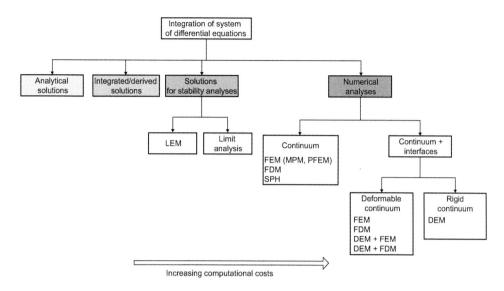

Figure 9.2 Available calculation methods.

heterogeneity/discontinuities at the 'mega'-scale. The second distinction is relative to the aim of approaching the problem in either small or large displacements. In case of small displacements, the most common numerical codes are based on either Finite Element Method (FEM, Section 9.1.5.1; Zienkiewicz et al., 2005), or Finite Difference Method (FDM, Section 9.1.5.2). In the last decades, dealing with large displacements in the framework of continuum approaches is possible by employing (i) Particle Finite Element Method (PFEM, Onate et al., 2004) and Material Point Method (MPM, Sulsky et al., 1994), introduced as extensions of FEM, and (ii) Smoothed Particle Hydrodynamics (SPH, Pastor et al., 2009) method, inspired by fluid mechanics. In the framework of discontinuum mechanics (rigid blocks interacting along deformable interfaces), the most common numerical method is the Discrete Element Method (DEM, Section 9.1.5.3, Cundall & Strack, 1979), particularly suitable for simulating the mechanical response of rock masses.

9.1.2 Analytical solutions

The system of non-linear differential equations can be solved analytically in very few circumstances, characterized by extremely simple geometries and loading conditions. Despite the strong assumptions made, the use of analytical solutions is still very popular for the preliminary design stage and as a reference for the validation of results obtained by performing more complex numerical simulations. Moreover, in some situations the considered schemes are even sufficiently realistic, as discussed in the following paragraphs (Sections 9.2.4.2.2 and 9.2.5.2.2).

The distribution of both stresses and strains around a tunnel assumed to be infinitely long (the problem of cylindrical cavity), and the related displacement field, can be easily derived for a one-phase medium obeying either an elastic or an elastic–perfectly plastic constitutive law if the following hypotheses are introduced: (i) circular tunnel, (ii) uniform state of stress ('hypothesis of deep tunnel', for which the gravity is disregarded), (iii) plane strain conditions (considering tunnel sections far from the tunnel face) and (iv) initial isotropic state of stress. These conditions are characterized by axial symmetry, and the state of stress/strains of any point around the tunnel depends only on the radial distance r from the tunnel centre. The hypotheses listed above can be released at the cost of more complicated analytical formulations (Section 9.2.5.2.2).

As far as the tunnel face is concerned, the mechanical response, typically three-dimensional, is commonly assessed by using the solution obtained for a spherical cavity and under hypotheses (ii) and (iv) listed above (Yu, 2013).

Some analytical solutions also exist for the predictions of settlements induced by shallow tunnels (e.g. Kelvin, 1848; Sagaseta, 1987; Verruijt & Booker, 1996), but, as discussed in Chapter 10, semi-empirical approaches are generally preferred for a better prediction of tunnelling-induced subsidence.

For the assessment of tunnel performance under seismic loads is common the employment of 'uncoupled approaches', according to which the wave propagation and structural problems are solved separately. The problem of one-dimensional wave propagation in stratified media constituted of either elastic or elastic-viscoplastic materials can analytically be solved in the frequency domain. These solutions are implemented in codes (e.g. Schnabel et al., 1972). To take into account the non-linearity of the mechanical behaviour of soils, codes based on the iterative use of the analytical solutions are also available (Phillips & Hashash, 2009).

9.1.3   Integrated/derived solutions

Closed-form solutions, derived from either parametric numerical or stability analyses, are also available and fruitfully adopted in the engineering practice. In Chapter 10, the evolution of radial displacements with face distance predicted on the basis of numerical axisymmetric simulations is provided, whereas the characteristic curve for the tunnel face under both undrained and drained conditions is given in Sections 9.2.4.1.1 and 9.2.4.2.1.

9.1.4   Solutions for stability analyses

Different approaches can be adopted for the preliminary assessment of the stability conditions of the tunnel face and the cross section. Since deformations are neglected and a rigid plastic behaviour is considered, the problem is governed by a reduced number of differential equations. Typically, in addition to the strength criteria of the involved materials, equilibrium and compatibility are conveniently imposed. In the following, two of the most diffuse methods (LEM and limit analysis) in the context of tunnelling design are briefly recalled. These two approaches cannot consider the eventual reduction in strength (i.e. the fragility in the material response) and the phenomenon of progressive failure, since a unique factor of safety is assumed along the entire failure surface.

9.1.4.1 Limit equilibrium method

The limit equilibrium method is a simplified approach aimed at the assessment of a scalar quantity (the safety factor) related to the probability for the system to fail. It is a function of geometry, mechanical properties and hydraulic conditions. The first step of the LEM is the definition of a set of failure mechanisms, consisting in rigid blocks interacting along interfaces, where plastic strains develop. For each mechanism, the factor of safety is determined by imposing the balance of momentum and by introducing the safety factor as a reduction coefficient to the material shear strength necessary to reach failure conditions. Among all these calculated values, the best approximation of the system factor of safety is the minimum one.

In tunnel engineering, this method has found numerous practical applications in evaluating the stability of unreinforced/reinforced tunnel faces (Section 9.2.4.1.1) and cross sections (Section 9.2.5.1.1).

9.1.4.2 Limit analysis

The limit analysis theory is a more rigorous approach aimed at assessing the factor of safety. According to limit analysis theory, the collapse conditions of a system may be determined without the need for simulating the deformation process and the relative stress/strain history, given the hypothesis of a rigid perfectly plastic constitutive law. The limit analysis consists in two methods, in which either (i) an admissible stress field and the material failure condition (static method), or (ii) an admissible velocity field and the material failure condition (kinematic method) are imposed. In case (i) the failure locus is convex and (ii) the flow rule is assumed to be associated (Drucker et al., 1952), the solutions obtained by means of the static and the kinematic methods are lower and upper bounds for the real solution, respectively. The unique case in which the flow rule may be assumed to be associated and the material mechanical behaviour to be perfectly plastic is that of a fine-grained soil under undrained conditions. For this reason, in general, the convergence to the solution is not ensured.

An interesting application of the limit analysis theory (Sloan, 1988, 1989) consists in obtaining the failure mechanism and the related geometry/stress field by employing the finite element method (Section 9.1.5.1).

9.1.5 Numerical methods

9.1.5.1 FEM

In the last decades, the use of FEM codes has become very common. For hydro-mechanical geotechnical applications, these are employed to assess: (i) displacements induced by either a mechanical or hydraulic perturbation, (ii) the spatial distribution and time evolution of stresses and pore water pressure, (iii) the stability of a system in case of complex failure mechanisms or when failure conditions cannot be dealt with standard LEM or limit analysis approaches (Section 9.1.4), since either hydro-mechanical coupling plays a fundamental role, or material fragility has to be accounted for.

The obtained solution is inevitably approximated, because the medium is discretized into a number of points and the numerical convergence is achieved according to a weak instead of a strong formulation.

FEM codes require the definition of (i) the spatial domain discretization, (ii) the temporal discretization (in case physical time is not accounted for, the discretization concerns the perturbation applied and is generally indicated with the term 'time step'), (iii) initial and (iv) boundary conditions, (v) integration scheme, (vi) constitutive relationships, (vii) evolution of geometry and change of material properties, (viii) structural elements and (ix) large displacements. Each item is discussed below separately:

(i)   Either the refinement of the spatial discretization or the use of higher-order polynomial shape functions provides the convergence to the solution, but they cause an increase in the computational costs. Suitable strategies to refine the mesh where either strains or stress gradients concentrate are usually employed. Particular attention has to be paid in case the material constitutive relationship is characterized by a softening behaviour, since the convergence to the solution is not always granted or this latter depends on the adopted discretization. In these cases, 'high-order' constitutive relationships (non-local approaches and gradient models) have to be preferred to avoid mesh dependence of the numerical solutions.

(ii)   In geotechnical computational analyses, the physical time is generally accounted for (i) when hydro-mechanical coupling is taken into consideration, (ii) seismic/dynamic perturbations are applied to the system, and (iii) the constitutive relationship is rate dependent. In all the other cases, the computational step has to be tailored to reduce the error associated with the finite integration of the incrementally linearized numerical equations system. In case of non-linear numerical analyses, the solution may be neither unique nor existing and when the numerical convergence is not achieved, that is the numerical error does not reduce by increasing the number of incremental steps, the causes of the numerical instability have to be investigated. In many cases, the numerical instability derives from very localized mechanical processes not governing the global response. In these cases, the analyst should find 'ad hoc' numerical strategies to solve the problem.

For time-dependent problems, the numerical solution is also affected by the time discretization. In particular, the solution depends on (i) the quality of the description of temporal evolution of the perturbation and (ii) the ratio between the dimension of the elements of the spatial discretization and the time increment (Section 9.4.3).

(iii) In non-linear numerical analyses, the solution is affected by initial conditions. As previously mentioned, the analyst has to mimic, as better as possible, the geological history of the domain, by adopting the constitutive relationships employed in the following stages of the simulation. The most popular simplified strategies consist in the (1) definition of the final model geometry and activation of a pre-defined state of stress (for simple geometries as in the case of horizontal stratigraphy and topography) and alternatively in the (2a) increase the material weight layer by layer (simulating the deposition process) or (2b) progressively increase the material weight in the whole soil domain.

(iv)   Analogously, the definition of boundary conditions is crucial for the reliability of the numerical results. In case of hydraulic, mechanical and hydro-mechanical coupled problems, boundary conditions need to be defined according to the case for (i) stresses or alternatively displacements (not necessarily nil), or for (ii) pore pressure or alternatively water flux. In some special cases, the finite element numerical code allows us to define elastic/frictional/cohesive/damping interfaces at the boundaries. Similarly, for the hydraulic conditions at the face and the cavity of a tunnel excavated in a saturated continuum, the outgoing water flux has to be allowed, but not the inflow, since internally the tunnel is not submerged.

(v)   The temporal solution of the system of incremental equations is usually obtained by employing explicit integration schemes. On the contrary, to both reduce the computational costs and improve the solution accuracy, the spatial integration of both constitutive relationships and field equations is performed by employing implicit integration schemes (Gens & Potts, 1988; Borja, 1991; Potts & Ganendra, 1994; Tamagnini & Viggiani, 2002).

(vi)   Depending on the employed constitutive relationship, FEM analyses can be either linear or non-linear. In general, numerical analyses are performed by taking into account a non-linear material mechanical behaviour. The analyst has to carefully choose the constitutive relationship, which has to be calibrated on the available experimental data. In case of tunnel excavation, particular attention should be paid since most of the material points in the soil domain are expected to follow unloading stress paths. Moreover, the definition of the plastic potential (the dilatancy law) is also crucial, since it governs both the pre-failure and the critical material response. It is worth mentioning that in all soil–structure interaction (SSI) problems, the system response is expected to be significantly affected not only by the material strength, but also by material deformability parameters (Chapter 10), that should be defined on the basis of the expected average strain level.

In case of seismic analyses (Section 9.4.3), common is the use of viscoelastic/elastic-hysteretic constitutive models that are, however, not suitable for reproducing the dissipation associated with the accumulation of irreversible strains expected during intense seismic excitations.

In case of hydro-mechanical coupled analyses (Section 9.2.1), the numerical solution is also likely to be significantly influenced by the development of irreversible strains before getting the critical conditions. In this case, the employment of constitutive relationships, conceived in the framework of the critical state theory and characterized by the presence of a 'volumetric cap' (e.g. Cam Clay model) or by the use of multi-surface plasticity models, is suggested.

(vii)   In tunnelling problem, the simulation of material removal is essential. This is usually introduced by progressively decreasing both the stiffness and weight of the excavated material. At the end of the process, the corresponding elements can be 'switched off'. On the contrary, to simulate the construction of the lining, elements are activated and their mechanical properties (stiffness/weight) are progressively increased up to their final value. Imposing a variation in the material properties is also typical in case a grouting process is simulated.

(viii) Frequently, the schematization of the structural elements (e.g. the lining, the steel ribs and bolting) by means of solid finite elements is not possible owing to unacceptable computational costs and possible localized mesh distortions. An alternative consists in adopting structural finite elements, e.g. truss, beam, plate and shell elements.

(ix) In principle, FEM is capable of dealing with large displacements. However, the mesh distortion may cause numerical instabilities. To overcome this problem, alternative numerical approaches have recently been proposed: (i) PFEM codes, based on a continuous remeshing around the 'particles' that carry material information (state variables) and whose position is progressively updated during the numerical simulation, and (ii) MPM codes, based on the use of 'material points' carrying material information (state variables) and moving in a spatially not evolving mesh. At present, although their employment could be useful for simulating grouting and excavation techniques, for practical applications in tunnel engineering they are not yet used.

9.1.5.2   FDM

From a theoretical point of view, the finite difference method is the simplest approach proposed to solve systems of differential equations. According to this method, derivatives may be approximated by incremental ratios. This allows us to obtain linear systems of equations. Owing to the simplicity of implementation and the small computational costs associated with every single time step, this method is very popular. In most of the cases, FDM equations are explicitly integrated. The achievement of the numerical solution is in general provided without any pathological convergence problem. Nevertheless, the accuracy and the reliability of the numerical results are more difficult to be assessed.

Analogous to FEM, also FDM requires the definition of (i) the spatial domain discretization, (ii) the temporal discretization, (iii) initial and (iv) boundary conditions and (v) constitutive relationships and allows us to deal with (i) an evolving geometry, (ii) a change in material properties, (iii) large displacements and (iv) the introduction of structural elements. For the discussion of these points, the reader is invited to refer to Section 9.1.5.1.

9.1.5.3   DEM

In the discrete element method, discontinuous media are represented as an assembly of discrete blocks, while discontinuities are modelled as deformable contacts. The blocks may be rigid or deformable and may experience rotations and displacements. An overlap of limited extent between blocks, controlled by a penalty function, is possible: as a matter of fact, contact between blocks is present only when two blocks are overlapped. Normal and tangential forces generated at contacts depend on the contact law. The definition of the contact law and the calibration of its parameters is crucial. As it was previously mentioned, tunnel excavation is essentially an unloading process and, therefore, the adopted contact law has to correctly reproduce the material behaviour under unloading stress paths.

DEM numerical algorithms are based on the explicit integration at each time step of the dynamic equations of motion for each block. Displacements and velocities of each block are thus updated, and owing to the temporal explicit integration, accelerations are calculated. The convergence of the analysis is numerically favoured by adding viscous dampers at the contact, whose physical/mechanical meaning is commonly considered to be questionable. At any rate, when quasi-static problems are solved, the solution obtained is independent of the viscous parameter employed.

At each time step, the block position is updated, allowing without any computational difficulty the solution to the mechanical problem under large displacements. This makes, for instance, the method particularly suitable for simulating the local detachment of rock blocks.

In some commercial codes:

* special bonds among blocks may be defined in order to simulate the presence of nails. Their constitutive relationship has to be carefully calibrated on the local structural response of the stiff inclusion taken into account;
* the blocks are not rigid, and their strain field is calculated by locally solving a continuum mechanic problem under a known geometry and an assigned stress distribution on the block boundaries.

9.1.5.4   Safety factor numerical estimation

A simplified approach, not taking into consideration the correct material stress/strain history, (known in the literature as c–φ reduction method), consists in defining an initial stable condition and progressively decreasing the material strength parameters until the numerical instability is achieved. The instability is identified as an uncontrolled increase in displacement of a set of 'check points'. In many cases, the numerical instability derives from very localized mechanical processes not governing the global response. This method is employed to numerically evaluate the system stability when the Mohr–Coulomb failure criterion is adopted. The value of strength parameters corresponding to the system failure provides a safety factor estimation.

9.2   ANALYSIS OF THE EXCAVATION PHASE

In this paragraph, two specific aspects of the excavation process in soils/rock masses, the hydro-mechanical coupling (Section 9.2.1) and the evolution of geometry (Section 9.2.2), are initially addressed. By following a standard approach in tunnel engineering, the stability of the tunnel face (Section 9.2.4) and the cavity (Section 9.2.5) by using LEM, limit analysis theory or analytic/numerical solutions based on continuum mechanics is discussed. Finally, the stability (sliding and detachment) of a single rock block is analysed (Section 9.2.6).

9.2.1   Hydro-mechanical coupling

Since tunnels are frequently driven below the groundwater table and the excavation implies a change in the pore water pressure in the spatial domain, the hazards related to the face/cavity stability may be assessed by solving a hydro-mechanical coupled problem. Among other representative examples of the effects of hydro-mechanical (HM) coupling, we mention the influence of face advance rate on tunnel stability and on tunnelling-induced subsidence.

In a quite general setting, a fully coupled numerical formulation of the HM problem is usually obtained by using fluid mass balances and linear momentum balance for the porous solid. The governing equations include a set of constitutive laws, relating stresses and fluid mass contents, in the case of partially saturated soils, to solid skeleton strains and pressures of interstitial fluids. The coupling between these equations can be described by means of a Biot's coupling tensor (Callari & Abati, 2011), whose proper particularization allows the applicability of the computational method to tunnelling in soils and in fractured rock masses treated as (not necessarily isotropic) equivalent continua (Section 9.1.1). Fluid flow equations are also required (e.g. of the Darcy's kind), and further constitutive laws can be employed to model the dependence of permeability on saturation degree. Furthermore, in intensely jointed rock masses treated as equivalent continua, a strong influence of strain on permeability tensor is frequently observed. Such a further source of HM coupling can be modelled by assuming the fluid flow to occur only in rock mass discontinuities. In this way, the permeability tensor of the equivalent continuum can be calculated by means of the additive decomposition of the contributions of each family of planar joints, whose opening, in general, may vary during excavation. These contributions can be calculated by using models based on the so-called 'cubic law' for permeability (Callari & Abati, 2009).

However, in some cases of interest, the numerical simulation of tunnelling below the groundwater table can be performed by means of uncoupled approaches, assuming either a 'perfectly drained' or 'perfectly undrained' response of the saturated ground. For example, in high-permeability soils and rock masses, the final steady-state flow field is often calculated as the solution of an uncoupled seepage problem, i.e. neglecting the ground deformability, once proper boundary conditions are imposed, including the inner boundaries of the tunnel. In a second calculation step, an uncoupled analysis of the solid skeleton mechanical response is developed in terms of effective stresses, submerged unit weight and seepage forces, where the latter forces are provided by the aforementioned seepage solution.

Conversely, in low-permeability soils and rocks, an undrained response to excavation is often assumed (the volume is locally imposed to be constant and the pore water pressure is calculated), equivalent to considering an infinite face advance rate. After such an uncoupled calculation, a coupled numerical analysis is usually performed by imposing proper conditions at the cavity boundary to describe the time evolution of the post-excavation response. This consolidation analysis is finalized to the assessment of the 'long-term' post-excavation response.

Both the aforementioned uncoupled approaches should not be used in the intermediate cases where comparable characteristic times of consolidation and excavation can be expected. In these cases, indeed, if the solid skeleton behaviour is path dependent, the simplifying assumption of an instantaneous (i.e. undrained) excavation followed by a consolidation process can lead to final results in significant disagreement with the actual response. Analytical assessments of both consolidation and excavation times are reported by Callari (2004) and Callari and Casini (2005) as a function of tunnel problem and ground properties.

In HM coupled analyses of tunnelling, care must be given to the proper setting of the computational domain, in terms of (usually variable) geometry and boundary conditions on displacements/stresses and pore pressures/flows. Examples of 2D and 3D modelling of full-face excavation of a shallow circular tunnel in a saturated soil deposit are presented in Figure 9.3a and b, respectively. Compared to the uncoupled case of tunnelling above the groundwater table, a significantly wider domain is usually required to minimize the effects of the lateral boundaries on flow and displacement fields. In particular, far-field conditions of zero displacements are imposed on such boundaries, in view of their considerable distance from the cavity axis: 10D in (a) and 7.7D in (b). As far as hydrostatic pore pressure (u_{w0}) is concerned, larger dimensions of the numerical model could be more suitable.

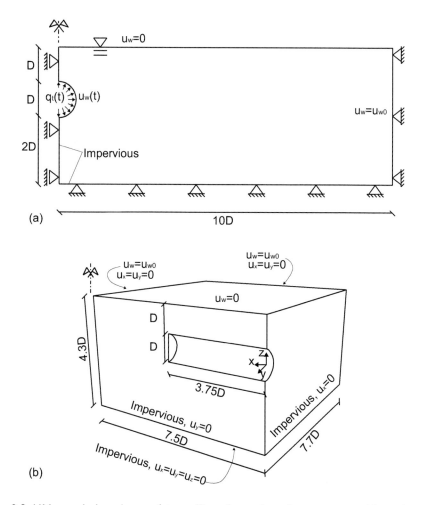

Figure 9.3 HM coupled analyses of tunnelling. Examples of geometry and boundary conditions on displacements/stresses and pore pressures/flows: (a) 2D problem geometry with applied BCs, including 'excavation' forces q_t and 'excavation' pore pressures $u_w(t)$ on the cavity boundary; (b) 3D problem geometry with assumed BCs.

9.2.2   The excavation process as an evolving geometry problem

9.2.2.1   Three-dimensional analyses

In this section, the main aspects to be considered in numerical analyses aimed at determining the influence of the excavation process on the stress and strain fields are summarized. For the sake of clarity, conventional tunnelling (Section 9.2.2.1.1) and mechanized tunnelling (Section 9.2.2.1.2) are discussed separately.

9.2.2.1.1   CONVENTIONAL TUNNELLING

When the tunnel is excavated, the evolution of geometry of (i) soil domain, (ii) first-phase support, (iii) invert and (iv) final lining (Figure 9.4) has to be simulated. Since non-linearities dominate the system response, the numerical solution depends on the stress/strain history and, therefore, on the excavation phases.

Usually, the numerical analyses have to simulate (i) the material removal, (ii) the first-phase support installation, (iii) the invert excavation and (iv) the final lining cast (Figure 9.4) and, eventually, the insertion of reinforcement systems. The soil is progressively removed (Section 9.1.5.1), whereas lining and reinforcement system are progressively activated. In case a soil improvement technique is adopted, within the soil domain some zones are progressively made stiffer and more resistant (grouting and freezing). In this case, crucial is the role of volumetric strains induced by the soil improvement and the simulation of the local change in stresses.

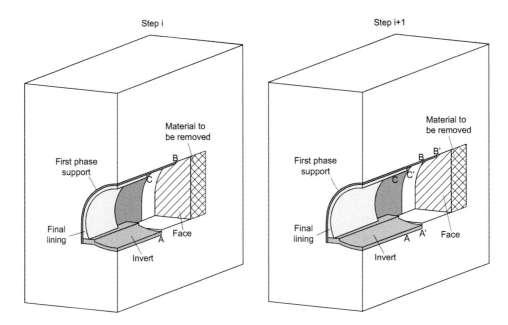

Figure 9.4 Scheme of conventional tunnelling numerical simulations.

In fully 3D and HM coupled simulations of tunnelling, the actual face advance rate can be stepwise approximated by the progressive deactivation of subdomains of proper length (Section 9.2.2). The complete deactivation procedure is the iteration of a two-step sequence (Figure 9.5): (i) a practically instantaneous face advance is simulated by the deactivation of a single region (i.e. a constant volume excavation) and proper pore pressure values are imposed at the new cavity boundary surface, including the excavation face, and (ii) the tunnel face location remains fixed for the time interval required to approximate the given advance rate; then, for the given excavation step, also the region modelling the lining can be activated, after a proper initialization of its strains (Callari & Casini, 2006; Callari et al., 2017). The permeability of the lining material should be realistically modelled, and proper conditions have to be set on pore pressures at the inner boundary of the lining. In case waterproofing is inserted between first-phase and final linings and drainage systems collecting the water flux are positioned there, the inner hydraulic boundary conditions have to be modified to numerically simulate the time history of the hydraulic perturbation induced by the excavation.

The numerical solution to the excavation problem allows the evaluation of the influence of the imposed variation in geometry on the system response. This can be described by considering a fixed control point (P in Figure 9.6a) positioned on the boundary of the tunnel cross section and computing the variation in radial stresses σ_r and displacements u_r, while the tunnel face is progressively moving.

As an example, the 3D FEM non-linear numerical analysis results, relative to an unlined tunnel reported in Flessati (2017), are plotted in Figure 9.6. The analysis was

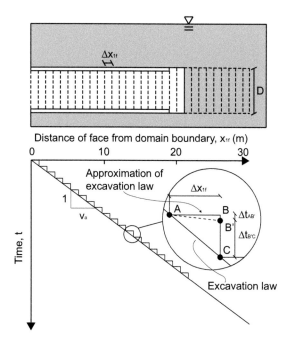

Figure 9.5 Simulation of advancing excavation with rate v_a and lining construction with given unsupported span.

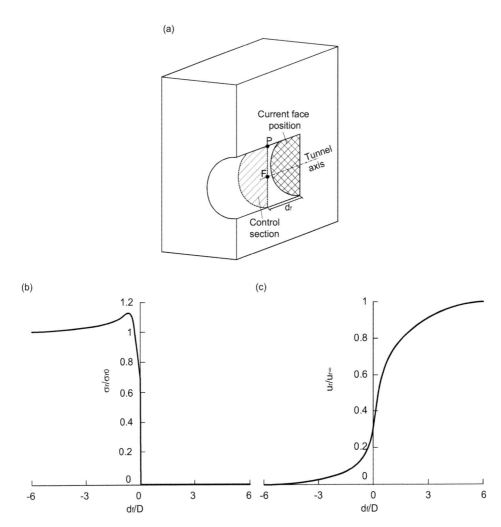

Figure 9.6 (a) Definition of the fixed control points; (b) variation in radial stress and (c) in radial displacements with the distance from the face. (From Flessati 2017.)

performed by considering a homogeneous soil, whose undrained behaviour is modelled by means of an elastic–perfectly plastic constitutive law (relating total stresses and strains). The failure condition is given by the Tresca criterion, and the flow rule is assumed to be associated (nil dilatancy angle value). In Figure 9.6, σ_r (normalized with respect to the geostatic one σ_{r0}) and u_r (normalized with respect to the final value $u_{r,\,\infty}$) are plotted versus the current position of the tunnel face d_f (normalized with respect to D). Negative d_f values mean that the material in the position of point P is not excavated and an increase in d_f corresponds to the advance of the tunnel face.

   As is evident from Figure 9.6, the face advance induces a variation in radial stresses (Figure 9.6b) and an accumulation of displacements (Figure 9.6c) in the control section.

For negative values of d_f, the stress variation is characterized by a peak (also experimentally observed by Nomoto et al., 1999; Shin et al., 2008; Berthoz et al., 2012), testifying the development of an 'arching effect' in the soil domain. For $d_f=0$, radial stress becomes nil and, since the cavity is unlined, for $d_f>0$ remains constant. On the contrary, displacements are continuously increasing with d_f, meaning that they are induced by a variation in radial stresses in points far from the control one. This non-local effect is a consequence of the compatibility conditions to be satisfied in the whole continuum domain. The value of (non-dimensional) displacement accumulated when d_f becomes nil is approximately equal to 30% (from a theoretical point of view this value is expected to be influenced by the material mechanical properties). The remaining 70% is accumulated after the soil is removed at the control point (positive d_f values).

On the contrary, if point F in Figure 9.6a is taken into consideration (in this case the control point is on the tunnel axis), the variations in horizontal stresses σ_h (normalized with respect to the initial geostatic value σ_{h0}) and displacements u_h (normalized with respect to the final value $u_{h\infty}$) are different from zero and physically meaningful (Figure 9.7) for $d_f<0$. For $d_f=0$, the control point belongs to the tunnel face, and for $d_f>0$, stresses and displacements cannot be defined.

9.2.2.1.2 MECHANIZED TUNNELLING

For the sake of brevity, hereafter only the case of tunnel boring machines (TBMs, Chapter 8), in which the shield is present and the tunnel face is supported, is considered. In this case, as it will be detailed in Chapter 10, (i) the material removal, (ii) the TBM shield, (iii) the segmental lining, (iv) the pressure applied on the tunnel face, (v) the grouting pressure, (vi) the grout pumped at the lining extrados and (vii) the jack forces applied to the lining (Figure 9.8) have to be accounted for. The TBM shield is commonly simulated as a temporary lining, progressively moving during

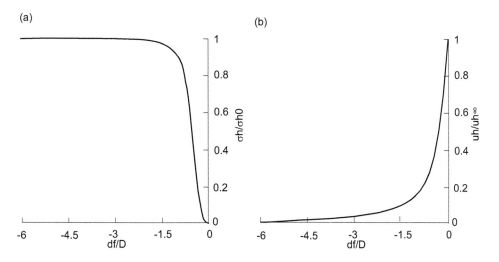

Figure 9.7 (a) Variation in horizontal stress and (b) in horizontal displacements with the distance from the face. (From Flessati 2017.)

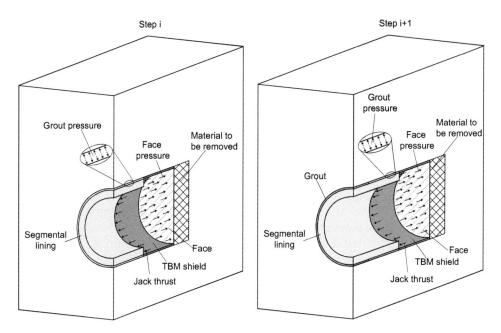

Figure 9.8 Scheme of mechanized tunnelling numerical simulations.

the excavation process. At each excavation phase, the applied pressure distributions (Figure 9.8) have to be spatially updated while the excavation advances: for instance, at every step the face pressure needs to be applied on the elements defining the current face position.

In case of fully HM coupled analyses, the procedure suggested at the end of Section 9.2.2.1 has to be adapted to the construction phases listed here above.

9.2.2.2   Axisymmetric approach

From a computational point of view, the 3D analyses reported in Section 9.2.2.1 are quite expensive. For this reason, in the current engineering practice, simplified approaches are preferred. A very popular strategy consists in describing the geometry evolution by assimilating the problem to an axisymmetric one and by assuming: (i) the tunnel and lining geometry to be cylindrical, (ii) the gravity to be disregarded and (iii) the initial state of stress to be isotropic (Figure 9.9). This approach is commonly used to simply determine the evolution with the face distance of stresses, strains and displacements. These are employed for simplified 2D fixed geometry analyses (Section 9.2.2.3 and Chapter 10).

9.2.2.3   2D fixed geometry analyses

From a computational point of view, numerical analyses reproducing the progressive evolution of the system geometry are very expensive. For this reason, in many practical cases, the tunnel excavation is reproduced by considering a fixed geometry and

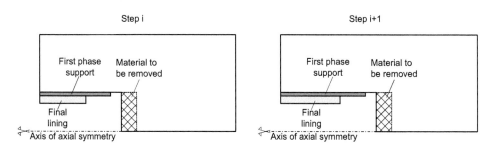

Figure 9.9 Scheme of axisymmetric tunnelling numerical simulations.

by simulating the material removal as a progressive reduction in a (fictitious) pressure distribution applied on a part of the subdomain (i.e. the tunnel face or the tunnel cross section).

In case of tunnel faces, the reduction in horizontal stresses is adopted not only in the context of numerical analyses, but also for analytical solutions (Section 9.1.2), integrated/ derived solutions (Section 9.1.3) and solutions for stability analyses (Section 9.1.4). These will be discussed in Section 9.2.4.

For the cavity, this approach is applied by considering a 2D plane strain geometry, corresponding to a transversal cross section, and simulating the progressive tunnel excavation by reducing the fictitious internal pressure applied at the boundary. Under this assumption, (i) the response of all the cross sections is coincident, although temporarily out of phase, and (ii) an increment in displacements is possible only if radial stresses decrease (if the considered material is rate independent), implying that the previously cited non-local effect (Figure 9.6 in Section 9.2.2.1.1), consisting in an accumulation of radial displacements while the local radial stresses are constant, cannot be simulated. The relationship between applied stresses and radial displacements is commonly named ground reaction curve (GRC). The GRC can be obtained either numerically or analytically (Section 9.2.5.2.2).

As it is evident, the simplified 2D approach cannot take the face advance explicitly into account. In a simplified way, using the elastic or the elastic–plastic results (e.g. Panet & Guenot 1983; Vlachopoulos & Diederichs, 2009) under the hypothesis of homogeneous isotropic medium and axisymmetric conditions (Section 9.2.2.2), the convergence of the cavity is related to the distance from the tunnel face. Both the Panet and Guenot (1983) solution and GRCs are essential for the simplified approaches of lining design (Chapter 10).

This method can also be extended to the 2D HM coupled simulation of tunnelling by considering, in addition to the release of stresses applied on the cavity, also properly defined 'excavation pore pressures' resembling the drainage conditions induced by the advancing excavation face (Figure 9.3a). The formulation and 3D validation of such a procedure is presented in detail by Callari and Casini (2005).

9.2.3   Modelling of support and reinforcement systems

In principle, the mechanical response of lining, invert and reinforcement systems (anchors, nails and bolts) may be simulated by using solid elements. Nevertheless, this is not usually possible for the associated computational costs. In practice, both support and reinforcement systems are modelled by employing structural elements (Section 9.1.5.1): in particular, for supports the employment of plate elements is common, whereas for nails and anchors, beam or truss elements are adopted, respectively. Structural elements not only allow abruptly reducing the number of degrees of freedom, but also permitting the direct estimation of generalized stresses (axial and shear forces and bending moments) acting in them.

Interfaces are commonly adopted to simulate the contact between soil and structural elements. These interfaces are rigid under compression and characterized by a nil tensile strength. Along the tangential direction, a friction angle (nullifying in the limit case of smooth interface), representing the roughness of the contact, has to be defined.

To further reduce computational costs, the use of a substructuring approach is common, as discussed in Chapter 10. For both nails and anchors, (i) the employment of 'embedded elements' (Sadek & Shahrour, 2004), not sharing the nodes with the solid elements in which they are positioned, or (ii) the substitution of the reinforcements and the adjacent soil with a homogenized material, characterized by an additional 'equivalent cohesion' (Grasso et al., 1989), is common.

9.2.4   Simplified approaches for estimating the mechanical response of tunnel face in soils

9.2.4.1   Shallow tunnels

9.2.4.1.1   SOLUTIONS FOR STABILITY ANALYSIS

When the cover-to-tunnel diameter ratio value is sufficiently small (i.e. lower than 4), the behaviour of the tunnel face cannot be longer analysed using the hypothesis of spherical symmetry (Section 9.2.4.2.2) and, as such, no analytical solutions exist. In this case, the ground surface influences the face response and 'chimney-like' failure mechanisms take place.

As far as undrained conditions are concerned, by following the simplified approach cited in Section 9.2.2.3, numerous both upper and lower limit analysis solutions, inspired by the pioneering work of Davis et al. (1980), have been proposed. These solutions are considered to be reliable when the cover-to-diameter ratio value is smaller than 3 (Davis et al. 1980).

As far as drained conditions are concerned, the mechanical response of tunnel faces was studied by both the LEM and the limit analysis theory (Section 9.1.4).

The LEM was employed by Horn (1961), who introduced a failure mechanism consisting in a wedge positioned in the proximity of the face and a parallelepiped-shaped chimney reaching the ground surface. More recently, the same failure mechanism has also been employed in case of mechanized tunnels to estimate the pressure to be applied on the face to avoid the collapse (Balthaus, 1988; Anagnostou & Kovari, 1996; Broere, 2001) and to study the influence of seepage forces (Perazzelli et al., 2014).

Using the limit analysis approach, the minimum pressure to be applied on the face was estimated for the first time by Leca and Dormieux (1990) by using lower and upper bound solutions.

9.2.4.1.2 INTEGRATED/DERIVED SOLUTIONS

For drained conditions, Vermeer et al. (2002), by assuming (i) the tunnel excavation to be modelled as a progressive reduction in a geostatic pressure initially applied to the face (Section 9.2.2.3), (ii) the initial stress distribution to be linearly increasing with depth (the self-weight is considered) and (iii) the lining to be rigid, numerically calculated the minimum pressure to be applied on the face to prevent its collapse. This limit pressure value depends on the tunnel diameter, soil unit weight, cohesion and friction angle, but is not influenced by tunnel cover, dilatancy angle value and material elastic properties. The expression found by Vermeer et al. (2002) is in a very satisfactory agreement with both limit analysis solutions (Leca & Dormieux, 1990) and small-scale model test results (Kirsch, 2010).

9.2.4.2 Deep tunnels

9.2.4.2.1 INTEGRATED/DERIVED SOLUTIONS

Under drained conditions, the previously cited expression, proposed by Vermeer et al. (2002) for shallow tunnels (Section 9.2.4.1.2), is still valid for deep tunnels.

As far as undrained conditions are concerned, di Prisco et al. (2018) numerically analysed the response of tunnel faces by imposing assumptions very close to those listed above (Section 9.2.4.1.2) referring to the Vermeer et al. (2002) solution. In particular, the authors have introduced a couple of non-dimensional variables (one for the stress applied on the face and one for the face displacement, both in the horizontal direction), allowing to derive a unique non-dimensional GRC, independent of the material mechanical properties, initial state of stress and tunnel diameter. They proposed an analytical expression, inspired by the analytical solution for spherical cavities (Section 9.2.4.2.2) and also valid in case the initial state of stress is not isotropic, for reproducing GRC:

$$
q_f = \begin{cases} Q_f, & Q_f < a_f \\ a_f e^{Q_f/a_f - 1}, & Q_f \geq a_f \end{cases} \tag{9.1}
$$

where

$$
Q_f = \frac{\sigma_{f0} - \sigma_f}{S_u} \tag{9.2}
$$

$$
q_f = \frac{u_f}{u_{fr,el}} \frac{\sigma_{f0}}{S_u} \tag{9.3}
$$

where σ_f is the average stress applied on the face, σ_{f0} the initial geostatic value of σ_f, S_u the undrained strength, and u_f the average face displacement, this latter given by:

$$
u_{fr,el} = \frac{\sigma_{f0}}{K_{el} E_u} D \tag{9.4}
$$

Here, E_u is the undrained Young's modulus, K_{el} is a non-dimensional coefficient depending on the cover-to-diameter ratio (according to di Prisco et al., 2018, $K_{el} = 3$ for cover-to-diameter ratio larger than 5), and a_f is a non-dimensional coefficient depending on the initial stress anisotropy factor $\bar{k}$ (i.e. the geostatic ratio between total horizontal and total vertical stresses).

The numerical results reported in di Prisco et al. (2018) also put in evidence that the dimension of the plastic subdomain continuously increases while the face is unloaded, implying the absence of occurrence of collapse mechanisms. This means that (i) both the LEM and the limit analysis theory are not suitable for analysing this case and (ii) the proposed GRC expression has to be employed in the framework of a displacement-based design perspective.

9.2.4.2.2   ANALYTICAL SOLUTIONS

The mechanical response of deep tunnel faces has been assimilated to the one of a spherical cavity where the excavation process is modelled as a progressive reduction in the internally applied pressure, from the in situ value to zero (Section 9.2.2.3). In case (i) the initial state of stress is isotropic and uniform (i.e. the self-weight is disregarded), (ii) the tunnel depth is infinite (the tunnel cover-to-diameter ratio is sufficiently large so that the effect of the ground surface is negligible), and (iii) the material is uniform and isotropic, the problem is characterized by a spherical symmetry and can be solved in case the material mechanical behaviour is either elastic or elastic–perfectly plastic. The analytical expressions of the GRC for the elastic case and for the elastic–plastic case (both for the drained case with the Mohr–Coulomb failure criterion and for the undrained case with the Tresca failure criterion) are reported in Yu (2013). When the material is elastic–plastic, the analytical solutions for both drained and undrained conditions are characterized by (i) an indefinite hardening (that is the plastic radius can continuously increase in size) and (ii) the absence of a collapse mechanism. This unphysical result is a consequence of neglecting the body forces (gravity).

9.2.5   Simplified approaches for the analysis of the tunnel cross section

As observed in Section 9.2.2.3, stress–strain and stability analyses of the tunnel cross section can be conveniently performed by adopting a plane strain approach, in which the 3D conditions and the related 'face effect' are fictitiously reproduced by an internal radial pressure q applied at the tunnel boundary. In case of shallow tunnels, limit analysis solutions are available for the evaluation of the cavity stability (Section 9.2.5.1.1), whereas in case of deep tunnels, analytical solutions are quite popular to define the GRC of the cavity (Section 9.2.5.2.2).

9.2.5.1   Shallow tunnels

9.2.5.1.1   SOLUTIONS FOR STABILITY ANALYSIS

If 2D plane strain conditions are adopted (Section 9.2.2.3) and the tunnel axis is located at a depth equal to H below a ground surface loaded by a uniform surcharge

q_s, the minimum pressure to be applied on the cavity to avoid its collapse (q_t) under drained conditions can be evaluated by employing the following equation, inspired by the well-known expression for bearing capacity of shallow foundations:

$$q_t = \frac{1}{2}\gamma D Q_\gamma + q_s Q_s + c' Q_c \tag{9.5}$$

with γ being the soil unit weight and

$$Q_c = (Q_s - 1)\cot\phi' \tag{9.6}$$

whereas the non-dimensional coefficients Q_γ and Q_s, depending on the cover-to-diameter ratio and on the friction angle, are calculated by employing the limit analyses theory (D'Esatha & Mandel, 1971; Atkinson & Potts, 1977; Ribacchi, 1978; Mühlhaus, 1985).

Similar solutions have been introduced for assessing the stability of tunnels in fine-grained soils under undrained conditions, for which the Tresca strength criterion is adopted (Davis et al., 1980).

9.2.5.1.2 CAVITY RESPONSE IN INCLINED STRATA

Tunnel excavation in inclined strata induces stress and strain changes in the surrounding domain significantly different from those occurring when the ground surface is horizontal. This affects both stresses acting on the lining and the settlement at the ground surface (Section 9.3.2). This SSI problem is further complicated by the possible presence of an already active landslide body intersected by the tunnel (Section 9.4.4), with different scenarios depending on the relative orientation of the slope to that of the tunnel axis.

A more comprehensive understanding of the physical phenomena can be obtained by using 3D elastic–plastic numerical analyses considering the real geometry of the slope and simulating the progressive tunnel excavation, provided that the pre-existing state of stress in the slope is correctly estimated (Urciuoli, 2002). However, to provide a first insight into the interaction problem, 2D numerical analysis results can be employed.

For example, for the ideal case of an infinite slope and circular tunnel, D'Effremo (2015) demonstrated that the plastic zone around the cross section increases more than linearly with slope inclination.

9.2.5.2 Deep tunnels

9.2.5.2.1 SOLUTIONS FOR STABILITY ANALYSIS

Equation 9.5 (Section 9.2.5.1.1) introduced for shallow tunnels to evaluate q_t can also be adopted for deep tunnels. The limit analysis solutions (D'Esatha & Mandel, 1971; Atkinson & Potts, 1977; Ribacchi, 1978; Mühlhaus, 1985) put in evidence that for sufficiently large cover-to-diameter ratio values (larger than 5), Q_s becomes nil, whereas Q_γ assumes a constant value, since for deep tunnels, a local (not involving the ground surface) failure mechanism is expected.

9.2.5.2.2 ANALYTICAL SOLUTIONS

Under the hypothesis of an infinitely long cylindrical cavity, excavated within an infinite homogeneous elastic medium, subject to an anisotropic state of stress independent of depth (Figure 9.10), Kirst (1898) introduced a solution in terms of stresses, strains and displacements along radial and circumferential directions. In terms of displacements (u_r and u_θ are the displacements along the coordinates r and θ defined in Figure 9.10),

$$u_r = \left[\frac{(S_v + S_h)}{2} - q_t\right]\frac{r_0^2}{2Gr} - \left[\frac{(S_v - S_h)}{2} - q_{dev}\right]\left(4 - 4v - \frac{r_0^2}{r^2}\right)\cos 2\theta\, \frac{r_0^2}{2Gr} \quad (9.7)$$

$$u_\theta = \left[\frac{(S_v - S_h)}{2} - q_{dev}\right]\left(2 - 4v + \frac{r_0^2}{r^2}\right)\sin 2\theta\, \frac{r_0^2}{2Gr} \quad (9.8)$$

with S_v and S_h the stresses applied on the boundaries here imposed infinitely distant from the cavity axis, q_t a loading variable defining the isotropic part of the stress applied on the tunnel cavity boundary ranging between $(S_v + S_h)/2$ (initial condition) and 0, q_{dev} a loading variable defining the deviatoric part of the stress applied on the tunnel cavity boundary ranging between $(S_v - S_h)/2$ (initial condition) and 0, G the elastic shear modulus, and v the Poisson ratio.

For an isotropic state of stress ($S = S_v = S_h$), $u_\theta = 0$ and the solution does not depend on the anomaly θ:

$$u_r = \frac{1 + v}{E}(S - q_t)r_0 = \frac{S - q_t}{2G}r_0 \quad (9.9)$$

In terms of stresses, the most severe stress condition is obtained at the tunnel boundary where $\sigma_r = q_t$ and $\sigma\theta = 2S - q_t$.

The GRC is also analytically obtained in case an infinitely long cylindrical cavity, subject to an isotropic state of stress independent of depth and the medium, is assumed to be either elastic–plastic or elastic–perfectly fragile with the Mohr–Coulomb strength criterion and a residual not nil shear resistance (Ribacchi & Riccioni, 1977). In this latter case, the characteristic curve can be expressed as:

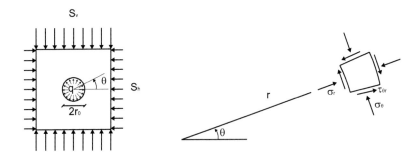

Figure 9.10 Infinitely long cylindrical cavity.

$$u_r = \frac{1+v}{E}\left[\frac{R^{N_\psi+1}}{r_0^{N_\psi}}\left(S + c'_p\cot\phi'_p\right)\sin\phi'_p + \left(S + c'_r\cot\phi'_r\right)(1-2v)\left(\frac{R^{N_\psi+1}}{r_0^{N_\psi}} - r_0\right) \right.$$
$$\left. + \frac{1 - N_\psi N_r - v\left(N_\psi + 1\right)\left(N_r + 1\right)}{\left(N_r + N_\psi\right)r_0^{N_r-1}}\left(q_t + c'_r\cot\phi'_r\right)\frac{R^{N_\psi+1}}{r_0^{N_\psi}} - r_0^{N_r} \right] \tag{9.10}$$

with

$$N_r = \frac{1 + \sin\phi'_r}{1 - \sin\phi'_r} \tag{9.11}$$

$$N_\psi = \frac{1 + \sin\psi}{1 - \sin\psi} \tag{9.12}$$

In Equation 9.10, E is the Young's modulus, ϕ'_p, c'_p and ϕ'_r, c'_r are the friction angle and cohesion at peak and residual conditions, respectively, ψ is the dilatancy angle, and the plastic radius (R_p) is given by:

$$\frac{R_p}{r_0} = \left(\frac{\dfrac{2S - f_p}{N_p + 1} + \dfrac{f_r}{N_r - 1}}{q_t + \dfrac{f_r}{N_r - 1}} \right)^{\frac{1}{N_r-1}} \tag{9.13}$$

with

$$N_p = \frac{1 + \sin\phi'_p}{1 - \sin\phi'_p} \tag{9.14}$$

$$f_p = 2c'_p\sqrt{N_p} \tag{9.15}$$

$$f_r = 2c'_r\sqrt{N_r} \tag{9.16}$$

In case of isotropic elastic–perfectly plastic material, the solution is obtained by substituting in Equations 9.10 and 9.13 $\phi' = \phi'_p = \phi'_r$ and $c' = c'_p = c'_r$.

In the elastic case, the above restrictions can be released at the cost of more complicated analytical formulations. For example, gravity can be accounted for by considering a radial volume force equal to γ pointing towards the tunnel centre in the upper half of the section and outwards in the lower part (Fenner, 1938). The influence of pore water pressure in two-phase media can be accounted for analytically only by following an uncoupled approach (Section 9.2.1) according to which the hydraulic problem is separately solved by assuming a purely radial flow.

Moreover, Detourney and Fairhurst (1982) and Wong and Kaiser (1991) analysed the problem by considering an anisotropic natural state of stress and put in evidence the change in the shape of the plastic zone as a function of the stress anisotropy ratio and applied pressure q_t. At a low stress ratio, plastic zone tends to separate and to form two ear-shaped zones developing from the spring lines upwards.

9.2.6   Stability of rock wedges

Tunnels excavated in fractured rock masses can be affected by the collapse of rock blocks at the tunnel face or in the cross sections, depending on the relative orientation of the excavation to that of the joint sets.

A frequent geometry is given by a tetrahedral wedge delimited by three surfaces belonging to three different joint sets within the rock mass and by the excavation surface itself. Once the rock wedges kinematically free to move are identified by the analysis of joint orientation (Chapter 3), their stability is assessed by using the LEM under the following hypotheses:

- Joints are considered persistent and ubiquitous.
- Lithostatic state of stress is neglected, since the weight of the rock block is the main driving force together with those possibly related to (i) water pressure, (ii) seismic actions and (iii) actions due to the mitigation measures (Chapter 7).

Wedges of different dimensions, depending on joint spacing and maximum size compatible to that of the tunnel cross section diameter, are analysed.

The nature of the possible collapse mechanism is governed by the vectoral composition of the previously cited forces: in case contact along at least one surface is maintained, a sliding mechanism takes place, whereas in case the contact with all the surfaces is loss, fall/detachment mechanisms occur.

In special circumstances, the influence of the natural state of stress cannot be neglected, especially in deep tunnels for which very high circumferential stresses are reached at the tunnel boundary as a consequence of stress redistribution after excavation. In these cases, an alternative consists in calculating the state of stress evaluated in the absence of joints. Then, the local stress conditions (i.e. normal and shear stresses) at the joint level are applied and a discontinuum approach is employed to assess the wedge stability (e.g. Goodman, 1995).

A more general analysis can be carried out by adopting the block theory proposed by Goodman and Shi (1985), capable of assessing the most critical blocks to stability, characterized by any number of faces and thus shape. The collapse of the first block at the excavation boundary could then open the way to the instability of further blocks located inside the rock mass, and for this reason, its identification is of primary relevance.

As it was discussed in Section 9.1.5.3, DEM analyses are today possible and the assessment of most unfavourable blocks to stability is an automatic result of the computation. In fact, also in this case, strength parameters of joints can be progressively reduced (according to the c–ϕ reduction procedure introduced in Section 9.1.5.4) in order to investigate the occurrence of failure conditions.

9.3   INTERFERENCES

This paragraph concerns the effects of the tunnel excavation on the surrounding environment, i.e. according to an extended definition in the 'far field'. In particular, in Section 9.3.1 potential disturbances on the hydro-geological system in which the tunnel is excavated are considered, while Section 9.3.2 is dedicated to the induced subsidence; finally, in Section 9.3.2.3, the vibrations induced by blasting are analysed.

9.3.1   Tunnel water inflow and influence of tunnelling on groundwater regime

The influence of tunnel excavation on the natural groundwater regime can be analysed by using different calculation methods. A fully coupled HM analysis (Section 9.2.1) is generally required to simulate the transient regime induced by tunnel excavation and the time evolution of stresses acting on the support systems. For the prediction of the long-term groundwater regime and the average steady-state inflow into the tunnel, uncoupled flow analyses are sufficient.

The choice between coupled and uncoupled approaches depends on both geometrical and hydro-mechanical factors: tunnel diameter and cover depth, mechanical and hydraulic properties of the ground, and hydraulic conditions imposed at the tunnel wall as well as at outer limits of the ground volume (boundary conditions).

The coupled approach is generally needed for the analysis of relatively shallow tunnels in soft ground deposits. In this case, the key features to establish a satisfactory modelling approach are: (i) the relatively small extent of the ground volume affected by the drainage effect of the tunnel; (ii) well-defined and predictable hydro-geological conditions; and (iii) temporary support systems (of tunnel face and wall), final lining (pervious or impervious) and water pressure control systems explicitly represented in the model. The last requirement is particularly important for modelling closed-face mechanized excavation in urban areas. In this case, 'near field', near the face and around the tunnel lining including the ground support, cannot be modelled separately from the 'far field', e.g. water table drawdown and other far-reaching effects of tunnel-induced drainage.

On the contrary, the uncoupled flow approach can be considered for many situations of deep tunnels in fractured rock masses. For the simple case of a rock mass of homogeneous permeability, under steady-state flow conditions, the hydraulic head distribution near the tunnel is approximately represented by a logarithmic law (Fernandez & Alvarez, 1994; Ribacchi et al., 2002). The drainage effect on the surrounding rock mass will be maximum if tunnel walls can be considered fully permeable (water pressure $u_w = 0$).

As far as the hydraulic outer boundary conditions are concerned, the main distinction is between the case of confined seepage, i.e. with a fixed water table, and that of unconfined seepage, i.e. with a possible drawdown of the water table (Figure 9.11). The lowering of the water table primarily depends on the infiltration rate ε from the surface, and possibly on the lateral recharge from nearby aquifers, the drawdown being particularly important for a ground of relatively high permeability. In the first case, a constant piezometric head $h(t) = $ cost is imposed at the boundary, while in the second case, the piezometric surface (where the water pressure is equal to zero) is a priori unknown and time-varying, until the final stationary conditions are reached.

If the water table lowers down to the tunnel wall, the stationary flow conditions in a vertical cross section are provided by the classical Dupuit solution, originally

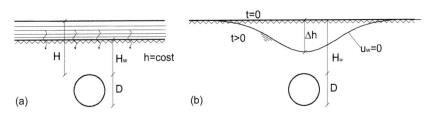

Figure 9.11 Typical hydraulic boundary conditions for confined (a) and unconfined (b) seepages.

proposed for the horizontal flow towards a draining trench (Harr, 1962). More sophisticated numerical approaches can take into account also the presence of a partially saturated zone above the water table and the dependency of the ground permeability on the saturation degree.

For deep tunnel projects, the detection of water-bearing zones, faults and karstified rock zones ahead of the tunnel face is of paramount importance. The analysis of transient inflow rates recorded near the tunnel face can be instrumental to 'update' major design parameters, such as transmissivity and storativity of the aquifer, back-calculate the characteristic sizes of the water-bearing zone intercepted by the excavation and infer the spatial distribution water pressures. Comparison of measured data with simplified solutions for the transient flow phase, e.g. the 'draining box model', based on the solution of Federico (1984), has proven particularly useful in the tunnelling practice (e.g. Ribacchi et al., 2002).

9.3.2   Induced subsidence and interaction with pre-existing underground and surface structures

The analysis of the interference between tunnel excavation and surrounding structures is mainly aimed at assessing any possible damage to existing underground structures and superstructures.

The subsidence induced by the tunnel construction depends on the chosen construction methods and is mainly related to: face and cavity (unsupported and supported) deformations due to 3D stress release, consolidation due to the dissipation of pore water pressure in clayey soils and long-term deformations associated with the development of viscous strains in weak rocks.

Mechanized excavation is generally assumed to cause small and negligible local reductions in stresses, in particular in case of shield excavation and especially with pressurized TBM. At the same time, other movements around a TBM can be caused by some extra cutting, due to the typical conical shape of the shield (Chapter 7) or to the larger diameter of the cutterhead, both necessary to reduce the friction between the advancing shield and the ground or to the inevitable shield pitching, rolling and yawing. These latter contributions are more important for large shields and may become not negligible in case of not well-trained workmanship. Furthermore, additional ground deformations naturally develop while closing the gap between the lining and the shield tail. This fraction can be reduced by a careful backfilling.

The problem can be solved by either using either a SSI uncoupled (Section 9.3.2.1) or a SSI coupled approach (Section 9.3.2.2). In the former case, 'geotechnical' and structural problems are solved separately. In the latter case, the spatial domain is not substructured and the solution is fully coupled.

9.3.2.1   SSI uncoupled approaches

In this case, the goal of the preliminary analysis consists in calculating the displacement field induced by the tunnel excavation. In case of superstructures, the goal is to calculate the ground displacement field induced in greenfield, that is neglecting the presence of structures (Chapter 10). In some cases, an upgraded approach consists in taking into consideration the presence of existing buildings by substituting these with equivalent pressure loads locally applied. The ground displacement field is then imposed to a refined model of the building to calculate the potential damages (Chapter 10). This solution is considered to be very conservative since it disregards the structure stiffness in the calculation of settlements of foundations.

Different methods can be used to estimate the ground movements induced at the ground surface and in depth by tunnelling: empirically based methods (Chapter 10), and theoretical methods based on analytical (e.g. Sagaseta, 1987; Verruijt & Booker, 1996) or numerical models (see Aversa et al., 2017).

In the last 40 years, many authors (Al Tabba & Wood, 1989; Stallebrass et al., 1994a, b; Stallebrass & Taylor, 1997; Addenbrooke et al., 1997; Grammatikopoulou et al., 2006, 2008) have discussed the high non-linearity of the soil mechanical behaviour on the evaluation of induced settlements.

9.3.2.2   SSI coupled approaches

Nowadays, it is possible to perform fully coupled soil–structure 3D numerical interaction analyses in which tunnel, soil domain and existing structures/superstructures are considered. In this case, the tunnel–soil–structure interaction is largely influenced by the accuracy in modelling the building structures. The properties of the materials, sometimes old and not classified, the geometry of the structural members, not always easy to be determined, and the overall shear and bending stiffness of the structural components should be investigated before any possible model assembling.

This fully coupled approach (e.g. Burd et al., 2000; Liu et al., 2001; Mroueh & Shahrour, 2003; Fargnoli et al., 2015) can be, in many cases, too complex and too long to be handled according to a typical design schedule. Simplified approaches have been suggested in order to reduce the computational costs based on suitable lumping techniques of the superstructure (e.g. Maleki et al., 2011; Losacco et al., 2014).

9.3.2.3   Inclined strata

In this case, the relative inclination of tunnel axis with respect to the maximum inclination slope direction governs the settlements' spatial distribution. For a tunnel perpendicular to the slope dip direction, settlements and horizontal displacements in a transversal cross section are no longer symmetric with respect to the tunnel axis. For their assessment, a graphical approach, based on the projection to an inclined plane of the classical semi-empirical curve valid for a horizontal ground surface (Chapter 10), was proposed by Whittaker and Reddish (1989) and Shu and Bhatacharya (1992). For increasing inclination angles, the subsidence trough expands upstream and contracts downstream.

A more comprehensive understanding of the system response is derived by the analyses of either 2D or 3D elastic–plastic numerical analyses results, once the pre-existing state of stress in the slope is accurately simulated (Urciuoli, 2002). For the ideal case of an infinite slope and a circular tunnel (D'Effremo, 2015), the plastic zone around the tunnel cross section increases more than linearly with slope inclination (Section 9.2.5.1.2). Thus, displacements at the ground surface become asymmetric and downhill directed.

The excavation of a tunnel in a dormant landslide is of particular interest, since human-induced changes in stresses may reactivate the movements. In this case, only elastic–plastic analyses, accounting for the presence of the failure surface with its estimated strength properties, are capable of simulating the phenomenon. D'Effremo (2015), again considering an infinite slope and a circular tunnel, demonstrated that the affected portion of failure surface increases with slope inclination and decreases with tunnel cover and tunnel–slip surface relative distance.

In terms of slope factor of safety, position and orientation of tunnel, induced by tunnel excavation, can be analysed by means of limit equilibrium (Section 9.1.4.1) and numerical methods (Section 9.1.5) and strongly depend on the initial characteristics of the state of stress, tunnel depth and slope inclination (Picarelli et al., 2002; D'Effremo, 2015).

9.3.3   Vibrations induced by drill and blast tunnel excavation

As described in Chapter 7, numerous are the techniques for tunnel excavation. Among these, when the tunnel is excavated in high-quality rock masses (Chapters 3 and 4), the drill and blast technique is still very popular. One of the main hazards of this technique is the damage of the nearby pre-existing under- or above-ground structures due to ground vibrations generated by explosions.

Numerical simulations are aimed at investigating the response of the structure at risk under the dynamic load induced by blasting. The reliability of the numerical solution is governed by the suitability of the numerical model in terms of spatial and time discretization and boundary condition imposed (Section 9.4.3) and depends on the correct definition of the dynamic input (explosion). The numerical strategy, despite its complexity, has the advantage of allowing a direct assessment of ground vibrations and internal forces inside the structure, subsequently used for structural verifications.

In most of the numerical works (e.g. Cho & Kaneko 2004; Ma & An, 2008; Jommi & Pandolfi, 2008; Sazid & Singh, 2013; Yan et al., 2015; Perazzelli et al., 2019), blasting is simulated by applying a pressure (P_{bh}) time evolution law at the boundary of the blast hole, as, for example, the one proposed by Cho and Kaneko (2004):

$$P_{bh} = P_p \xi \left(e^{-\alpha t} - e^{-\beta t} \right) \tag{9.17}$$

where $\xi = 1/(e^{-\alpha t0}-e^{-\beta t0})$, $t_0 = (1/(\beta-\alpha))\ln(\beta/\alpha)$, P_P is the peak pressure, t_0 is the time at which the pressure peak is reached, and α and β are empirical parameters.

If the aim of the analysis is the simulation of the so-called 'far-field' effects of the explosion (i.e. far from the blasting hole), rock behaviour can be modelled by employing a simple elastic relationship with the addition of a Rayleigh damping (Perazzelli et al., 2019). In contrast, near-field conditions (i.e. near the blasting hole), characterized by fracture propagation and rock comminution, can only be reproduced by means of sophisticated constitutive relationship based on either fracture (Jommi & Pandolfi, 2008) or damage mechanics (Ma & An, 2008).

A preliminary assessment of the model response under free-field conditions and its comparison with empirical solution is strongly suggested.

The simplified empirical approaches consist in the assessment of the so-called 'peak particle velocity' (PPV): the local maximum velocity of a point distant s from the charge point. Once the distance of the structure from the charge point is known, PPV defines a measure of disturbance level and has to be compared with empirical threshold values.

According to the literature (e.g. Kumar et al., 2016), PPV (expressed in mm/s) depends on the charge per delay Q_d (expressed in kg) and distance from the centre of gravity of the charge s (expressed in m). Numerous are the empirical equations available (Singh & Vogt, 1998); for instance, the one proposed in USBM (Duvall and Petkof, 1959) reads as follows:

$$\text{PPV} = p_1\left(\frac{s}{\sqrt{Q_d}}\right)^{-p_2} \tag{9.18}$$

where empirical parameters p_1 and p_2 are determined on site-specific in situ test results.

9.4 LONG-TERM TUNNEL PERFORMANCES

9.4.1 HM coupled processes

The construction of tunnels in saturated grounds induces a modification of both total stresses and pore pressure around the tunnel (mechanical and hydraulic perturbations). The spatial diffusion of such a perturbation within the soil/rock mass domain takes place with time and is governed by both mechanical and hydraulic properties of the materials involved. Hydro-mechanical coupled problems (Section 9.2.1) may also be analysed by taking into consideration the evolution of hydraulic properties, such as the material permeability, with the progressive deformation of the solid skeleton and, in case, of the discontinuities.

If the soil/rock permeability value is low, the excavation of the tunnel and the installation of the support system can generally be modelled as an undrained process. The changes in the pore pressure generated in the ground in the short-term situation and the modified boundary conditions (drainage effects of the excavation if no waterproofing systems, see Chapter 10, are installed) initiate a consolidation process in which time-dependent deformations occur, followed by changes in the lining loads, until the long-term steady-state flow conditions are attained.

A very special case is that occurring in an elastic medium under the hypothesis of axisymmetry (see Section 9.2.5.2.2). Under undrained conditions, the excavation, represented as a stress relief at the tunnel boundary, does not change the mean stress inside the ground ($\Delta p = 0$, with p being the total average pressure) and no pore pressure variation Δu is generated at $t = 0$. During the consolidation stage, the pore pressure perturbation propagates from the tunnel wall inside the ground causing an inflow seepage and a change in effective stresses; however, the convergence remains the same and the lining loads do not change. In case $\bar{k} \neq 1$, the effect of the average and deviatoric stress components can be superimposed. In this case, a time-dependent loading of the lining occurs, but only related to the latter stress component, which causes a purely flexural deformation of the lining (ovalization).

In the elastic–plastic case, the consolidation process induced by the mechanical and hydraulic perturbations is characterized by a modification in the tunnel convergence, even in the axisymmetric case, and thus by a corresponding time-dependent change in lining loading.

In weak rock masses, such as stiff argillaceous formations, the stress change caused by excavation may cause opening of pre-existing fissures and new fracturing within the 'plastic' annulus around the tunnel, i.e. the 'excavation-damaged zone (EDZ)'. The loosened fissure network increases the overall permeability of the ground mass (often represented as an equivalent continuum medium) and favours the desaturation process of the EDZ. A modelling approach in which the negative pore pressure is limited or, more roughly, the pore pressure is nullified in EDZ can better represent this situation.

In fractured hard rock mass, the water flow is mainly localized within a network of hydraulically interconnected discontinuities, often characterized by varying aperture, loose infilling materials and intersection with high-transmissivity zones such as karstic cavities and fault zone. This highly localized flow conditions tend to produce local shearing and non-uniform load distributions on the lining, which can be better analysed by discontinuous medium models.

9.4.2   HTCM coupled processes

As is well known, high pressures and high temperatures may induce viscous phenomena in rock-like materials, but time-delayed irreversible strains may develop even in materials whose mechanical behaviour is dominated by hydro-chemo-mechanical coupling. For instance, this is the case of salt rocks and gypsum. The velocity of the evolution of the microstructure processes defines a sort of characteristic time for the material, and in case of chemical processes, this is influenced by temperature, water content and water chemistry. A correct description of time-delayed irreversible processes in geomaterials is a very difficult task because in situ conditions may affect the process evolution (e.g. scale effect). The tunnel excavation induces a hydro-mechanical perturbation, whose consequences in some cases are delayed with time, causing an evolution of stresses on tunnel structure (increase in loads applied to the lining). A rigid separation between purely mechanical induced delayed strain rates and hydro-chemo-mechanical (squeezing and swelling) is almost impossible, since at the microstructural level irreversible strains (for instance microcrack formations) may facilitate chemical reactions. In the literature, numerous are the constitutive relationships conceived to take into account viscous effects, based on the original model of Perzyna (1963) (vis-coplastic approaches) and inspired to the Maxwell–Kelvin–Voight rheological scheme (Ghaboussi & Gioda, 1977). More recently, even constitutive relationships capable of taking into consideration the effect of temperature (Cecinato et al., 2011; Alonso et al., 2016) and chemistry (Nova et al., 2003; Ciantia & di Prisco, 2016) have been proposed and employed to solve boundary value problems, once implemented in finite element numerical codes.

All these approaches may justify, from a qualitative point of view, the phenomena affecting the tunnel structure, but their use is quite complex and the calibration of the constitutive parameter is necessarily done by back-analysing in situ data.

9.4.3   Full SSI dynamic analyses for tunnels under seismic actions

The design of the tunnel considering the performance of the system under seismic ac-tions will be approached in Chapter 10. Hereafter, the main peculiarities of dynamic numerical analyses are recalled.

At the preliminary design stages, the effects of seismic actions on tunnel lining are generally calculated by using either analytical solutions or simplified numerical methods. For a better prediction of the seismic behaviour of tunnels, however, full dynamic numerical SSI analyses have to be carried out (FHWA, 2009; ISO 23469, 2005), by using either Finite Element Method (Section 9.1.5.1) or Finite Difference Method (Section 9.1.5.2) for soils/intact rock masses, and, sometimes in case of frac-tured rocks, Discrete Element Method (Section 9.1.5.3) codes. These are needed to describe both kinematic and inertial aspects of the SSI. In such analyses, complex structural layout, ground conditions and interfaces can be taken into account. In ad-dition, the non-linear soil behaviour can be modelled, by using suitable constitutive relationships. However, there are a few issues to be carefully addressed to guarantee the reliability of the dynamic numerical results: choice of the dynamic input, appropri-ate refinement of the numerical mesh, minimization of boundary effects and control of energy dissipation in the numerical integration algorithm used in time domain analy-ses. Due to the linear extension of the tunnel, to capture the asynchronous motion of the tunnel along its longitudinal axis, the dimensions of the 3D numerical model have to be suitably tailored. Dynamic analyses of long structures such as tunnels, much longer than the significant wavelengths of the seismic ground shaking, should take into account the effect of travelling seismic waves. Non-uniform or asynchronous seismic excitations generally produce larger tunnel deformations compared with the uniform or synchronous loading case. In fact, during earthquakes, different segments of a long tunnel are not subject to the same ground motion at the same instant. Numerical mod-els generally assume different portions of the tunnel to be subject to the same signal with different arrival times (coherence). The seismic input is typically modelled as a plane wave. Wave scattering and three-dimensional propagation are neglected, even if these phenomena can determine a variation in stresses and strains along the tunnel axis. Actually, the results of three-dimensional full dynamic analyses, under simplified assumptions (e.g. Fabozzi et al., 2018), show that the asynchronous tunnel motion may affect the structural response of the transverse section of the tunnel lining, in terms of dynamic increment in the normal/shear forces and bending moments.

The extension of the calculation domain (domain size) should be large enough to allow the reproduction of the propagation of the maximum wavelength, corresponding to the lower frequency of the input signal. Furthermore, the position of the mesh boundaries and the adopted mechanical constraints should allow energy transmission outwards the computation domain, thus preventing the propagation of spurious reflected waves from the boundaries within the domain. Silent boundaries are often used, ensuring the energy arriving at the boundary to be totally or partially absorbed (Lysmer & Kuhlemeyer, 1969; Kramer, 1996; Ross, 2004). Free-field boundaries (Cundall et al., 1980) and tied degrees of freedom (Zienkiewicz et al., 1988) may also be used along vertical side boundaries, for level ground conditions and horizontal layering. More complex techniques are based on the Domain Reduction Method (Bielak et al., 2003).

The dynamic response of a numerical method is influenced by the space/time discretization. The size of the elements (mesh coarseness) should be selected to allow the largest frequency content of the input motion to be transmitted through the domain. This implies that the element size must be lower than the minimum wavelength, $\lambda_{min} = \alpha_\lambda \cdot V_S/f_{max}$ (V_S is the shear wave velocity in the ground, and f_{max} is the maximum frequency of the ground motion). The coefficient α_λ typically ranges between 1/10 and 1/8, depending on the number of nodes of the element along the direction of wave propagation (Kuhlemeyer & Lysmer, 1973). The use of elements with a too high aspect ratio (generally, with length/width ratio lower than 5) is not recommended. The size of the time step is also limited to a certain value (i.e. by multiplying the smallest fundamental period of vibration by the same factor α_λ).

9.4.4 Tunnel–landslide interaction

When the tunnel crosses an active landslide, displacements accumulating with time may significantly change the stress applied to the lining. This stress variation is expected to be very large where the tunnel intersects the landslide slip surfaces. As highlighted by previous studies on pipelines interacting with landslides (O'Rourke & Lane, 1989; Spinazzè et al., 1999; Calvetti et al., 2004; Maugeri et al., 2004; Cocchetti et al., 2009a, b; Wang, 2010), damage patterns are governed by the relative orientation of the construction to that of the slope movement. Analysis of convergence measurements can provide useful information on the tunnel portion affected by the landslide phenomena (Bandini et al., 2015).

The limit analysis approach can provide the assessment of the possible maximum loads developing in the lining, by assuming the landslide plastically flow. Under the assumption of a landslide developing in an infinitely long stratum characterized by a thickness H_s and of a circular tunnel of diameter D crossing orthogonally the landslide body, Scandale (2015) numerically showed that the internal actions acting in the lining are a function of both the cover–diameter ratio (H/D) and H_s/D.

Numerical SSI analyses allow us to evaluate the relationship between landslide movements and the internal actions in the lining. In particular, these latter show a tendency to reach a maximum value, corresponding to that calculated by the limit analysis, no longer affected by a further increase in the slope movements. These calculations may be usefully calibrated on monitoring data, obtained by either topographic or inclinometer measurements.

9.4.5 Thermo-mechanical coupling

To correctly model the tunnel behaviour under fire, the initial conditions, in terms of stress distribution in the lining, have to be known. In order to obtain such a stress distribution, one of the numerical strategies mentioned in Section 9.2.1 is used.

Fire is applied assuming the air temperature to vary with time according to the selected fire scenario. The fire action is usually defined as a temperature versus time curve applied in the air volume where the fire is expected to develop. The initial temperature is usually assumed to coincide with the ambient temperature (20°C).

The effect of fire is modelled by performing partially coupled thermo-mechanical analyses. In the thermal analysis, the convection–diffusion equation is solved and the corresponding temperature field is used in the successive mechanical analysis to determine the mechanical response of the system. Both thermal conductivity and specific heat for concrete have to be considered as temperature dependent.

In the fire analysis, the thermal dilatation of concrete should be taken into account, since, due to the soil constraints, the dilation may induce a significant increment in stresses in the lining. Furthermore, the reduction in lining stiffness (due, for example, to concrete cracking phenomena) can be considered, in order to correctly estimate the actions on the lining due to the thermal dilatation effects. The zone of lining exposed to fire needs to be correctly accounted for: in fact, due to the read level, the lower part of the tunnel is not directly exposed to fire and, as a consequence, the thermal dilatation effect is not symmetric, inducing lining ovalization and bending actions.

The analysis has to follow the evolution of the lining behaviour during the overall time scenario (considering also the cooling phase, if required). According to this type of structural analysis (ultimate analysis under extreme loads), local verifications are not required and if limited convergence is numerically obtained, the verifications are satisfied.

In case spalling is expected, this can be numerically modelled by removing concrete layers at different time instants. A typical value of spalling rate is 2.5 cm/15 minutes. Accordingly, the thermal boundary condition has to be updated.

AUTHORSHIP CONTRIBUTION STATEMENT

The chapter was developed as follows. di Prisco and Boldini: chapter coordination and §9.1, 9.2, 9.3, 9.4; Bilotta: §9.3, 9.4; Callari: §9.2; Desideri: §9.3, 9.4; Flessati: §9.1, 9.2, 9.4; Graziani: §9.3, 9.4; Meda: §9.4;. Russo: §9.3. All the authors contributed to chapter review. The editing was managed by Boldini, di Prisco and Flessati.

REFERENCES

Addenbrooke, T.I., Potts, D.M. & Puzrin, A.M. (1997) The influence of pre-failure soil stiffness on the numerical analysis of tunnel construction. *Géotechnique*, 47(3), 693–712.
Alonso, E.E., Zervos, A. & Pinyol, N.M. (2016) Thermo-poro-mechanical analysis of landslides: from creeping behaviour to catastrophic failure. *Géotechnique*, 66(3), 202–219, https://doi.org/10.1680/jgeot.15.LM.006.

Al-Tabbaa, A. & Wood, D.M. (1989) An experimental based 'bubble' model for clay. *International Conference on Numerical Models in Geomechanics* (eds. A. Pietruszczak and G. N. Pande), pp. 91–99. Rotterdam: Balkema.

Anagnostou, G. & Kovari, K. (1996) Face stability conditions with earth-pressure-balanced shields. *Tunnelling and Underground Space Technology*, 11(2), 165–173.

Atkinson, J.H. & Potts, D.M. (1977) Stability of a shallow circular tunnel in cohesionless soil. *Geotechnique*, 27(2), 203–215.

Aversa, S., Bilotta, E., Ferrero, A.M. & Migliazza, M. (2017) Settlements induced by TBM excavation Rock Mechanics and Engineering Volume 4: Excavation, Support and Monitoring, pp. 701–723.

Balthaus, H. (1988) Standsicherheit der flüssigkeitsgestützten Ortsbrust bei schildvorgetriebenen Tunneln. In Festschrift, pp. 477–492. Springer Berlin Heidelberg.

Bandini, A., Berry, P. & Boldini, D. (2015) Tunnelling-induced landslides: the Val di Sambro tunnel case study. *Engineering Geology*, 196, 71–87.

Berthoz, N., Branque, D., Subrin, D., Wong, H. & Humbert, E. (2012). Face failure in homogeneous and stratified soft ground: Theoretical and experimental approaches on 1g EPBS reduced scale model. *Tunnelling and Underground Space Technology*, 30, 25–37.

Bielak, J., Loukakis, K., Hisada, Y. & Yoshimura, C. (2003) Domain reduction method for three-dimensional earthquake modelling in localized regions. Part I: theory. *Bulletin of the Seismological Society of America*, 93(2), 817–824.

Borja, R.I. (1991) Cam-Clay plasticity, part II: implicit integration of constitutive equation based on a nonlinear elastic stress predictor. *Computer Methods in Applied Mechanics and Engineering*, 88(2), 225–240.

Broere, W. (2001) Tunnel face stability & new CPT applications. PhD thesis, Delft University of Technology, Delft.

Burd, H.J., Houlsby, G.T. & Augarde, C.E. (2000) Modelling tunnelling-induced settlement of masonry buildings. *Proceedings on ICE Geotechnical Engineering*, 143, January, pp. 17–29.

Callari, C. (2004) Coupled numerical analysis of strain localization induced by shallow tunnels in saturated soils. *Computers and Geotechnics*, 31(3), 193–207.

Callari, C. & Abati, A. (2009) Finite element methods for unsaturated porous solids and their application to dam engineering problems. *Computers and Structures*, 87, 485–501.

Callari, C. & Abati, A. (2011) Hyperelastic multiphase porous media with strain-dependent retention laws. *Transport in Porous Media*, 86(1), 155–176.

Callari, C., Alsahly, A. & Meschke, G. (2017) Assessment of stand-up time and advancement rate effects for tunnel faces below the water table. *4th ECCOMAS Conference on Computational Methods in Tunneling and Subsurface Engineering (EURO:TUN 2017)*, Innsbruck, April 2017.

Callari, C. & Casini, S. (2005) Tunnels in saturated elasto-plastic soils: three-dimensional validation of a plane simulation procedure. *Mechanical Modelling and Computational Issues in Civil Engineering*, Springer, 2005, 143–164.

Callari, C. & Casini, S. (2006) Three-dimensional analysis of shallow tunnels in saturated soft ground. *Geotechnical Aspects of Underground Construction in Soft Ground – Fifth International Symposium*. TC28, 2006, pp. 495–501. Balkema.

Calvetti, F., di Prisco, C. & Nova, R. (2004) Experimental and numerical analysis of soil-pipe interaction. *Journal of Geotechnical and Geoenvironmental Engineering*, 130, 1292–1299.

Cecinato, F., Zervos, A. & Veveakis, E. (2011) A thermo-mechanical model for the catastrophic collapse of large landslides. *International Journal for Numerical and Analytical Methods in Geomechanics*, 35(14), 1507–1535.

Cho, S.H. & Kaneko, K. (2004) Rock fragmentation control in blasting. *Materials Transactions*, 45(5), 1722–1730.

Ciantia, M.O. & di Prisco, C. (2016) Extension of plasticity theory to debonding, grain dissolution, and chemical damage of calcarenites. *International Journal for Numerical and Analytical Methods in Geomechanics*, 40(3), 315–343.

Cocchetti, G., di Prisco, C. & Galli, A. (2009b) Soil-pipeline interaction along unstable slopes: a coupled three-dimensional approach. Part 2: numerical analyses. *Canadian Geotechnical Journal*, 46(11), 1305–1321.

Cocchetti, G., di Prisco, C., Galli, A. & Nova, R. (2009a) Soil-pipe interaction along unstable slopes: an uncoupled/coupled 3D approach – part 1: theoretical formulation published. *Canadian Geotechnical Journal*, 46(11), 1289–1304.

Cundall, P.A., Hansteen, H., Lacasse, S. & Selnes, P.B. (1980) NESSI- Soil Structure Interaction Program for Dynamic and Static Problems. Report 51508-9 Norwegian Geotechnical Institute.

Cundall, P.A. & Strack, O.D. (1979) A discrete numerical model for granular assemblies. *Geotechnique*, 29(1), 47–65.

Davis, E. H., Gunn, M. J., Mair, R. J., & Seneviratine, H. N. (1980). The stability of shallow tunnels and underground openings in cohesive material. *Geotechnique*, 30(4), 397–416.

D'Effremo, M.E. (2015) Meccanismi di riattivazione di movimenti di versante indotti dallo scavo di una galleria. Tesi di dottorato. Dipartimento di Ingegneria Strutturale e Geotecnica – Sapienza Università di Roma.

D'Esatha, Y. & Mandel, J. (1971) Profondeur critique d'éboulement d'un souterrain, C.R. Acad. Sc. Paris, 273 Ser. A., 470–473.

Detourney, E. & Fairhurst, C. (1982) Generalization of the ground reaction curve concept. *Proceedings – Symposium on Rock Mechanics*, pp. 924–934.

di Prisco, C., Flessati, L., Frigerio, G. & Lunardi, P. (2018) A numerical exercise for the definition under undrained conditions of the deep tunnel front characteristic curve. *Acta Geotechnica*, 13(3), 635–649.

Drucker, D.C., Prager, W. & Greenberg, H.J. (1952) Extended limit design theorems for continuous media. *Quarterly of Applied Mathematics*, 9(4), 381–389.

Duvall and Petkof. (1959) Spherical propagation of explosion generated strain pulses in rock. U.S. Department of the Interior, Bureau of Mines.

Fabozzi, S., Bilotta, E., Yu, H. & Yuan, Y. (2018) Effects of the asynchronism of ground motion on the longitudinal behaviour of a circular tunnel. *Tunnelling and Underground Space Technology*, 82, 529–541.

Fargnoli, V., Gragnano, C.G., Boldini, D. & Amorosi, A. (2015) 3D numerical modelling of soil–structure interaction during EPB tunnelling. *Geotechnique*, 65(1), 23–37.

Federico, F. (1984) Il processo di drenaggio da una galleria in avanzamento. *Rivista Italiana di Geotecnica*, 4, 191–208.

Fenner, R. (1938) Untersuchungen zur Erkenntnis des Gebirgsdruckes. Gluckauf.

Fernandez, G. & Alvarez, T.A. (1994) Seepage-induced effective stresses and water pressures around pressure tunnel. *Journal of Geotechnical Engineering* (ASCE), 120(1), 108–128.

FHWA (Federal Highway Administration) (2009) Technical manual for design and construction of road tunnels-Civil elements. Publication No. FHWA-NHI-10–034, Department of Transportation, Federal Highway Administration, Washington D.C., U.S.

Flessati, L. (2017) Mechanical response of reinforced deep tunnel fronts in cohesive soils: experimental and numerical analyses, Ph.D. Thesis, Politecnico di Milano.

Gens, A. & Potts, D.M. (1988) Critical state models in computational geomechanics. *Engineering Computations*, 5, 178–197.

Ghaboussi, J. & Gioda, G. (1977) On the time-dependent effects in advancing tunnels *International Journal for Numerical and Analytical Methods in Geomechanics*, 1(3), 249–269.

Goodman, R.E. (1995) Block theory and its application. *Géotechnique*, 45(3), 383–423.

Goodman, R.E. & Shi, G.H. (1985) *Block Theory and Its Application to Rock Engineering*. Englewood Cliffs, NJ: Prentice-Hall.

Grammatikopoulou, A., Zdravkovic, L. & Potts, D.M. (2006) General formulation of two kinematic hardening constitutive models with a smooth elasto-plastic transition. *International Journal of Geomechanics*, ASCE, 6(5), 291–302.

Grammatikopoulou, A., Zdravkovic, L. & Potts, D.M. (2008) The influence of previous stress history and stress path direction on the surface settlement trough induced by tunnelling. *Géotechnique*, 58(4), 269–281.

Grasso, P., Mahtab, A. & Pelizza, S. (1989) Reinforcing a rock zone for stabilizing a tunnel in complex formations. *Proceedings of International Congress on Progress innovation in Tunnelling*, Toronto, 2, 671–678.

Harr, M.E. (1962) *Groundwater and Seepage*. New York: McGraw-Hill.

Horn, N. (1961) Horizontaler erddruck auf senkrechte abschlussflächen von tunnelröhren. Landeskonferenz der ungarischen tiefbauindustrie, 7–16.

ISO (International Organization for Standardization) (2005) *ISO 23469: Bases for Design of Structures – Seismic Actions for Designing Geotechnical Works*, International Standard ISO TC98/SC3/WG10. Geneva, Switzerland: International Organization for Standardization.

Jommi, C. & Pandolfi, A. (2008) Vibrations induced by blasting in rock: a numerical approach Rivista Italiana Di Geotecnica, 2, 77–94.

Kelvin, L. (1848) Note on the integration of the equations of equilibrium of an elastic solid. *Cambridge and Dublin Mathematical Journal*, 3, 87–89.

Kirsch, A. (2010) Experimental investigation of the face stability of shallow tunnels in sand. *Acta Geotechnica*, 5(1), 43–62.

Kirst, G. (1898) Die theorie der elastizität und die bedürfnisse der festigkeitslehre. *Zeitschrift des Vereines Deutscher Ingenieure*, 42(28), 797–807.

Kramer, S.L. (1996) *Geotechnical Earthquake Engineering*. Upper Saddle River, NJ: Prentice Hall, Inc., 653 p.

Kuhlemeyer, R.L. & Lysmer, J. (1973) Finite element method accuracy for wave propagation problems. *Journal of the Soil Mechanics and Foundation Division*, 99(SM5), 421–427.

Kumar, R., Choudhury, D. & Bhargava, K. (2016) Determination of blast-induced ground vibration equations for rocks using mechanical and geological properties. *Journal of Rock Mechanics and Geotechnical Engineering*, 8(3), 341–349.

Leca, E. & Dormieux, L. (1990) Upper and lower bound solutions for the face stability of shallow circular tunnels in frictional material. *Géotechnique*, 40(4), 581–606.

Liu, G., Houlsby, G.T. & Augarde, C.E. (2001) 2-dimensional analysis of settlement damage to masonry buildings caused by tunnelling. *Structural Engineering*, 79(1), 19–25.

Losacco, N., Burghignoli, A. & Callisto, L. (2014) Uncoupled evaluation of the structural damage induced by tunnelling. *Géotechnique*, 64(8), 646–656.

Louis, C. (1969) Etude des dcolements d'dau dans les roches fissurdes et de leur influence sur la stabilitd des massifs rocheurs. *BRGM. Bulletin de la Direction des t~tudes et R6cherches*, Serie A. (3), 5–132. (These presentee a l'Univ, de Karlsruhe).

Lysmer, J. & Kuhlemeyer, R.L. (1969) Finite dynamic model for infinite media. *ASCE Journal of Engineering and Mechanical Division*, 859–877.

Ma, G.W. & An, X.M. (2008) Numerical simulation of blasting-induced rock fractures. *International Journal of Rock Mechanics and Mining Sciences*, 45(6), 966–975.

Maleki, M., Sereshteh, H., Mousivand, M. & Bayat, M. (2011) An equivalent beam model for the analysis of tunnel-building inter-action. *Tunnelling Underground Space Technology*, 26(4), 524–533.

Maugeri, M., Casamichele, P. & Motta, E. (2004) Non-linear analysis of soil-pipeline interaction in unstable slopes. *XIII World Conference on Earthquake Engineering*, Vancouver, Canada.

Mroueh, H. & Shahrour, I. (2003) A full 3-D finite element analysis of tunneling–adjacent structures interaction. *Computers and Geotechnics*, 30(3), 245–253.

Mühlhaus, H.B. (1985) Lower bound solutions for circular tunnels in two and three dimensions. *Rock Mechanics and Rock Engineering*, 18(1), 37–52.

Nomoto, T., Imamura, S., Hagiwara, T., Kusakabe, O. & Fujii, N. (1999) Shield tunnel construction in centrifuge. *Journal of Geotechnical and Geoenvironmental Engineering*, 125(4), 289–300.

Nova, R., Castellanza, R. & Tamagnini, C. (2003) A constitutive model for bonded geomaterials subject to mechanical and/or chemical degradation. *International Journal for Numerical and Analytical Methods in Geomechanics*, 27(9), 705–732.

Onate, E., Idelsohn, S., Del Pin, F. & Aubry, R. (2004) The particle finite element method. An overview. *International Journal of Computational Methods*, 1(2), 267–307.

O'Rourke, T.D. & Lane, P.A. (1989) Liquefaction hazard and their effects on buried pipelines. Technical Report NCEER-89-0007, National Center for Earthquake Engineering Research, State University of New York at Buffalo.

Panet, M. & Guenot, A. (1983) Analysis of convergence behind the face of a tunnel: tunnelling 82. *Proceedings of the 3rd international Symposium*, Brighton, 7–11 June 1982, pp. 197–204. London: IMM, 1982. International Journal of Rock Mechanics and Mining Sciences & Geomechanics Abstracts, 20.

Pastor, M., Blanc, T. & Pastor, M.J. (2009) A depth-integrated viscoplastic model for dilatant saturated cohesive-frictional fluidized mixtures: application to fast catastrophic landslides. *Journal of Non-Newtonian Fluid Mechanics*, 158(1–3), 142–153.

Perazzelli, P., Leone, T. & Anagnostou, G. (2014) Tunnel face stability under seepage flow conditions. *Tunnelling and Underground Space Technology*, 43, 459–469.

Perazzelli, P., Soli, C. & Boldini, D. (2019) Empirical and numerical analysis of the blast-induced structural damage in rock tunnels. *Tunnels and Underground Cities: Engineering and Innovation meet Archaeology, Architecture and Art – Proceedings of the WTC 2019 ITA-AITES World Tunnel Congress 2019*, pp. 4159–4168.

Perzyna, P. (1963) The constitutive equations for rate sensitive plastic materials. *Quarterly of Applied Mathematics*, 20(4), 3321–3326.

Phillips, C. & Hashash, Y.M.A. (2009) Damping formulation for nonlinear 1D site response analyses. *Soil Dynamics and Earthquake Engineering*, 29(2009), 1143–1158.

Picarelli, L., Petrazzuoli, S. & Warren, C.D. (2002) Interazione tra gallerie e versanti. Le opere in sotterraneo in rapporto con l'ambiente. IX Ciclo di conferenze Torino.

Potts, D.M. & Ganendra, D. (1994) An evaluation of substepping and implicit stress point algorithms. *Computer Methods in Applied Mechanics and Engineering*, 119(3–4), 341–354.

Ribacchi, R. (1978) L'influenza delle forze di volume sullo stato di sollecitazione intorno ad una galleria circolare. *Rivista italiana di Geotecnica*, 12, 33–47.

Ribacchi, R., Graziani, A. & Boldini, D. (2002) Previsione degli afflussi d'acqua in galleria ed influenza sull'ambiente. *Le Opere in Sotterraneo e il Rapporto con l'Ambiente, IX Ciclo di Conferenze di Meccanica e Ingegneria delle Rocce*, Torino 26–27 November 2002 (eds. Barla & Barla), pp. 143–199.

Ribacchi, R. & Riccioni, R. (1977) Stato di sforzo e di deformazione intorno ad una galleria circolare. *Gallerie*, 5, 7–20.

Ross, M. (2004) Modeling methods for silent boundaries in infinite media, ASEN 5519-006: Fluid-Structure Interaction, University of Colorado at Boulder.

Sadek, M. & Shahrour, I. (2004) A three dimensional embedded beam element for reinforced geomaterials. *International Journal for Numerical and Analytical Methods in Geomechanics*, 28(9), 931–946.

Sagaseta, C. (1987) Analysis of undrained soil deformation due to ground loss. *Géotechnique*, 37, 301–320.

Sazid, M. & Singh, T.N. (2013) Two-dimensional dynamic finite element simulation of rock blasting. *Arabian Journal of Geosciences*, 6(10), 3703–3708.

Scandale, S. (2015) Effetti dei movimenti di versante sui rivestimenti di una galleria. Ph.D. Thesis, Dipartimento di Ingegneria Strutturale e Geotecnica – Sapienza Università di Roma.

Schnabel, P.B., Lysmer, J. & Seed, H.B. (1972) SHAKE: A computer program for earthquake response analysis of horizontally layered sites. EERC Report 72-12, University of California, Berkeley.

Shin, J.H., Choi, Y.K., Kwon, O.Y. & Lee, S.D. (2008) Model testing for pipe-reinforced tunnel heading in a granular soil. *Tunnelling and Underground Space Technology*, 23(3), 241–250.

Shu, D.M. & Bhattacharyya, A.K. (1992) Modification of subsidence parameters for sloping ground surfaces by the rays projection method. *Geotechnical and Geological Engineering*, 10, 223–248.

Singh, P.K. & Vogt, W. (1998) Ground vibration: prediction for safe and efficient blasting. *Erzmetall*, 51(10), 677–684.

Sloan, S.W. (1988) Lower bound limit analysis using finite elements and linear programming. *International Journal for Numerical and Analytical Methods in Geomechanics*, 12(1), 61–77.

Sloan, S.W. (1989) Upper bound limit analysis using finite elements and linear programming. *International Journal for Numerical and Analytical Methods in Geomechanics*, 13(3), 263–282.

Spinazzè, M., Bruschi, R. & Giusti, G. (1999) Interazione tubazione-terreno in pendii instabili. *Rivista Italiana di Geotecnica*, 1, 65–70.

Stallebrass, S.E., Jovicic, V. & Taylor, R.N. (1994a) The influence of recent stress history on ground movements around tunnels, S. Shibuya, T. Mitachi, S. Miura (Eds.), *Pre-Failure Deformation of Geomaterials*, vol. 1. Rotterdam: Balkema, pp. 615–620.

Stallebrass, S.E., Jovicic, V. & Taylor, R.N. (1994b) Short term and long term settlements around a tunnel in stiff clay. *Proceedings on 3rd International Conference on Numerical Methods in Geotechnical Engineering*.

Stallebrass, S.E. & Taylor, R.N. (1997) The development and evaluation of a constitutive model for the prediction of ground movements in overconsolidated clay. *Géotechnique*, 47(2), 235–253.

Sulsky, D.L., Schreyer, H. & Chen, Z. (1994) A particle method for history-dependent materials, *Computer Methods in Applied Mechanics and Engineering*, 118(1), 179–196.

Tamagnini, C. & Viggiani, G. (2002) Constitutive modelling for rate-independent soils: a review. *Revue française de génie civil*, 6(6), 933–974.

Urciuoli, G. (2002) Strains preceding failure in infinite slopes. *International Journal of Geomechanics*, 2(1), 93–112.

Vermeer, P.A., Ruse, N. & Marcher, T. (2002) Tunnel heading stability in drained ground. *Felsbau*, 20(6), 8–18.

Verruijt, A. & Booker, J.R. (1996) Surface settlements due to deformation of a tunnel in an elastic halfplane. *Géotechnique*, 46(4), 753–756.

Vlachopoulos, N. & Diederichs, M.S. (2009) Improved longitudinal displacement profiles for convergence confinement analysis of deep tunnels. *Rock Mechanics and Rock Engineering*, 42(2), 131–146.

Wang, T.T. (2010) Characterizing crack patterns on tunnel linings associated with shear deformation induced by instability of neighboring slopes. *Engineering Geology*, 115, 80–95.

Whittaker, B.N. & Reddish, D.J. (1989) *Subsidence Occurrence Prediction and Control*. Amsterdam: Elsevier.

Wong, R.C.K. & Kaiser, P.K. (1991) Performance assessment of tunnels in cohensionless soils. *Journal of Geotechnical Engineering*, 117(12), 1880–1901.

Yan, B., Zeng, X. & Li, Y. (2015) Subsection forward modeling method of blasting stress wave underground Mathematical Problems in Engineering, art. no. 678468.

Yu, H.S. (2013) Cavity *Expansion Methods* in *Geomechanics*. Springer Science & Business Media.

Zienkiewicz, O.C., Bianic, N. & Shen, F.Q. (1988) Earthquake input definition and the transmitting boundary condition. *Conference on Advances in Computational Non-linear Mechanics* (ed. St. Doltnis I.), pp. 109–138.

Zienkiewicz, O.C., Taylor, R.L. & Zhu, J.Z. (2005) The *Finite Element Method: Its Basis* and *Fundamentals*. Elsevier.

Chapter 10

Assessment of excavation-related hazards and design of mitigation measures

C. di Prisco
Politecnico di Milano

C. Callari
Università degli Studi del Molise

M. Barbero
Politecnico di Torino

E. Bilotta and G. Russo
Università di Napoli Federico II

D. Boldini and A. Desideri
Sapienza Università di Roma

L. Flessati
Politecnico di Milano

A. Graziani
Università degli Studi Roma Tre

A. Luciani
SWS Engineering

A. Meda
Università degli Studi di Roma Tor Vergata

M. Pescara
Tunnel Consult

G. Plizzari and G. Tiberti
Università degli Studi di Brescia

A. Sciotti
Italferr SpA

DOI: 10.1201/9781003256175-10

CONTENTS

10.1 INTRODUCTION

As mentioned in Chapters 2 and 7, tunnelling process has to be inspired by risk management strategies. For each hazard and in each tunnelling phase (planning/feasibility study, engineering or construction), risk is individuated/assessed/evaluated. When the assessed risk is considered to be acceptable, mitigation measures are not necessary; when not, according to the tunnelling phase, mitigation measures have to be defined/designed/implemented. This chapter is devoted to the presentation of the design approaches to the most common mitigation measures (Chapter 8).

In the current engineering practice, the mitigation measures are defined in the pre-liminary design according to the strategy described in Chapter 7, where the definition of the residual risk acceptability thresholds is crucial. The quantitative risk assessment highlights the role played by (i) the soil–structure interaction, (ii) excavation and con-struction stages and (iii) interaction with groundwater.

In the engineering phase, the design for the construction of mitigation measures is quantitatively done by employing the calculation methods described in Chapter 9 and by assigning thresholds to key performance indicators (Chapter 2). In most of the cases, these thresholds are defined in design codes (Section 10.1.1), but in other cases, they must be tailored for the project-specific environment/constraints/hazards.

A generally valid approach cannot be defined for tunnelling, since the three items mentioned below increase the complexity of the relevant problem:

i.   All the problems of soil–structure interaction, as it is in highly redundant struc-tures, are quite difficult to be dealt with, since calculating loads/actions in the structural elements without assigning material properties is practically impossi-ble. For instance, the values of internal actions in tunnel linings are significantly influenced by both soil/rock mechanical properties and lining geometry/mechani-cal properties.

ii. In the case of tunnels, the problem solution is more difficult, since the induced perturbations are mainly related to geometry (excavation and construction stages), to the new conditions to be imposed at the new boundaries (e.g. at tunnel face and walls) and also to the artificially induced changes in soil/rock mechanical properties as a consequence of reinforcement and/or improvement techniques.

iii. The previously mentioned changes in geometry (excavation/construction) may also cause the evolution of pore water pressure distribution that, owing to hydro-mechanical coupling, affects with time the displacement field and the stress distri-bution in the whole domain.

Therefore, the designer should follow an organic multidisciplinary approach to the whole geotechnical–structural problem, taking into consideration all the significant interactions, instead of introducing a sharp separation between 'geotechnical' and 'structural' assessments. Since every structural element ('substructure') contributes to the mechanical response of the whole system, the designer has to clarify the function of every structural element introduced and to justify all the design choices (e.g. typology and the number of structural elements to be employed) in the light of a quantitative evaluation of the global mechanical response.

As mentioned in Chapter 2, the hazards can essentially be subdivided into two categories: the ones taking place during the tunnel construction (excavation hazards, H_e) and the ones occurring during the tunnel lifetime (operational hazards, H_o), po-tentially affecting the tunnel's long-term performance. The requirements for the mit-igation measures to reduce the likelihood/impact of the two hazard categories are different and, therefore, are discussed separately in Sections 10.2 and 10.4. In tunnel-ling, the presence of uncertainties/unknowns is unavoidable, whose management is based on the employment of the observational method (Section 10.3).

10.1.1 Codes and partial factors

The current design codes are based on the employment of both the ultimate limit state (ULS) approach and partial factors of safety, accounting for the uncertainties associated with both material parameters and applied actions. The limit states to be considered are related to both soil/rock failure (e.g. tunnel face collapse) and structural element failure.

As far as the partial factors for geomaterials are concerned, the design codes only consider partial factors on the parameters defining material strengths. As far as the determination of geomaterials' constitutive parameters is concerned, the strategy suggested in Chapter 4 taking into account the structure to be built and the construction techniques has to be followed. However, in most of the geotechnical problems and, in particular, in case of soil–structure interaction problems, due to the static redundancy, the internal actions in the structural elements are influenced not only by material strengths, but also by soil–structure relative stiffnesses. Moreover, in case of hydro-mechanical coupled problems, the internal actions also depend on excavation rate and material permeability. From a theoretical point of view, in these cases, additional material partial coefficients, not considered by the design codes, should be adopted (Flessati & di Prisco, 2020).

The ULS approach was originally conceived for structural design where, in most of the cases, actions applied on structural elements (e.g. self-weight and wind) can be treated as input data. On the contrary, in most of the geotechnical problems and, in particular, in tunnelling, the actions applied on structural elements (e.g. stress distribution applied on lining) are calculated by solving a soil–structure interaction problem. As a consequence, internal actions in structural elements (i.e. axial and shear forces and bending moment) are to be determined by employing 'coupled approaches' (either simplified, such as the convergence confinement method (Section 10.2.1.6), or numerical, Chapter 9) taking into consideration both the soil domain and the structural elements. Only in some special cases, for instance when a single detached rock block interacts with the lining, internal actions in structural elements can be calculated without solving a soil–structure interaction problem ('uncoupled approaches').

The peculiarity of the mechanical problems related to excavation consists in the nature of the perturbation imposed: not an external action but an imposed progressive change in geometry.

The design codes (e.g. EC7), for the verification of structural elements in soil–structure interaction problems, allow the introduction of partial coefficients on the effect of the actions, i.e. on the internal actions in the structural elements. This implies (i) the internal actions in the structural elements are evaluated by employing either coupled or uncoupled approaches without the introduction of partial factors of safety on both material strength properties and actions, and only subsequently (ii) the internal actions are increased by using partial factors. The structural verification is performed by also applying partial factors on structural material strength.

On the contrary, for the evaluation of ULS only related to soil/rock mass failures (for instance unsupported cavity/face collapse), numerical analyses can be performed reducing the characteristic soil strength properties (for instance adopting a c–φ reduction method, Chapter 9, Section 9.1.5.4) and by not imposing any amplification of actions.

10.2 EXCAVATION PHASE

In Chapters 8 and 9, different excavation techniques have been described; these are sub-divided into two categories: conventional tunnelling (Section 10.2.1) and mechanized tunnelling (Section 10.2.2). The choice of the designer of the excavation technique, governed by tunnel geometry and layout, material mechanical properties, environmental conditions and expected hazards, changes abruptly the framework within which the designer has to conceive the mitigation measures to be implemented. For instance, for the lining design, in case of conventional tunnelling, both temporary (shotcrete shells often combined with steel ribs) and cast-in-place permanent concrete linings are employed, whereas in mechanized tunnelling using shielded TBMs, only precast permanent segmental linings are installed. For the segmental lining, each segment has to be designed not only to bear stresses transferred by the soil, but also to bear the stresses developing (i) during handling and transportation to the TBM and (ii) when the TBM pushes on the installed lining for advancing. The intermediate case of open TBM is not approached separately since, in this case, the temporary and the permanent linings practically coincide with those used in the conventional tunnelling (Chapter 8).

In both conventional and mechanized excavations, although with different goals and methods, the soil/rock is often improved/reinforced (Section 10.0) and the lining extrados is waterproofed (Section 10.2.4).

As it was previously mentioned (Chapter 9), the excavation process not only induces a perturbation in the proximity of the tunnel ('near field'), but also influences the 'far field'. In Section 10.2.5, settlements induced at the ground surface are discussed in case of horizontal ground surfaces, whereas in Section 10.3.1, the case of inclined slopes is discussed. Far-field hydraulic consequences due to the tunnel excavations are omitted in this chapter, for the sake of brevity, but they have already been addressed in Chapter 9.

10.2.1 Conventional tunnelling

In conventional tunnelling, in addition to the excavation means, drill and blast or mechanical (described in Volume 2, Chapter 2), both supports and auxiliary excavation methods (Chapter 8) are adopted using either a full-face excavation approach or a partialized excavation (sequential excavation) approach. This subsection is concerned with:

- Full-face excavation.
- Sequential excavation (Section 10.2.1.1).
- Pre-support by means of umbrella arch (Section 10.2.1.2).
- Pre-confinement by means of face reinforcements (Section 10.2.1.3).
- Dewatering and drainage (Section 10.2.1.4).
- Design of pre-improvement techniques for tunnelling (Section 10.2.1.5).
- First-phase and final supports (Section 10.2.1.6).

The choice of describing auxiliary methods for excavation before considering first-phase and final supports is justified by the temporal sequence of installation/construction. Very often, simplified lining design approaches do not take into account the presence of pre-support, but when numerical analyses are performed, the actual temporal sequence of intervention has to be simulated in order to obtain reliable results. In many cases, computational costs are balanced by a significant reduction in construction costs.

As described in Chapter 8, full-face excavation is generally performed for standard tunnel cross section areas up to $200\,\mathrm{m}^2$ and is recommended to be adopted in soil and bad rock conditions. For a correct design of the support, all the construction phases and the adopted improvement/strengthening techniques should be numerically modelled (Chapter 9) step by step (Chapter 8), from the pre-confinement of the tunnel face to the final lining cast. The correct simulation of the evolving geometry and the design of the mitigation measures employed (Sections 10.2.1.2–10.2.1.6) are crucial for the safe (Chapter 5) implementation of the operational procedures.

10.2.1.1   Sequential excavation

Sequential excavation (Chapter 8), particularly common for the excavation of large tunnels (ITA WG 19, 2009) or for reducing the risks associated with rock burst (Zhang et al., 2012), consists in splitting the face excavation into different phases. The most popular strategy consists in excavating: (i) the crown first, (ii) subsequently the bench and (iii) finally the invert, but alternative excavation sequences were proposed in the past (Callari & Pelizza, 2020).

As is evident, sequential face excavation is associated with a significant increase in the complexity of the management of the construction site. These logistic issues are not addressed in this section (Chapter 6 of Volume 2). In this section, only some issues associated with the design are discussed.

For a correct design of supports, the sequential excavation should be modelled by considering the evolution of the system geometry. The problem is associated with a progressively evolving three-dimensional geometry, which cannot be assumed as axisymmetric (Chapter 9, Section 9.2.2.2), thus making difficult a proper definition of the 'excavation stress' release needed for a plane strain analysis (Chapter 9, Section 9.2.2.3). For the same reason, the design of supports cannot be based on simplified approaches (e.g. analytical solutions, integrated–derived solutions and solutions for stability analyses, Sections 9.1.2–9.1.4 of Chapter 9) and numerical methods (Section 9.1.5 of Chapter 9) have to be employed. As already discussed in Chapter 9, the choice of the numerical method essentially depends on the characteristic size of the representative elementary volume/discontinuity spacing. If the elementary volume/discontinuity spacing is significantly smaller than the tunnel diameter, a homogenized continuum approach can be employed (finite element or finite difference method), whereas if tunnel diameter and discontinuity spacing are comparable, the use of a discontinuum approach (discrete element method) is necessary. Independent of the adopted numerical method, all the stages associated with the face advance (various excavations and support installation) must be reproduced (Section 9.2.2 of Chapter 9).

10.2.1.2   Pre-support by means of forepoling techniques

Forepoling techniques, consisting of either steel or pipe umbrellas, are often employed to pre-support the tunnel vault (see Chapter 8). As illustrated in Figure 10.1a, the surface

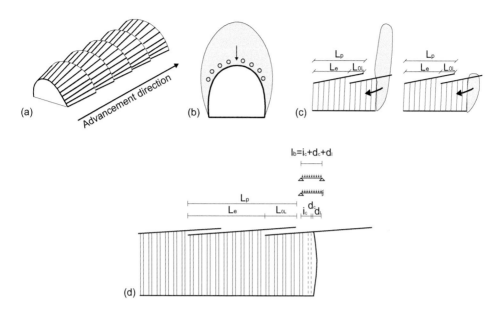

Figure 10.1 (a) 3D view of a sequence of steel pipe umbrellas; (b, c) schematic illustration of the stabilizing action of pipes on vault and face; (d) longitudinal section with indication of construction design lengths L_p, L_e, L_{OL} and typical structural schemes used for the assessment of single pipes.

enveloping the pipes of the umbrellas is an iterated sequence of troncoconical surfaces. However, umbrellas must not be considered as arch structures. On the contrary, each single pipe must be modelled as a separated longitudinal beam, since along their (moving) unsupported span, extending from the last installed rib to the advancing excavation face (Figure 10.1d), the pipes are not structurally connected in the hoop direction.

As for tunnel face nailing (Section 10.2.1.2), also in the design of steel pipe umbrellas, a crucial role is played by the setting of the following design parameters: pipe length L_p and distance L_e between two consecutive executions of the umbrellas ($L_p > L_e$). The difference $L_{OL} = L_p - L_e$ is the so-called 'overlapping length', whose function is to ensure stability during the execution of the next umbrella. Indeed, such a construction step is characterized by the highest vulnerability (minimum pipe length) and exposition to hazard (maximum concentration of workers and machines at the stationary excavation face). Therefore, the design assessment must be focused on the overlapping length L_{OL}. The maximum value of pipe length L_p is limited by technological requirements, e.g. the compliance with the designed cavity boundary, so the setting of L_e should also comply with a L_{OL} large enough to ensure the support of the soil above the vault, even in cases of local instabilities at the face (Figure 10.1c). In view of the typical sizes of face collapse mechanisms, a preliminary sizing of L_{OL} can be defined as a fraction of the characteristic size D of the tunnel face (e.g. $L_{OL} > 0.4D$).

Since each pipe of the umbrella must be modelled as a separated longitudinal beam, no form of the so far available 2D models (analytical or numerical) is suitable for its design assessment. Therefore, in the last decades, an increasing number of 3D numerical analyses (Chapter 9, Section 9.1.5) have been developed for research and design (Sterpi & Moro, 1995; Galli et al., 2004; Volkmann & Schubert, 2007; Shi et al., 2017). However, in standard design practice, a simplified approach can be used, which considers the instant immediately preceding the installation of a new rib as the most unfavourable condition for each pipe (Figure 10.1d). In this situation, indeed, the pipe is unsupported for a length not less than $(i_c + d_c)$, where i_c is the spacing between the ribs and d_c is the distance of the rib from the face at the instant of its installation. Hence, the assessment calculation is carried out by assuming for the single pipe a simplified structural scheme consisting of a beam of length $l_b = i_c + d_c + d_l$, where d_l is the depth of pipe insertion in the face where pipe displacements are negligible (in the absence of other data, $d_l = 0.5\,\text{m}$). The action of the overburden on the steel pipe can be modelled as a uniform vertical load $q_b = \sigma_v\, i_p$, where i_p is the spacing between adjacent pipes (typically 0.40 m) and σ_v can be calculated, for example, using the limit vertical load obtained by Terzaghi (1943). As far as constraints conditions at the beam boundaries are concerned, the beam is assumed to be hinged at the two ends or hinged at the rib and fixed at the other end (see Figure 10.1d). For both these cases, the maximum moment is $M_{\max} = q_b\, l_b^2/8$. Assessment calculations should be repeated for different values of overburdens and of pipe spacing. Especially for tunnelling under extremely difficult conditions, e.g. in granular incoherent soils below the groundwater table, umbrellas of jet grouting columns are preferred, as illustrated in Section 10.2.3.

10.2.1.3 Pre-confinement by means of face reinforcements

To support the tunnel face, the introduction of linear inclusions in the advance core is very popular (Chapter 8). To model the presence of reinforcements, a popular approach is based on the concept of 'equivalent cohesion' (Grasso et al., 1989; Lunardi, 2008). According to this approach, the effect of reinforcements is converted into a pseudo-cohesion for the homogenized reinforced material.

Recently, different design approaches, based on the limit equilibrium method (in particular on the failure mechanism proposed by Horn (1961)), have been proposed by Anagnostou and Perazzelli (2015) and Perazzelli and Anagnostou (2017). In particular, in Anagnostou and Perazzelli (2015), the approach proposed is suitable for granular soils, whereas in Perazzelli and Anagnostou (2017), cohesive soils are accounted for. These approaches are reliable only in case of shallow tunnels (Chapter 9, Section 9.2.4.1), where failure mechanisms develop.

In contrast, they are not reliable in case of deep tunnels excavated in ductile materials (Chapter 9). In this case, a displacement-based design approach is preferable. A first step towards this direction is the analytical approach proposed by Wong et al. (2000, 2004), (i) where the tunnel face response is assimilated to the one of a spherical cavity excavated in a heterogeneous infinite soil domain subject to a uniform and isotropic state of stress and (ii) the presence of reinforcements is simulated by defining a suitable homogenized material.

Recently, di Prisco et al. (2020), by interpreting a series of FEM numerical results, have proposed a displacement-based design approach for deep tunnel faces under undrained conditions. The numerical results (Figure 10.2a), in terms of tunnel face GRC (ground reaction curve), plotted in the non-dimensional Q_f–q_f plane (Chapter 9, Equations 9.2 and 9.3) put in evidence that: (i) the initial slope of the characteristic

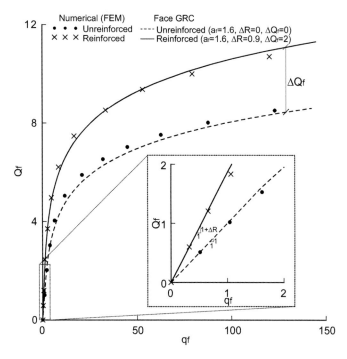

Figure 10.2 Unreinforced and reinforced tunnel face GRC. (Adapted from di Prisco et al. 2020.)

curve of the reinforced face is larger than the unreinforced one, (ii) the amplitude of the initial linear branch of the reinforced curve is larger, and (iii) the final branches of two curves are almost parallel. These differences were also experimentally observed by di Prisco et al. (2018a), who performed a series of 1 g small-scale model tests on unreinforced and reinforced tunnel faces in cohesive soils.

The previously mentioned numerical results (di Prisco et al., 2020) also allow us to describe the increase in normal stresses applied on the lining and their evolution during the unloading. The use of face reinforcements allows us to reduce such an increase, which also depends on the lining stiffness.

The design approach proposed by di Prisco et al. (2020) is based on a modified tunnel face GRC, where the increase in the curve stiffness (ΔR of Figure 10.2) and the translation of the final branch (ΔQ_f of Figure 10.2) are assumed to depend on the reinforcement number, diameter and spacing and relative reinforcement stiffness (di Prisco et al., 2019). In di Prisco et al. (2020), two expressions for calculating ΔR and ΔQ_f are provided. The comparison between the numerical results and the curves corresponding to the reinforced GRC are reported in Figure 10.2.

The approach can be employed in displacement-based design approaches, for instance, to choose the number of reinforcements necessary to obtain a given system performance in terms of face displacement (Flessati & di Prisco, 2020; Flessati et al., 2020). Moreover, the approach can also be employed to calculate the maximum axial force in the most loaded reinforcement.

10.2.1.4 Dewatering and drainage

Tunnelling at atmospheric pressure below the groundwater table can induce water flux mainly oriented towards the excavation face and the unsupported tunnel span. In medium- to high-permeability soils and rock masses, such a configuration of flow and of related seepage forces can be quickly reached, leading to abundant water inrushes and, especially for incoherent soils, to instabilities of tunnel face and/or vault. Even in those cases where catastrophic failures are not induced by water inrushes (e.g. in strong enough rock masses), the presence of abundant water can be detrimental for construction site safety and efficiency.

To mitigate these hazards in tunnels with moderate overburden, a possible design solution can be the dewatering from the surface using deep wells. The purpose of such a solution is to lower the groundwater table (or piezometric surface), thus decreasing pore pressure and, more importantly, minimizing the piezometric gradient and the related seepage forces in the soil close to tunnel. Among the advantages offered by this solution, the independence between pumping operations, performed from the surface, and tunnel excavation stages can be mentioned. Indeed, such a separation of activities can result in a reduction in construction times, with positive effects also on workers' safety. Furthermore, the effects of dewatering from surface can significantly decrease the need of other (more expensive) improvement techniques (Section 10.2.1.5).

For the considered dewatering strategies, the assumed design parameters can be validated by employing 3D numerical formulations of the mechanically coupled and variably saturated seepage problem (Callari & Casini, 2005, 2006; Callari, 2009), which would also provide results in terms of characteristic dewatering times and predictions about possible dewatering-induced subsidence (see Section 9.2.1 of Chapter 9). However, also some simplified analytical methods are available for a rapid validation and management of the proposed design solution, which can be particularly useful for a prompt application of the observational method during construction (Section 10.3). In the following, as an example of simplified approach, the well-known 'method of image wells' (Leonards, 1962; UFC, 2004) is mentioned.

As an example of dewatering strategy, Figure 10.3 illustrates a 350-metre-long alignment section of a highway tunnel consisting of two single-way tubes. The alignment section interferes with water-bearing fine sand deposits, including pressurized (artesian) persistent layers, with piezometric heads 20 m higher than tunnel invert. For the pumping operations, the design considers 34 wells aligned along two directions, parallel and normal to the tunnels. Along both these directions, the average spacing between wells is approximately 30 m. The wells are excavated up to 10 m below the tunnel invert (about 50 m of maximum depth) and equipped with a submersible pump placed at each well bottom. For their portion close to the tunnel, the 30-centimetre-large boreholes are equipped with perforated casings and proper filters.

In the considered alignment section, the tunnel face advancement starts from the south portals (on the RHS of Figure 10.3), first in the north-bound and then in the south-bound tubes (faces highlighted in red and black, respectively), ensuring a proper reciprocal distance (e.g. not less than 50 m). In this way, during excavation advancement, steady-state pumping is ensured in not less than four cross sections (i.e. 12 wells), which follow one another in order to continuously identify a 90-metre-long alignment section (continuous blue rectangle), including both the excavation faces of the two

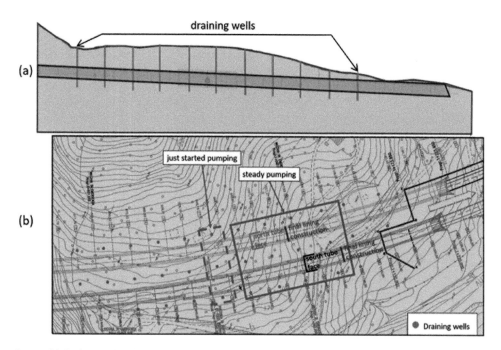

Figure 10.3 Example of a dewatering design solution based on pumping from the surface using deep wells: (a) surface and tunnel longitudinal profiles with location of wells; (b) plan view with tunnel tubes, draining wells, advancing faces in north-bound and south-bound tubes; 90-metre-long alignment section of 12 steady pumping wells (continuous rectangle) and cross section where the pumping starts (dashed rectangle).

tubes. Furthermore, at 30 m in advance with respect to the north tube excavation face, the pumping starts in the cross section immediately further north (dashed blue rectangle), to ensure that steady-state conditions are reached before the face approaches. In each pumping section, after the assessment of a safe response to excavation, pumping can be gradually stopped after the passing of the permanent lining installation section (marked in green), following the face at a maximum distance of 30 m.

The assessment of dewatering design requires geotechnical profiles describing the relations between tunnels and aquifers. In the considered example, the excavation face of the north-bound tube advances in a persistent sand layer with a maximum thickness of about 5 m, which, according to the estimated piezometric surface elevation, is subject to a piezometric head of about 15 m, measured from the sand layer bottom. The excavation face of the south-bound tube advances in a water-bearing sand layer with a maximum thickness of about 12 m. Furthermore, these sand layers are included between low-permeability layers (silty clays), and the interstitial water is thus confined and pressurized. In view of this hydrogeological situation, the calculations can be carried out under the conservative assumption of a 5- to 12-metre-thick sand layer under pressure, with piezometric heads of 25 m measured from the bottom layer, located just 1 m below the tunnel invert. The calculations are repeated by assuming two different values for permeability k of the sand layers: 10^{-5} and 10^{-4} m/s. Furthermore, the predictions of piezometric surface lowering and of pumped flows are developed under the following simplifying assumptions: (i) the pervious sand layers are infinitely long, homogeneous, isotropic and of constant thickness t_{sl}; (ii) the undisturbed initial piezometric surface is horizontal, at elevation H_u; (iii) N_w wells are arranged along two horizontal directions, parallel (y) and normal (x), to the assumed-straight tunnel axis; (iv) three wells are in each cross section (Figures 10.3b and 10.4), and the average spacing between wells is about 30 m; (v) the undisturbed initial piezometric surface H_u is recovered along an infinitely long line parallel to tunnel axis (y) and located up-stream of y at horizontal distance L_i from well 'i'; (vi) consistently with the designed pumping operation, the calculation is performed by assuming pumping to be acti-vated in 12 wells located in four cross sections ($N_w = 4 \times 3 = 12$ wells); (vii) the pumped flow is constant, and the seepage flow is steady; (viii) the aquifer is artesian (with the conservative assumption of an undisturbed piezometric surface H_u of the sand layers, which is higher than elevation $z_{t, sl}$ of the top of the layers); (ix) the wells cross the en-tire aquifer thickness $t_{sl} = z_{t, sl} - z_{b, sl}$, i.e. the elevation of the bottom of wells is lower than the elevation $z_{b, sl}$ of the bottom of sand layers; and (x) the piezometric gradient is uniform along the vertical direction, and the flow vectors are horizontal.

In view of these assumptions, the piezometric head h at a generic point $P(x, y)$, not coincident with any of the wells, can be calculated by using the relation:

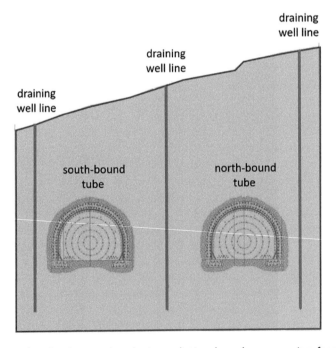

Figure 10.4 Example of a dewatering design solution based on pumping from the surface using deep wells: cross section of the two tunnel tubes with location of the three deep well lines employed for a dewatering design solution based on pumping from the surface.

$$h(x,y) = H_u - \frac{F(x,y)}{2\pi k t_{sl}} \quad \text{with} \quad F(x,y) = \sum_{i=1}^{N_p} Q_{wi} \ln\left[\frac{s_i(x,y)}{r_i(x,y)}\right] \tag{10.1}$$

for flow $Q_{w,i}$ pumped from well i, and distances r_i and s_i of the generic point $P(x, y)$ from real well i and its conjugated 'image well', respectively. Thus, distances r_i and s_i depend also on length L_i, whose choice is based on ground surface shape in the tunnel neighbourhood and on the maximum influence length L_{si} of a single well 'i', as is obtained from the empirical relation (Leonards, 1962; UFC, 2004):

$$L_{si} \cong c(H_u - h_{wi})\sqrt{k} \tag{10.2}$$

with k expressed in 10^{-4} cm/s and L_{si}, H_u and h_{wi} in feet. The value $c = 3$ can be assumed to be constant. The applicability of the considered calculation method is verified to be reliable if $L_i < L_{si}/2$.

As a first step of design assessment, the calculation method described above can be used to evaluate the flow rates that should be pumped to minimize the piezometric surface elevation. To this purpose, a realistic minimum value for the piezometric height $h(x, y)$ is set in Equation 10.1 and the problem is solved with respect to flow rate Q_w, assumed to be the equal for all the wells ($Q_{wi} = Q_w$ for $i = 1...N_p$).

The effects on the piezometric surface induced by pumping with the considered $N_p = 12$ wells can be predicted by using values of single-well flow rate Q_w not larger than those given by Equations 10.1 and 10.2. The 3D view of the depressed piezometric surface (Figure 10.5a) as well as a plan view of the contours of piezometric head h (Figure 10.5b) allow capturing the predicted effects of well pumping. The piezometric head profiles calculated along the two main directions of the well system, i.e. parallel (y) and normal (x) to the tunnel axis, are shown in Figure 10.5c and d, respectively.

Consistently with the observational method (Section 10.3), to assess the modelling of the pumping design and the effectiveness of the dewatering technique, the first drilled wells should be used to perform pumping field tests, including the survey of the piezometric head decrease induced for different flow rates. These tests might also be used to estimate the permeability of the sand layers. As confirmed by both field observations and hydro-mechanical coupled numerical studies, the face stability below the groundwater table typically increases for increasing face advancement rates (Callari et al., 2017; di Prisco et al. 2018a) and decreases during standstills (Callari, 2015), with the relevant stand-up time decreasing for increasing permeability values (Anagnostou et al., 2016). For both conventional and mechanized tunnelling, a frequently considered solution to increase the stability of the face is the pre-drainage from the face by means of subhorizontal drain pipes (see Chapter 8). For an advancing excavation face, the design parameters, in terms of drain pipe length L_d, execution length L_e and overlapping length L_{OL}, should be optimized (see Section 10.2.1.5.3) with the purpose of obtaining a significant and stable decrease in the piezometric gradient close to the face (i.e. a significant reduction in the destabilizing seepage forces), as it can be assessed by using a numerical formulation of seepage flow. The full assessment of tunnel face stability requires a hydro-mechanical coupled method.

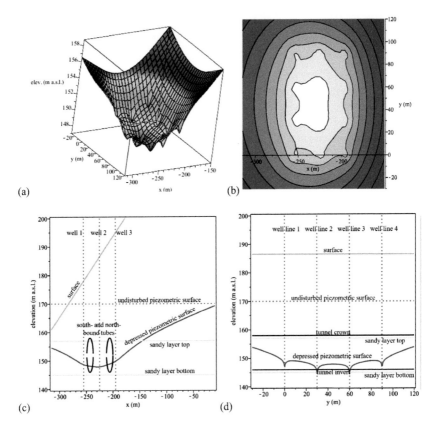

(a)   (b)   (c)   (d)

Figure 10.5 (a) 3D view from below of the depressed piezometric surface; (b) x–y plane contour plot of piezometric surface elevation. Profiles along the tunnel direction y(c) and x(d) of the depressed piezometric surface. Results obtained for the arrangement of pumping wells (Figure 10.3) and $k = 10^{-4}$ m/s; single-well flow rate $Q_w = 5.5$ L/s, sand layer thickness $t_{sl} = 12$ m; and undisturbed piezometric head $H_u = z_{b,sl} + 25$ m.

10.2.1.5 Design of pre-improvement techniques for tunnelling

In addition to dewatering and pre-drainage strategies, presented in Section 10.2.1.4, several other techniques are often implemented in tunnelling design to improve the properties of both soils and rock masses. A detailed description of the most frequently employed improvement techniques is already available in Chapter 8, which also includes indications about the relevant application fields (Table 8.4 of Chapter 8) and several references. The present section deals with the design of these techniques in their typical applications in tunnelling. Although the pre-improvement techniques that will be illustrated in the following paragraphs can be applied for both conventional and mechanized tunnelling, they are mostly used for conventional tunnelling and for this reason, this subsection has been placed within the 'Conventional tunnelling' section.

10.2.1.5.1   PRE-IMPROVEMENT FROM SURFACE USING PERMEATION, COMPENSATION
GROUTING AND FREEZING

A geometric arrangement of pre-grouting from the surface of an urban environment is depicted in Figure 10.6a, which is an application of the traditional permeation grouting, a durable improvement technique obtained from the injection of proper mixes in soil pores or rock mass fractures at pressures ranging from 1 to 4 MPa. Therefore, grouting decreases the permeability and improves the mechanical parameters of the original porous/fractured mass, especially in terms of cohesion and stiffness. The solution adopted in Figure 10.6a is mainly aimed at pre-improving the soil volumes crossed by the tunnel face, and also at strengthening the volumes below the shallow footings of existing buildings.

However, the shallow subsoil in urban environment is often full of obstacles for the grouting pipe drillings (e.g. sewers, water pipes and cable ducts). In these cases, as an alternative solution in the same environment, Figure 10.6b illustrates the radial pre-grouting, from a pilot adit, of soil volumes in the neighbourhood of the tunnel face. The grout mixes in the form of suspensions, solutions or emulsions, can be either cement or chemically based (with the latter characterized by higher injectability) and are usually injected using sleeved grouting pipes.

Figure 10.7 illustrates the grouting from shaft as a further frequent alternative for pre-grouting above tunnel alignments and under the shallow footings of existing buildings in urban environments. It can be remarked that the shaft solution depicted in Figure 10.7 is also very often used for compensation grouting technique (Mair, 1994), which is typically aimed at compensating the ground losses due to mechanized tunnelling (Section 10.2.2). This compensation is obtained by means of high-pressure injections of cement-based and medium-to-high viscous grouts in soil layers lying between tunnel excavations and shallow foundations of existing buildings and structures. To achieve this goal, the grout should heave the ground by localized hydraulic fracturing. Hence, compensation grouting is an application of the so-called 'displacement grouting', as also stated by the European Standard EN 12715 (2000). The sleeved grouting pipes are installed before tunnel excavation, and the grouting operations typically

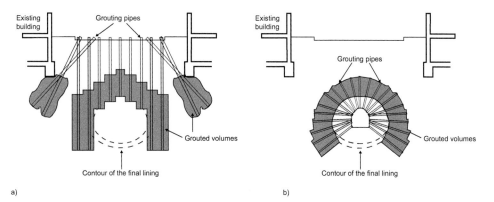

Figure 10.6 Examples of pre-grouting schemes typically used in tunnelling projects: (a) pre-grouting from surface in urban environment; (b) radial pre-grouting in urban environment from a pilot adit.

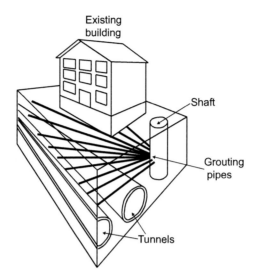

Figure 10.7 Example of pre-grouting of soil along tunnel alignments from a shaft in urban environment.

consist of three phases: (i) soil conditioning or pre-treatment by means of permeation grouting (especially for pervious and/or loose soils); (ii) heaving to compensate short-term settlements; and (iii) observational phase to mitigate long-term settlements. The pre-treatment phase (i) should be completed before tunnelling, so as to implement the compensation grouting phase (ii) during tunnel excavation (Littlejohn, 2003b; Kummerer, 2003). The control of compensation grouting procedures is based on an observational approach (Section 10.3). Therefore, a proper monitoring of building displacements must be planned (Chapter 11).

For pre-grouting from surface in urban environments, another available and more recent drilling technology is the horizontal directional drilling (HDD), schematically depicted in Figure 10.8.

A schematic example of the application of soil pre-freezing from the surface in urban environment is reported in Figure 10.9. Soil freezing is a temporary improvement technique that can be applied to water-saturated soils. The designed frozen ground

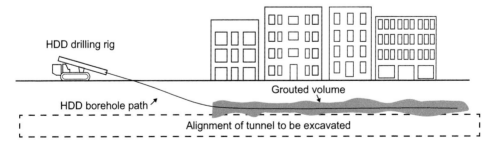

Figure 10.8 Pre-grouting of soil along tunnel alignment in urban environment by means of horizontal directional drilling from surface.

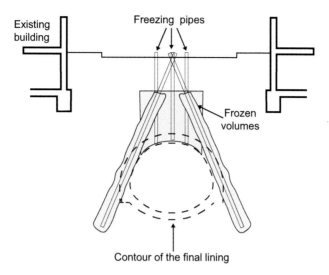

Figure 10.9 Example of pre-freezing of soil from the surface in urban environment.

volumes, if properly formed, are characterized by higher strength and reduced permeability values. The presence of seepage in soil volumes that should be frozen increases the time needed for freezing, because of the continuous supply of heat transported by the fluid flow. Freezing can be completely inhibited by large water flows, e.g. larger than 1–2 m/day (Shuster, 1972; Bell, 1993; Xanthakos et al., 1994; Andersland & Ladanyi, 2003), of relatively high temperatures. In view of its high costs, soil freezing is considered only in extremely difficult cases (Jessberger & Han, 1980; Pimentel et al., 2012).

10.2.1.5.2   PRE-IMPROVEMENT FROM THE TUNNEL FACE USING PERMEATION GROUTING AND JET GROUTING

Further pre-improvement approaches are based on grouting from the tunnel face. We consider the representative example of a tunnel whose alignment interferes with water-bearing fine sand deposits, including pressurized (artesian) persistent layers, with piezometric heads 20 m higher than tunnel invert. In similar difficult situations, the use of the so-called 'preventer' in drilling operation is suggested to contrast the inflow of drilling and injection fluids and to avoid seepage and solid transport through the drilling holes towards the tunnel face. As depicted in Figure 10.10, chemical permeation grouting from the face is combined with canopies of jet grouting columns to reduce soil permeability and to ease the formation of jet grouting columns. Jet grouting is a durable improvement technique, based on the high-velocity injection of one or more fluids (cement-based grout, air and water) into the soil, thus requiring high pressures (usually over 40 MPa). The fluids are injected through small-diameter nozzles placed on a pipe that is usually first drilled into the soil and is then moved towards the ground surface during jetting. Typically, the jets propagate orthogonally to the drilling axis, inducing a complex coupled fluid/mechanical phenomenon of soil remoulding and permeation, including some

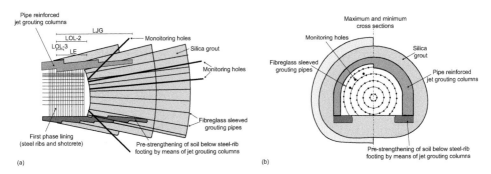

Figure 10.10 Example of pre-grouting from tunnel face (chemical grouts and jet grouting canopies): (a) longitudinal section; (b) minimum and maximum cross sections.

soil removal. The expected result is a cemented soil column with good mechanical properties and very low permeabilities. As in the case considered in Figure 10.10, the columns can be reinforced with steel or fibreglass pipes and geometrically assembled to form continuous elements of various shapes and sizes (Croce et al., 2014).

The available jet grouting techniques are based on single- (cement–water grout), double- (air and grout) or triple-fluid (grout and air and water) injection procedures. It must be remarked that only the single-fluid technique can be used to create subhorizontal jet grouting columns. Therefore, the single-fluid jet grouting technique is the most frequently used in tunnelling applications. Finally, as regards the jet grouting applications, it can be noted that a geometric scheme very similar to that depicted in Figure 10.6 might be used also for vertical jet grouting columns.

10.2.1.5.3   GEOMETRICAL AND EXECUTIVE PARAMETERS FOR PRE-GROUTING FROM TUNNEL FACE

In addition to the parameters related to each grouting technique, e.g. silica/cement contents, grout pressures and injected volumes (see Chapter 4 of Volume 2), further geometrical and executive parameters are needed, especially for the design of pre-grouting from the tunnel face. As is illustrated in Figure 10.10, these parameters can be identified with the grouting pipe length L_{GP}, the distance L_E between two consecutive grouting executions from the face, and the number $n_{GP, E}$ of grout pipes to be installed at each execution. A further parameter is $n_{E, LGP} = L_{GP}/L_E$, i.e. the integer number of grout pipe installations for each advancement L_{GP}. The design of the grouting scheme should be based on the consideration of several different combinations of L_{GP}, $n_{GP, E}$ and L_E parameter values, in order to find the optimal setting in terms of hazard mitigation, efficiency of construction site and minimization of costs. Furthermore, for each given parameter combination, the most vulnerable situation should be identified and subjected to hazard assessment.

As an application example, we consider the design illustrated in Figure 10.10, where the aforementioned parameters are set as $L_{GP} = 27$ m, $n_{GP, E} = 30$ and $L_E = 9$ m, thus leading to $n_{E, LGP} = L_{GP}/L_E = 3$. The effects of such a setting are depicted in the graph in Figure 10.11, where the stationary presence of a total number $n_{GP} = 3\, n_{GP, E}$ of grouting pipes is attained after $n_{E, LGP}$ grouting executions. More importantly, the

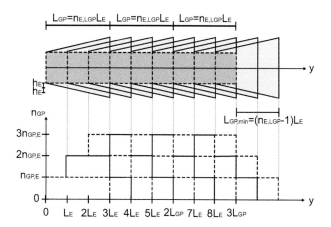

Figure 10.11 Example of design optimization of pre-grouting from tunnel face.

graph in Figure 10.11 clearly identifies the current location of the advancing face as the most vulnerable situation. Indeed, ahead of the excavation face, the pre-grouted volume is characterized by the minimum length $L_{GP, min} = (n_E, L_{GP-1})L_E$ and by the minimum available number of grouting pipes ($n_{GP} = 2 n_{GP, E}$ and $n_{GP} = n_{GP, E}$ for the first and the second execution distances LE, respectively). Furthermore, such a construction step is characterized by the highest exposition to hazard (maximum concentration of workers and machinery at the stationary excavation face due to repetition of grouting executions). Hence, the safety assessment of the design of the whole grouting scheme should be focused on such a situation. It can be also remarked from Figure 10.11 that the minimum excess thickness of the grouted volume with respect to the tunnel walls is $2h_E$, where h_E is the increase in thickness obtained for each L_E advancement.

The aforementioned parameters can be adapted also to the design of other pre-grouting (or pre-support, e.g. forepoling) schemes. For example, for the design of the jet grouting canopies in Figure 10.10, it is $L_{JG} = 20\,m$, $n_{JG, E} = 52$ and $L_E = 9\,m$, thus leading to $n_{E, LJG} = L_{JG}/L_E > 2$, i.e. to a double overlapping: $L_{JG, OL-2} = 11\,m$ and $L_{JG, OL-3} = 2\,m$ between two and three jet grouting columns, respectively. Hence, along such overlapping lengths, the total number of jet grouting columns is then $n_{JG} = 2n_{JG, E} = 104$ and $n_{JG} = 3n_{JG, E} = 156$, respectively, as depicted in Figure 10.10b. The highest vulnerability (minimum length of available jet grouting columns) and the highest exposition to hazard are again predicted at the advancing tunnel face. Therefore, the design assessment must be focused on the overlapping length L_{OL-2}.

10.2.1.5.4 TYPICAL HYDRO-MECHANICAL PROPERTIES OF IMPROVED SOILS

For techniques such as permeation grouting and jet grouting, with durable effects in terms of improved properties of treated soils, some of the strength and permeability data available in Xanthakos et al. (1994) are summarized as follows.

For permeation grouting, uniaxial compressive strength values ranging from 1.0 to 7.0 MPa and from 0.5 to 2.0 MPa have been measured for sandy soils treated with two-fluid and single-fluid sodium silicate mixes, respectively. For cement–water mixes, reductions up to 10^{-4} in the initial permeability values have been measured. Usually, a factor of 2 can be assumed to compare the cohesion of the treated soil to the virgin soil.

For jet-grouted soils, uniaxial compressive strength values between 1.5 and 10 MPa and between 10 and 30 MPa have been measured for fine-grained and coarse-grained soils, respectively. Jet-grouted silty sands exhibited permeability reductions by one or two orders of magnitude (Croce et al., 2014). An increased elastic modulus is also achievable with the jet grouting treatment: according to different soils, the values of the ratio between increased uniaxial compressive strength and increased elastic modulus are available in the scientific literature (Croce et al., 2014).

10.2.1.5.5   HEAVING HAZARD DUE TO GROUTING

In permeation grouting design, the maximum values of injection pressures should be high enough to ensure the full permeation of soil, but not so high to induce any uncontrolled foundation uplift in pre-existing structures and buildings. The same hazard might be also induced by jet grouting treatments. In the framework of the observational method (Section 10.3), the heaving hazard should be assessed in the design stage, e.g. using numerical methods able to evaluate the relevant threshold values in terms of induced movements and corresponding grouting parameters. Furthermore, an ad hoc monitoring of the displacements induced at the surface and in the subsoil should be planned in the design stage and carried out in the construction stage.

The first attempts of numerical simulation of the grout-induced heaving were mostly based on the assignment of inelastic extensional strains, as proposed by Nicholson et al. (1994), Canetta et al. (1996), Falk and Schweiger (1998), Schweiger and Falk (1998) and Nicolini and Nova (2000). In this approach, the calibration of magnitudes and orientations of the imposed inelastic strains usually requires a careful back-analysis of measured movements. An alternative approach was based on prescribed grout pressures, as originally proposed by Kummerer (2003), Wisser et al. (2005) and followed by Contini et al. (2007).

10.2.1.6   First-phase and final support design

The main objectives of a tunnel support consist in (i) confining ground deformations, (ii) keeping fluids in or out of the tunnel and (iii) preventing blocks and other fragments from loosening and falling from the rock mass into the excavation. The support systems are classified on the basis of their contribution to the tunnel stability: they can be modelled by assuming them to carry the load induced by the ground deformation (for example steel sets) or to increase the inherent strength properties of the ground (for example fully grouted rock bolts).

The design of a tunnel support is thus related to the ground/structure interaction, and consequently, the timing of its installation is an important variable. The time dependency of ground properties or hydro-mechanical coupled phenomena, leading to the well-known stand-up time, have to be taken into account. The ground/structure interaction analysis provides the state of stress induced into the support by the ground deformation which, in turn, depends on the soil/rock mechanical properties, possible additional loads (besides geostatic) and the advance of the tunnel face.

Usually, there are two classes of supports (Chapter 8): first-phase (temporary) and final (permanent) supports. The first-phase support is installed immediately after excavation and provides short-term tunnel stability. The final support is designed to guarantee the long-term stability of the cavity, by absorbing any deferred pressure (due to swelling, creep, water pressure, etc.), which might occur during the service life of the tunnel. In some cases, the final support is not installed and the long-term stability is provided by the first-phase support.

Different approaches are available for the design of first-phase and final supports, classified into empirical, analytical and numerical. While the empirical approach provides suggestions on the type and sizing of tunnel supports based on past experiences, analytical and numerical approached allow analysing the ground/structure interaction.

10.2.1.6.1 EMPIRICAL METHODS

Empirical methods are based on the assessments of precedent practice. Experience over many years has shown that the methods are generally successful when implemented by experienced tunnel engineers. Riedmuller and Schubert (1999) highlighted some disadvantages that should be taken into consideration when using these methods, among which are the following: the level of safety in the support design is unknown, there is little or no guidance on the timing of support installation, and the effects of natural or man-made adjacent structures are not considered. For these reasons, empirical approaches can be used as preliminary design of tunnel supports and successfully applied, provided that regular inspection and monitoring of the tunnel during construction are performed (observational method, Section 10.4). Several empirical methods are provided in the literature, for example the one based on RMR classification system (Bieniawski, 1989) and that provided by Barton and Grimstad (Barton et al, 1974; Grimstad, 2007) based on Q classification system (Chapters 3, 4 and 7).

Bieniawski, on the basis of a large number of case studies, provided guidelines for the selection of supports for tunnels as a function of the RMR class (Chapter 4) of the excavated rock mass. Only drilling and blasting procedures and horseshoe tunnel shape are considered. In 1983, Ünal proposed a simplified equation for the estimation of support load P as a function of RMR value and later as a function of GSI (Chapter 4) (Ünal, 1983; Osgoui & Ünal, 2009):

$$P = \frac{100 - \mathrm{RMR}}{100} \gamma B \tag{10.3}$$

$$P = \frac{100 - \left[\left(1 - \frac{D_c}{2}\right) \sqrt{\frac{\sigma_{\mathrm{cr}}}{100}} \mathrm{GSI} \right]}{100} C_s S_q \gamma D \tag{10.4}$$

where γ is the unit weight of rock, B the longest span of the opening, D_c the damage coefficient, $\sigma_{\mathrm{cr}} = S_r$ UCS (with S_r being a reduction factor lower than 1 in case of brittle rock behaviour), C_s the correction factor for the horizontal-to-vertical field stress ratio, and S_q the correction factor for the squeezing ground condition.

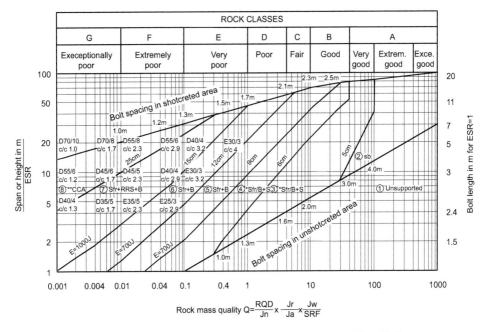

$$\text{Rock mass quality } Q = \frac{RQD}{Jn} \times \frac{Jr}{Ja} \times \frac{Jw}{SRF}$$

REINFORCEMENT CATEGORIES	4) Fibre reinforced shotcrete and bolting, 6-9cm, **Sfr+B**
1) Unsupported	5) Fibre reinforced shotcrete and bolting, 9-12cm, **Sfr(E700)+B**
2) Spot bolting, **sb**	6) Fibre reinforced shotcrete and bolting, 12-15cm, **Sfr(E700)+B**
3) Systematic bolting and unreinforced	7) Fibre reinforced shotcrete >15cm +
or fibre reinforced shotcrete, 5-6cm	reinforced ribs of shotcrete and bolting **Sfr(E1000)+RRS+B**
Sfg/B+S	8) Cast contrete lining **CCA or Sfr(E1000)+RRS+B**

The bolts are 20 or 25mm in diameter

E) Energy absorption in fibre reinforced shotcrete at 25mm bending during plate testing

D45/6 RRS with totally 6 reinforcement bars in double layer in 45cm thick ribs with centre to centre (c/c) spacing 1.7m
c/c 1.7 Each box corresponds to Q-values on the left-hand side of the box

*) Up to 10cm in large spans
) or **Sfr+RRS+B

Figure 10.12 The *Q* support chart, provided by Barton and Grimstad (Barton & Grimstad, 2014).

The *Q* system (Chapter 4) support recommendations were first developed in 1974 by Barton and updated over time. The *Q* support chart (Figure 10.12) is based on more than 1,250 base records by Barton and Grimstad (Grimstad, 2007). The support type and design characteristics are provided as a function of the *Q* value and the ratio between the tunnel span or height and a coefficient ESR, taking into account the safety level required by the final use of the tunnel considered (Table 10.1). Seven classes are defined on the basis of rock mass quality, and the corresponding support is suggested. In the area referred to bolt/fibre-reinforced shotcrete support (Sfr + B), the bolts spacing and the thickness of the shotcrete layer are suggested. In the area referred to rib reinforced shotcrete (RRS), the reported boxes contain the number of layers of reinforcing bars (double 'D' or single 'E'), thickness of ribs and the number of bars in each layer (first line) and centre-to-centre (*c/c*) spacing of each rib (second line).

Table 10.1 ESR values (modified from Barton and Grimstad (2014))

Type of excavation		ESR
A	Temporary mine openings, etc.	ca. 2–5
B	Permanent mine openings, water tunnels for hydropower (excluding high-pressure penstocks), pilot tunnels, drifts and heading for large openings and surge chambers	1.6–2.0
C	Storage caverns, water treatment plants, minor road and railway tunnels and access tunnel	0.9–1.1 storage caverns 1.2–1.3
D	Power stations, major road and railway tunnels, civil defence chambers, portals and intersections	Major roads and rail tunnels 0.5–0.8
E	Underground nuclear power stations, railway stations, sports and public facilities, factories and major gas pipeline tunnels	0.5–0.8

Relationships providing an estimation of the support loads (vertical and horizontal loads) and tunnel stand-up time S_{ut} as a function of Q and ESR are also available:

$$SUT = 2ESRQ^{0.4} \qquad (10.5)$$

$$P_v = \frac{0.2Q^{-\frac{1}{3}}}{J_r} \qquad (10.6)$$

where J_r is one of the parameters of Barton classification (Chapter 4). The horizontal support load is calculated with the same formula as P_v, where Q is modified in Q_p, equal to $5Q$ if $Q > 10$, $2.5Q$ if $0.1 < Q < 10$ and Q is $Q < 0.1$.

10.2.1.6.2   ANALYTICAL METHODS

Before either analytical or numerical modelling, the choice of employing a continuum- or discontinuum-based approach is mandatory (Chapters 4 and 9). This depends on the size of the grains/blocks constituting the ground with respect to the engineering dimension of the analysed work (for example diameter, span and height of underground excavation).

The analytical methods are generally two-dimensional idealizations assuming the ground to be a homogeneous continuum and the tunnel circular. One of the most popular is based on the GRC (convergence confinement method, Chapter 9) describing the response of the rock/soil domain during the tunnel excavation as well as the interaction between rock/soil and temporary or permanent support (Panet, 1995). Details on the GRC are given in Chapter 9. Under the hypothesis of deep circular tunnel, constructed by using full-face excavation techniques, under plane strain axisymmetric conditions, a hydrostatic fictitious pressure q_t is applied to the inner contour of a circular tunnel to account for the excavation advancement. It depends on the distance from the tunnel face and on the strength properties of the ground and can be defined as a percentage of the natural state of stress S, assumed isotropic, through the so-called 'stress release factor λ' (Figure 10.13). The presence of a circular support, installed at a certain distance from

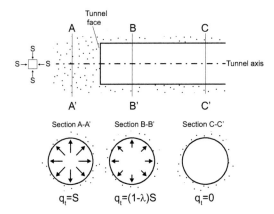

Figure 10.13 Stress release factor as a function of the distance from tunnel face.

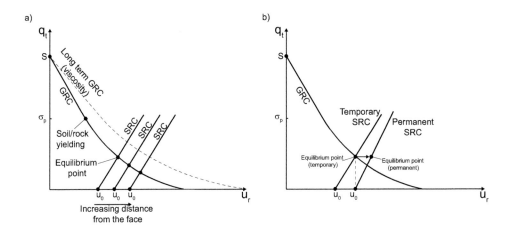

Figure 10.14 (a) Convergence-confinement method and (b) temporary and permanent linings.

the tunnel face and progressively loaded by the stress release of the ground by a pressure applied at its extrados, can be also accounted for. In the convergence–confinement method, two curves are defined (Figure 10.14a): the GRC, given by the relationship between the applied fictitious pressure q_t and the radial displacement u_r at the tunnel inner contour, and the support reaction curve (SRC), expressing the relationship between the pressure applied at the lining extrados and the corresponding radial convergence. The solution to the soil/structure problem is obtained in a simplified way by imposing equilibrium and compatibility between the two subsystems and is provided graphically by the intersection of the two curves in the plane q_t–u_r.

   The GRC is linear for an elastic medium on the contrary it is composed by a first linear elastic branch, followed by a non-linear one, associated with the development of irreversible strains, in case of elastic–plastic materials. The transition occurs for q_t equal to the yield pressure σ_R. A finite maximum value of tunnel convergence is obtained when $q_t = 0$ (the tunnel is fully excavated) if the medium displays a residual uniaxial compressive strength; otherwise, similarly to the plastic radius, it tends to infinity (i.e. the plastic radius tends to infinity if $f_r = 0$; see Equation 9.10 of Chapter 9).

For SRC, a typical linear relation is adopted:

$$q_t = k_s \left(\frac{u_r}{r_0} - \frac{u_0}{r_0} \right) \tag{10.7}$$

where k_s is the equivalent radial support stiffness, r_0 is the tunnel radius, and u_0 is the tunnel convergence that already occurred at the instant of time of lining installation. This latter quantity can be derived from adopting one of the models summarized in Chapter 9 (Section 9.2.2). The SRC line stops at the maximum support pressure, which is defined by the yield strength of the support system.

The typical expressions of a concrete ring or steel arch's stiffness are, respectively:

$$k_s = \frac{E_c}{1+v_c} \frac{t_r}{r_0} \frac{1-t_r/2r_0}{1-v_c-t_r/r_0\left(1-t_r/2r_0\right)} \tag{10.8}$$

where E_c = concrete Young's modulus, v_c = concrete Poisson's ratio, and t_r = the ring thickness, and

$$k_s = \frac{E_s A_s}{s_l r_0} \tag{10.9}$$

where E_s = the steel Young's modulus, A_s = the area of the transversal section of the steel arch, and s_l = the steel arch spacing. Equations 10.8 and 10.9 refer to an ideal case of a circular closed lining system that is not necessarily realized after the excavation. Very often, initially the invert is absent and is constructed far from the face. In these cases, initially the k_s values have to be suitably reduced.

Patterns of either active or passive anchorages can also be accounted for by employing the following expression:

$$k_s = \frac{\pi d^2 E_s}{4 L s_{b,t} s_{b,l}} \tag{10.10}$$

where L_b and d_b are the length and diameter of bolt, and $s_{b,\,t}$ and $s_{b,\,l}$ are the transversal and longitudinal bolt spacings, respectively.

When the support is constituted by more than one structural element, for example in case of steel sets and shotcrete, the stiffness is given by the sum of the stiffness of each element (the structural elements are in parallel). The maximum pressure is that of the element with the lowest critical elastic deformation (the one beyond which yielding occurs).

For composite linings, such as shotcrete and steel sets, the technique referred to as 'equivalent section' (taking into account not only the heterogeneity of the lining, but also the spacing among steel sets, Figure 10.15) can be used as suggested by Carranza Torres and Diederichs (2009), to which the reader is referred for details and formulas. Referring to the schematic representation in Figure 10.15 of a lining composed of steel

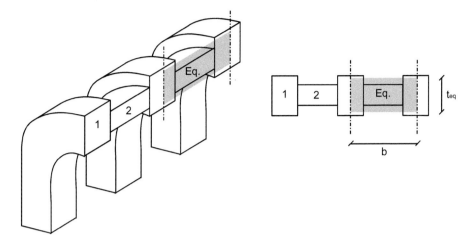

Figure 10.15 Schematic representation of a composite liner, constituted by materials 1 and 2 (for example, 1 is the steel set and 2 is the shotcrete). (Modified from Carranza Torres and Diederichs 2009.)

sets and shotcrete, the composite section can be regarded as an equivalent section of width b and thickness t_{eq}, characterized by an equivalent Young modulus E_{eq}. Both these equivalent parameters are calculated as a function of the elastic modulus, the area of the transversal section and the moment of inertia of the two materials involved.

The equilibrium of the tunnel/lining system is achieved when the two curves intersect. This equilibrium point provides the pressure acting hydrostatically on the lining extrados, which induces a global axial force (N) into the lining, when the lining is a concrete or steel sets and shotcrete ring, or tensile stresses into the bolts. As is evident in Figure 10.14a, the higher the stress release at the time of support installation (i.e. the higher the distance of the tunnel face when the support is installed), the lower is the pressure on the support and the higher is the final tunnel convergence. The stand-up time, the admissible convergence and the maximum pressure admissible on the support to guarantee it reacts elastically have to be accounted for when choosing the installation time.

When the stability problem is related to the presence of discontinuities systems in rock mass which can originate potential unstable rock blocks, the design of stabilizing support (i.e. rock bolts and shotcrete) can be based on a LEM analysis (Chapter 9, Section 9.1.4).

If the primary lining is followed by the installation of the final one, a new GRC is defined starting from the tunnel convergence already developed (Figure 10.14b). Obviously, the inclination of the straight line representing the permanent lining equivalent stiffness in Figure 10.9b is assumed to be larger than the one of temporary lining.

10.2.1.6.3 NUMERICAL METHODS

A popular structural simplified (the loads acting on the lining are assumed to be externally assigned according to empirical formulas) approach is the beam–spring model, where the lining is discretized in beam finite elements and the ground is modelled by means of radial and tangential springs. Although this approach is meaningless from a theoretical point of view, its popularity is because its numerical implementation is simple and the associated computational time is negligible. The stiffness of the springs can be calibrated on the basis of ground investigation test results. The loads acting on the beam are assumed to be vertical and uniform in the crown and horizontal and variable linearly with depth at the sidewalls. These are estimated on the basis of empirical formulas taking into account the ground mechanical properties (Terzaghi, 1943; Bieniawski, 1984; Barton et al., 1974, for example). An elastic behaviour is assigned to structural elements, whose stiffness is a function of lining thickness and material deformability. The radial springs are assumed to work only under compression. The subgrade reaction modulus along the normal direction (K_n) at sidewalls and crown is defined to be directly proportional to the deformability modulus of the ground (E_d) and inversely proportional to the equivalent tunnel radius (r_0) and Poisson's ratio (v), according to the following equation:

$$K_n = \frac{E_d}{r_0(1+v)} \tag{10.11}$$

At the invert and the bases of sidewalls, K_n is defined as:

$$K_n = \frac{E_d}{B_i\left(1-v^2\right)} \tag{10.12}$$

where B_i is the length of the base/invert.

The subgrade reaction modulus along the tangent direction K_t is usually assumed equal to 1/3 to 1/10 K_n and shear stresses at the contact between ground and structure have to be lower than the interface shear strength, usually defined as purely frictional strength (cohesion is assumed to be null).

To overcome the limits of analytical methods, numerical approaches can be used (FEM, FDM and DEM; Chapter 9) to explicitly model complex structures, geological conditions, constitutive behaviour and construction sequences (Chapter 9). These methods can simulate, under 2D and 3D conditions, the tunnel support by employing structural elements such as beams and shells located all around the excavation boundary. Numerical analyses provide the state of stress induced by the ground in the lining (summarized in terms of axial/shear forces and bending moments), taking rigorously the ground response to both boundary conditions and time effects into account.

In 2D analyses, the stress release at the time of support installation (both temporary and permanent linings) is obtained by using the GRC (Chapter 9). Different stages are considered, in which the stress release along the excavation boundary is simulated as a function of tunnel advance and excavation sequences. As previously reported, the presence of the first-phase lining is usually neglected when the final lining is installed and this is numerically simulated by deactivating the first-phase lining in the stage in which the secondary one is activated.

Whatever the method used for estimating the stresses induced in the support by the ground is, their compatibility with the mechanical characteristics of the support has to be verified according to the current regulations (Section 10.1.1). Capacity diagrams, commonly used for structural verifications, are useful tools for this purpose. Axial thrust and bending moment (M) or axial thrust (N) and shear force (V) values are plotted together with the support capacity diagrams (maximum values of axial thrust and bending moment or shear force that the lining can support).

The ground/support interaction analysis in case of composite linings is applied to this equivalent system, and thrust, bending moment and shear force are calculated. These have then to be distributed to each component involved, as shown in Figure 10.16, depending again on the elastic modulus, the area of the transversal section and the moment of inertia of the two materials. Once M, N and V are calculated for both the materials, their compatibility with the corresponding strength characteristics has to be verified by the usual relationships used for homogeneous materials.

The design of temporary or permanent supports has to be performed by using the more suitable methods for the type of support chosen, and the increase in the tunnel stability induced has to be clearly highlighted. For this purpose, whatever the method chosen is, the stability analyses must be conducted without and with the presence of support. The effectiveness of the support has to be verified by during-construction and in-operation monitoring (Section 10.3).

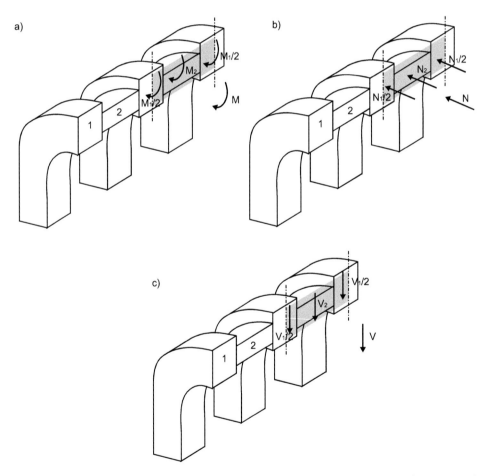

Figure 10.16 Distribution of (a) bending moment, (b) thrust and (c) shear force in each component of the liner. (Modified from Carranza Torres and Diederichs 2009.)

10.2.1.6.4 LINING DESIGN IN SQUEEZING MATERIALS

The large time-dependent deformations of rock masses during tunnelling are often referred to as squeezing behaviour (Terzaghi, 1943; Barla, 1995), a term introduced after the construction of the first base tunnels through the Alps. The amount of tunnel convergence and its rate are strictly related to the geotechnical properties of the rock mass (Chapters 3, 4 and 9), in situ state of stress and hydraulic conditions, but a primary role is also played by the adopted construction technique, which may prefigure the aim of limiting the ground tendency to deform with a stiff support or of allowing some deformation to occur with a flexible one, at the risk of a further weakening of the rock mass properties. Such a complex interaction requires a careful soil–structure interaction analysis carried out by using sophisticated numerical approaches (Chapter 9), in which different possible scenarios are preliminarily investigated and further specialized during the construction, once the first data from the monitoring systems are available. The direct observation and measure of squeezing rock behaviour during the excavation are in fact mandatory for a correct comprehension of the stress–strain behaviour of tunnel and support dimensioning.

A simplified approach taking into consideration delayed strains due to viscous effect, amplified in many cases by high temperatures, is based according to the literature on the use of the convergence-confinement method in Figure 10.14a. According to this approach, the mechanical properties (material stiffness) of the homogeneous soil/rock stratum are reduced with time, causing a progressive increase in stresses applied to the lining and in tunnel convergence. Nowadays, the evolution with time of stresses in the lining is calculated by performing either viscoelastic or elastic-viscoplastic numerical analyses, once experimental creep test results of the materials involved are available.

10.2.2 Mechanized tunnelling

In case of mechanized tunnelling, the design is generally focused on (i) the assessment of the support pressure to be applied in order to avoid unacceptable face displacements (geotechnical issue), (ii) the design of the thrust necessary to avoid the TBM jamming (geotechnical issue), (iii) the structural response of segmental linings (structural issue) and (iv) the durability of cutterhead or in general of all the excavation devices (technological/mechanical issue). In this chapter, only the first three topics will be addressed (Sections 10.2.2.1–10.2.2.3, respectively), and the fourth will be approached within Volume 2 (Chapter 2).

10.2.2.1 The design of the face support for 'earth pressure' (EPB) and 'slurry' shields (SS)

This section is focused on the two most common face support techniques implemented in soil TBMs, namely slurry support and earth pressure techniques. The applicability fields of these techniques are influenced by the hydro-chemo-mechanical properties of the involved materials. For this reason, the fields of application are expected to evolve with the use of new materials.

In slurry shields (SS), the support action exerted by the suspension strongly depends on the interaction between the bentonite slurry and the soil at tunnel face. Indeed, to stabilize the soil, the slurry pressure should not only balance the pore water pressure, but also be properly transferred to the solid skeleton, to balance effective stresses. Among the several available theories about such stress transfer (e.g. Anagnostou & Kovari, 1994; Broere & van Tol, 2000; Bezuijen et al., 2006; Zizka & Thewes, 2016; Zizka et al., 2020), the most frequently adopted in slurry-supported-face tunnelling is the approach originally proposed for diaphragm wall technology, summarized in DIN 4126.

This standard suggests that slurry pressure is transferred as sketched in Figure 10.17, where two different pressure distributions are considered along the horizontal axis.

a. Membrane-type transfer: the filtering of bentonite particles and their deposition on the face surface creates a filter cake, i.e. a quasi-impervious membrane on the tunnel face, which transfers the whole excess slurry pressure $\Delta p_s = p_s - u_w$ as a total pressure on the boundary of the soil domain. Therefore, such a transfer is fully localized at the (advancing) tunnel face (i.e. Δp_s is applied as a pressure jump).
b. Infiltration-type transfer: the slurry infiltrates the soil, and the excess slurry pressure is gradually transferred to the solid skeleton (in terms of effective stress variation) along the entire penetration length l_{max} by means of related seepage forces.

Hence, according to DIN 4126, the bentonite suspension can be modelled as a Bingham fluid, whose rheology is described by two parameters: the yield value τ_f of shear stress τ, i.e. the threshold below which the strain rates $\dot{\gamma}$ are nil, and the apparent slurry viscosity $\mu^* = \tau/\dot{\gamma}$, which governs the dependence of the shear stress τ on the shear strain rate $\dot{\gamma}$. In SS tunnelling τ_f plays a crucial role not only for tunnel face stabilization, but also for spoil conveyance, since the suspension is also used to convey the muck within the pipe system. To comply with this important function, τ_f and thus the apparent slurry viscosity ($\mu^* = \tau/\dot{\gamma}$) should be kept as lower as possible. As regards the role of τ_f in the stabilization of the tunnel face, the interaction of the slurry with the soil can be described by means of the support pressure gradient f_{s0}, defined as

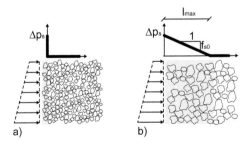

a)   b)

Figure 10.17 Transfer mechanisms of excess slurry pressure to the soil's solid skeleton according to DIN 4126: (a) membrane-type transfer, due to the formation of a 'filter cake'; (b) infiltration-type transfer, due to slurry penetration up to a length l_{max}. The symbol p_s denotes the slurry pressure, and $\Delta p_s = p_s - u_w$ is the excess slurry pressure with respect to pore pressure u_w. The relevant pressure gradient f_{s0} is also depicted. (Modified from Zizka et al. 2020.)

the decrease in slurry excess pressure per unit length of penetration into the excavation face. According to DIN 4126, the pressure gradient f_{s0} can be calculated as a function of both τ_f and the characteristic grain size d_{10} of the soil (Müller-Kirchenbauer, 1977):

$$f_{s0} = \frac{3.5\,\tau_f}{d_{10}} \tag{10.13}$$

As illustrated in Figure 10.17b, the existing pressure gradient f_{s0} can be experimentally determined in terms of imposed slurry excess pressure Δp_s and maximum measured penetration depth l_{max}, i.e.

$$f_{s0} = \frac{\Delta p_s}{l_{max}} \tag{10.14}$$

The support pressure gradient f_{s0} can be used to predict the most likely type of slurry pressure transfer (membrane- or infiltration-type; see Figure 10.17). According to DIN 4126, the infiltration-type transfer is attained if the support pressure gradient f_{s0} is lower than 200 kN/m³. In this case, as shown in Figure 10.18, if the slurry deeply infiltrates within the soil, the relevant stabilizing thrust vector must be obtained by integrating only the seepage forces acting within the unstable wedge (in dark grey) and not in the entire infiltration volume (in light grey), thus leading to a loss of support action with respect to the case of membrane-type transfer (Anagnostou & Kovari, 1994).

As illustrated in Figure 10.19a, the ideal range of application of bentonite slurry shields coincides with area A, since the grain size in this range is, at the same time, small enough to ensure the slurry face support action and large enough to reduce the effort needed for muck separation. Such a separation effort is higher, with likely clogging, for the fine-grained soils and clays falling in zone B. Zone C consists of very coarse-grained and possibly uniformly graded gravels, whose extremely high permeability could favour a deep penetration of slurry without stagnation, thus leading, even using a highly concentrated bentonite suspension, to an ineffective face support mechanism (see Figure 10.19a). Hence, for these materials, fillers should be added to the bentonite suspension to plug the larger soil pores.

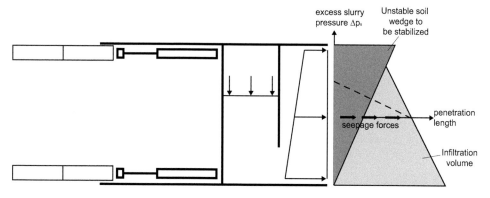

Figure 10.18 Support loss due to a slurry penetration volume larger than the soil wedge needing stabilization. (Modified from Zizka and Thewes 2020.)

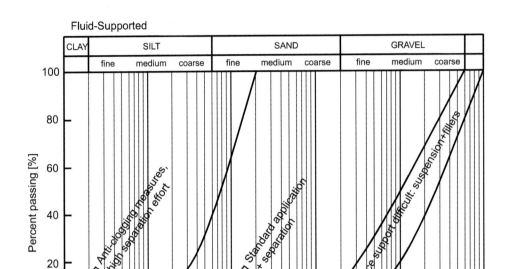

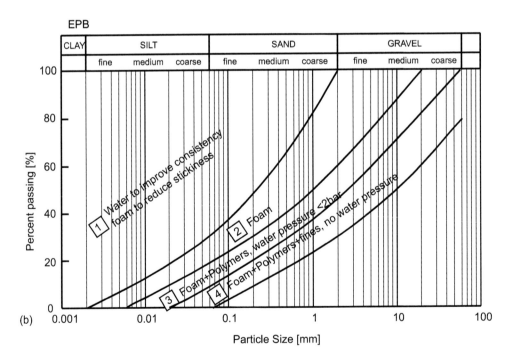

Figure 10.19 Ranges of application for: (a) slurry shields (SS) support; (b) earth pressure balance (EPB) support. (Modified from Thewes 2007.)

In earth pressure balance shields (EPBS), the soil excavated by the cutting wheel enters the excavation chamber, which is used to support the face. The support pressure is regulated by modifying the balance between the muck volume entering in the chamber (which can be regulated through the EPB advancement rate) and the muck volume extracted from the chamber by the screw conveyor, i.e. the so-called 'optimal advance rate' (Guglielmetti et al., 2008). To provide an efficient face support, the muck within the excavation chamber must satisfy some particular mechanical properties (Martinelli et al., 2019; Peila et al., 2019; Carigi et al., 2020a, b; Todaro et al., 2021). This goal can be obtained by mixing the muck with suitable additives, such as foam. During the excavation stages, the hydraulic jacks push on the bulkhead, which in turn transmits the related pressure to the muck contained in the excavation chamber. In this way, the support pressure is transferred to the face in terms of total stresses. Such a transfer is usually attained in cohesive soils with low permeability (the fine-grained soils in area '1' in Figure 10.19b), which are the optimal application range of EPB shields, where the muck can be effectively conditioned just injecting water in the excavation chamber. In the so-called 'extended EPB application range' (the coarser and non-cohesive soils in area '2' in Figure 10.19b), conditioning by injection of foam is used to reduce the permeability of the muck, thus ensuring a proper pressure transfer in the excavation chamber. For soils in areas '3' and '4', the muck in the excavation chamber must be conditioned by means of polymers and foam. Thanks to the progressive improvement in conditioning agents, the boundaries of the mentioned application areas are in continuous evolution.

In the EPB excavation chamber, the muck must satisfy several properties related to rheology, consistency, shear strength, stability, abrasiveness and clogging effects (Thewes & Budach, 2010; Galli & Thewes, 2014; Vinai et al. 2008). More information can be found in Chapter 2 of Volume 2. Hence, the earth muck parameters must comply with some proper value ranges. For the muck, different parameter ranges are required for the primary and the extended application ranges (areas '1' and '2' in Figure 10.19b, respectively). For the EPB primary application range, the proper parameter values, suggested to comply with given conditioning purposes, are suggested by (Thewes & Budach, 2010; Zizka & Thewes, 2016; Peila et al., 2019).

As regards the extended application range of EPB (area '2' in Figure 10.19b), the desired parameter properties can be achieved by conditioning the muck by means of foams, polymers or slurries of fines. Such additives are injected through nozzles located in front of the cutting wheel, in the excavation chamber and within the screw conveyor. The optimal amounts of conditioning additives to be injected can be determined by laboratory testing or based on practical experience. Several methods developed for the testing of additives and conditioned soils are cited by Zizka and Thewes (2016), Peila et al., (2019).

In the design of face supports, the considered ultimate limit states (ULSs) are typically those related to tunnel face collapse and to overburden break-up or support medium blowout. The assessment of these two ULSs requires the evaluation of a minimum and maximum value of support pressure (at face crown), respectively. For both ULSs, several calculation methods are available, including the analytical (limit analysis, global limit equilibrium and empirical relations) and the numerical formulations presented in Chapter 9. Further discussions on the analytical approaches can also be found in Broere (2001), Zizka and Thewes (2016), Anagnostou and Kovari (1994, 1996), among many others, and in Vu et al. (2016) for the blowout ULS.

In the assessment of ground failure ULSs, the EC7 code (C.E.N., 1997) requires the application of both the so-called 'resistance factor approach' (RFA) and 'material factor approach' (MFA), using the Design Approach 1 with Combination 1 (A1+M1+R1) and Combination 2 (A2+M2+R1) of partial factors, respectively. Hence, RFA is obtained by setting greater than one partial factors on actions (γ_F) and unit values for partial factors on material strengths (γ_M) and resistance (γ_R). Conversely, MFA is obtained by setting values greater than one for partial factors on strengths (γ_M) and unit values for partial factors on actions (γ_F) and resistance (γ_R).

The computational assessment of tunnel face ULSs is often based on the so-called 'strength reduction technique' (SRT), also in view of its consistency with the 'material factor approach' (MFA). By means of SRT, it is possible to evaluate the shear-strength-based safety factor

$$\eta_{sst} = \gamma_{lim}/\gamma_M \tag{10.15}$$

with γ_{lim} being the limit value for γ_M (Zienkiewicz et al., 1975). However, several doubts arise about the applicability of SRT to those frequent cases in which the tunnel stability is significantly dependent on the rate of the actions (Anagnostou et al., 2016). On the contrary, it is acknowledged that the unloading of tractions at cavity boundary is an action which realistically resembles the excavation effects. Hence, the tunnel face stability can be numerically assessed by reducing the pre-excavation pressures at the face, i.e. an 'unloading technique' (ULT). A comparison between the performances of SRT and ULT in the computational application of the MFA to the stability assessment of tunnel faces has recently been presented by Callari (2021). Furthermore, as an alternative to a commonly used, but faulty, loading-based safety factor for tunnel stability, Callari (2021) proposed and validated a new unloading-based safety factor:

$$\eta_{unl} := \frac{\lambda_{lim}}{\lambda_t} = \frac{1 - \sigma_{f,lim}/\sigma_{f0}}{1 - \sigma_f/\sigma_{f0}} \tag{10.16}$$

where σ_f is the (total) support pressure applied at the face, $\sigma_{f,lim}$ the limit σ_f value, and σ_{f0} a proper measure of the pre-excavation stress state at tunnel depth. Similarly to what is suggested in di Prisco et al. (2018b), the denominator λ_t of Equation 10.16 is representative of the real tunnelling 'action effect', i.e. the magnitude of the unloading from σ_{f0} to σ_f. The numerator λ_{lim} is the 'resistance', i.e. the limit value of the action effect. The factor η_{unl} can be evaluated by using any analytical method and the 'unloading technique' in numerical analyses.

In soil TBMs, the face support pressure is crucial not only for the aforementioned ULSs, affecting the face stability, but also for the limitation of volume losses at tunnel face. Indeed, a further purpose is to limit soil deformations, potentially damaging pre-existing both surface and underground structures. Hence, the assessment of this serviceability limit state (SLS) is required, not only in the design phase, but also during construction (see Section 10.3).

In EPB and SS, the SLS related to soil movements is also the main motivation for the pressurized mortar injection of the annular gap existing between lining and soil. The reader is referred to ITAtech (2014) and references therein for details on grout selection criteria (single- vs two-component grouts), advantages and disadvantages of different grout types, control of the grouting system (based on dual-stop criteria, i.e. on the control of injection pressure and of injected grout volume), pressure design based on geostatic calculations, transportation techniques, groundwater effects, early support of lining, required fluidity/pumpability, etc.

10.2.2.2   TBM jamming

In case of mechanized tunnelling, (i) the immobilization of the TBM during the excavation and (ii) the inability to resume TBM operations after a standstill are a main concern (Ramoni & Anagnostou, 2006). These two hazards essentially depend on the tangential stresses developing on TBM shield: if the resultant force of these tangential stresses is larger with respect to the maximum TBM thrust, the machine cannot move (TBM jamming). The resultant force of tangential stresses depends on (i) machine geometry, (ii) ground–TBM interface friction and (iii) the effective pressure radially applied on the TBM shield (Figure 10.20).

The main geometrical parameter is the ratio of shield length to diameter (Figure 10.20): by decreasing this ratio, the likelihood of jamming decreases. Other important geometrical parameters, strictly related to the pressure applied on the shield, are the difference between cutterhead and shield diameter (gap) and the shield conicity (Figure 10.20; Chapter 8; and Volume 2, Chapter 2).

The interface friction essentially depends on ground nature. By employing lubricants (e.g. bentonite), the friction between shield and ground can significantly be reduced (up to 50%, according to Gehring, 1996).

The effective radial pressure applied on the shield depends severely on both ground mechanical properties and TBM geometry. In addition, when delayed convergence is expected (e.g. when consolidation process takes place or when materials are characterized by a squeezing behaviour), (i) the excavation rate severely influences the effective pressure value (Ramoni & Anagnostou, 2007b) and (ii) during standstills an increase

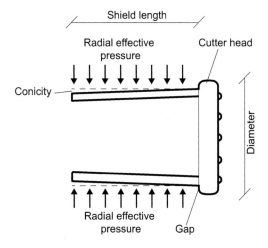

Figure 10.20 Sketch of TBM geometry.

in effective pressure is expected (Ramoni & Anagnostou, 2007a). For a reliable evaluation of the pressure applied on the shield, the numerical solution of a soil–structure interaction problem (Chapter 9), accounting for a time-dependent system response, is necessary.

For the preliminary design, to assess the minimum thrust force required in order to overcome shield skin friction, the abaci introduced by Ramoni and Anagnostou (2010), depending on both shield geometry and material properties, can be employed.

10.2.2.3 Design of precast segments

10.2.2.3.1 GENERAL OVERVIEW

The segmental lining can be defined as a sequence of precast concrete elements, the so-called 'segments', assembled in the rear shield of a shielded tunnel boring machine (shielded TBM).

The term 'shielded TBM' is generic and must be interpreted as any kind of TBM with a sequence of steel shields totally enveloping the area of the cutterhead motors, grippers if on-board, spoil evacuation system up to the beginning of the backup gantries. Therefore, reference is made to the single-shield rock TBM, the double-shield TBM, the EPB TBM, the hydroshield TBM and the more recent 'dual-mode' TBM (Chapter 8). Except for the double-shield TBM, all the other type of shielded TBMs advance by pushing themselves longitudinally against the segmental lining. The installation of a segmental lining in this type of machines is not an option, but it is integrated in the specific excavation system. Only the double-shield TBM might advance while driving in fair–good materials using the grippers (as on open TBM), and thus, the installation of the segmental lining is not strictly related to the advance of the TBM. In contrast, it becomes mandatory while driving in poor rock conditions where the grippers cannot properly work, and the machine is driven like a single-shield TBM. Nowadays, most of these machines are conceived to install segmental linings systematically, if used for the TBM advance. This testifies how much the use of such a lining is convenient.

From the above all the requirements for the overall dimensioning of the segmental lining derive, and thus, the segmental lining has to:

- follow the TBM along the alignment, and thus, it needs to accommodate horizontal and vertical curves;
- be strong enough to resist the loads coming from the TBM, without being damaged, since it must work as final lining, too (at least in most of the current applications);
- resist to any ground/rock load, at short and long terms, as it derives from the soil/structure equilibrium, since the installation time, which takes place at a distance from the face of about $1D$–$2.5D$;
- be watertight to avoid any water inflow (not allowed for different reasons depending on the project requirements) or to allow water inflow into the tunnel only from dedicated drainage points. This choice affects also the water pressure loads applied on the lining and therefore the lining mechanical response;
- allow the creation of openings to locate cross-passages or technical rooms;

- allow the installation of temporary equipment during the construction and permanent devices for the tunnel serviceability.

For all these reasons, it is evident that the design of TBM must proceed jointly with the design of the segmental lining.

In Volume 2, Chapter 3 is devoted to describing the specific geometry, material and accessories constituting the segmental lining, while in what follows, an insight into the main aspects of the structural dimension is proposed.

10.2.2.3.2 DESIGN PROCESS

In general terms, the segmental lining is a tunnel final lining and therefore must satisfy all the structural requirements already discussed in Section 10.1.1. Anyhow, the segmental lining is characterized by some particularities that need to be considered to get a proper structural dimensioning, deriving from its specific concept and geometry:

- The lining has a circular shape, but it cannot be considered as a continuous pipe.
- The lining is made of subsequent rings, and each of them needs to guarantee its integrity and stability.
- The rings are made of segments which are themselves, first of all, structural elements to be checked.
- From the number of segments composing a specific ring derives its flexural stiffness as an 'equivalent continuous ring', and this greatly influence its structural capacity.
- The loads coming from the rock mass or soil are transferred to the segmental lining throughout the annular gap filled in the zone of the rear of TBM with different materials and techniques influencing the short-term behaviour of the ring.
- The loads are transferred from one segment to the other inside a single ring through the longitudinal joints (segment-to-segment joint), characterized by a reduced contact thickness with respect to the general ring thickness; this contact area cannot resist tractions at any point.

Therefore, given the specific geometry of the segmental lining (see Volume 2, Chapter 3) and the points highlighted above, the design of segmental lining follows the following specific steps:

1. Structural check of the single segment at precast yard: extraction from mould (minimum required strength), first handling and storage (young concrete), long-term storage and transportation to TBM.
2. Installation of segments to create the lining at TBM rear shield location.
3. Push of the TBM to allow its advance.
4. Exit of ring from TBM where the lining is loaded by the longitudinal injection under pressure when this is the case, or from rock wedges with the risk of a not-complete confinement due to some unavoidable delay in filing the annular gap.
5. Long-term equilibrium (static, seismic and fire condition as applicable, Section 10.4).

Apart from guaranteeing the integrity of the single segments up to the installation point (steps 1 and 2), the steps 3 and 5 are structural dimensioning ones.

Step 3 is concerned with the phase during which the ring 'interacts' with TBM.

Step 4 is not a problem when the crew at TBM follows specific quality control procedures and at design level some assumptions for a not-complete confinement at crown are considered under different loading conditions.

10.2.2.3.3   LONG-TERM RING EQUILIBRIUM

As already anticipated, of importance is the identification of the flexural stiffness of the ring and to use it in the various calculations whether simple application of analytical formulas or complex FEM analyses are foreseen. The operational flexural stiffness of segmental lining may be assumed to range in between the full value of a continuous pipe with the given thickness and a pipe of the same thickness, but with longitudinal joints always in the same angular position, so that various hinges are recognized. In reality, the segmental lining is like a 'brick wall' (Figure 10.21) where all the segments are shifted and interlocked. Thus, the response of the system cannot be simulated by performing 2D numerical analyses.

For this reason, it is convenient to model the lining like an 'equivalent continuous pipe' and to consider the brick wall behaviour differently, that is reducing the stiffness of the lining according to the formulation proposed by Muir Wood (1975).

$$I_r = I_j + I_s \left(\frac{4}{n_s} \right)^2 \tag{10.17}$$

Figure 10.21 'Brick wall effect' of a segmental lining.

where I_r stands for the equivalent moment of inertia of the segmental lining; I_j the moment of inertia of the section of the joint depending on the thickness of the contact area at joint; I_s the moment of inertia of the segment cross section; and n_s the number of segments excluding key one (only in case of small key).

Once defined $\zeta = I_r/I_s$ and by employing Equation 10.17, the equivalent ring of the segmental lining will be simulated by assigning the real thickness, but reducing the modulus of the concrete by the same factor ζ. Thus, the flexural stiffness, associated with the product of the concrete Young's modulus (E_c) by the moment of inertia, is reduced accordingly:

$$E_c I_r = E_c \zeta I = E_{c(r)} I \tag{10.18}$$

Once the value of bending moment M is obtained by calculation, it is split into two terms: one assumed to be applied to the segment and one to the joint according to the scheme in Figure 10.21.

Since any joint can be in any angular position, it is mandatory to verify it for the maximum (positive/negative) bending moment with the concurrent axial force and the maximum axial force with the concurrent (positive/negative) bending moment, as reduced with the consideration above (Figure 10.22). In this check, the maximum (positive/negative) eccentricity must be captured.

The verification of the segment is a common structural verification of a steel reinforced concrete structure, while the verification of the joint is done according two steps:

- contact force verification, for example applying the rules given at Point 6.7 of Eurocode 2, which brings to a strong limitation of the acceptable eccentricity at joint;
- bursting effect inside a concrete volume when loaded on surface by a force acting on a reduced area, according to Point 6.5.3 of Eurocode 2, specifically developed to strength the area under the caps of tendons, but perfectly applicable to the specific case of the segments' joints.

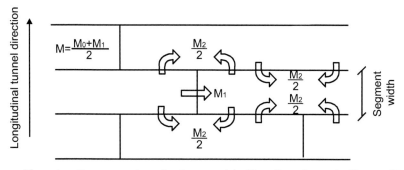

M = bending moment coming from model with uniform flexural stiffness $\zeta^* E_{c(r)}^* I$
M_0 = bending moment for segment calculation $M^*(1+\zeta)$
M_1 = bending moment for joint calculation $M^*(1-\zeta)$
M_2 = transferred bending moment due to staggered arrangement of joint ("brick wall" effect on the segmental lining) $M^* \zeta$

Figure 10.22 Mechanism to transfer load from joint to segment.

10.2.2.3.4   INTERACTION WITH THE TBM

Here below a checklist concerning the interaction of segmental lining with TBM is reported. Some of the following items involve segmental lining designer and TBM manufacturer, especially the geometrical aspects; others are numbers to be used as reference for both TBM and segmental lining design; and finally, there are constraints to be satisfied for the coupling of the two.

As far as geometry is concerned:

the internal radius of lining and specific geometry of the ring (thickness, length and taper of the universal ring) need to fit with the internal diameter of the rear shield to assure sufficient clearance to exit from TBM without risk to be damaged;

the length of ring together with the number of segments must be associated with weight/dimension that can be handled;

the number of longitudinal connections is recommended to be equal to the number of thrusting shoes;

the extra-stroke (the length of rams in excess to the length of the ring) of TBM must be adequate to allow the insertion of key segment with a reasonable tolerance on both sides (20–30 mm minimum);

the length of shield and its conicity must be adequate (to check the equilibrium between rock/soil and lining and thus define short- and long-term ground load).

As far as action transmitted by TBM is concerned:

• the maximum push for each ram is the reference for the structural verification together with the contact area of the thrusting shoes over the concrete;
• the maximum eccentricity of the ram on the middle of the joint has to be considered in structural verifications;
• the dimension of thrusting shoes has to be checked if the roll of the TBM is considered to overlap with the longitudinal joint.

As far as annular filling is concerned:

• the pressure and the numbers of longitudinal injection points have to be designed;
• the numbers of the injection lines for the radial injection have to guarantee full annular filling within the fourth ring out of the tail shield.

As far as backup is concerned:

• the maximum load transferred and the position of the supports has to be taken into consideration, especially if rubber wheels are the selected solution, which is a very punctual load with respect to the rail system.

With specific reference to the TBM push, the structural verification to be done coincides with those already introduced for the longitudinal joint verification because, conceptually, it is again a concrete surface on which a punctual load is applied on a

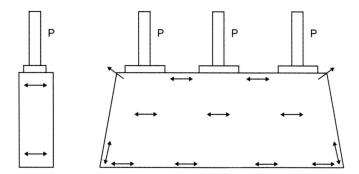

Figure 10.23 Zone of traction concentration along radial and circumferential directions when segment is loaded by the forces transmitted by TBM.

reduced area. So, accordingly, the verification of the contact stress and bursting effects must be performed.

In Figure 10.23, the zones where tensile stresses develop when the generic segment is loaded by the force P transmitted by the TBM rams are shown.

10.2.2.3.5 FIBRE REINFORCEMENT FOR USE IN PRECAST SEGMENTS

Fibre-reinforced concrete (FRC) is a material in which the crack bridging effect provided by the addition of fibres enhances the tensile behaviour after cracking, in terms of increase in residual tensile strength and ductility. In fact, after concrete cracking, fibres bridge the crack and allow transmission of forces between the crack planes (which are negligible in normal concrete).

FRC is a composite material and not the simple addition of fibres to a concrete matrix. Post-cracking concrete residual properties (toughness) depend on fibre characteristics (such as material properties, shape and aspect ratio), quantity (usually expressed by the volume percentage), distribution, orientation and the properties of the cementitious matrix surrounding the fibres. When referring to typical dosage adopted in precast tunnel segments (volume percentage generally lower than 1%), fibre reinforcement has a negligible effect on the elastic modulus, Poisson's ratio, compressive strength, electrical conductivity and porosity. Therefore, the concrete behaviour before cracking is not significantly modified by fibres.

The growing use of FRC as a structural material has been possible by the development of national and international standards concerning the design of FRC elements (ACI 544, 1996; RILEM TC 162-TDF, 2003). Among them, the *fib* Model Code 2010 (2013) has introduced post-cracking residual flexural strengths as performance parameters by adopting EN 14651 (2005) standard for classifying FRCs in terms of post-cracking residual strengths (f_{Ri}).

In the past two decades, FRC has been increasingly adopted in precast tunnel linings since FRC, with or without conventional reinforcement, enhances the general structural segment behaviour, together with the boost of the corresponding industrialized production. As a consequence, FRC represents a competitive material for tunnel segmental lining for the following main reasons:

- Fibres considerably improve the concrete post-cracking tensile behaviour, in the following defined as toughness.
- Fibre reinforcement enables a better crack control; hence, smaller crack openings are expected at SLS, which may result in an improvement in structural durability.
- Fibres provide higher resistance to impact loading.
- The industrial production process is improved, since a substitution (partial or complete) of conventional rebars can be achieved, allowing a time reduction in handling and placing curved rebars.
- A considerable reduction or elimination in storage areas for traditional reinforcement can be obtained.
- Fibre reinforcement is distributed everywhere in the segment, including the concrete cover which, in traditional reinforced concrete segments (RC segments), often needs to be considerably thick for the fulfilment of the fire protection and durability requirements.

Specific documents are currently available for the design of FRC precast tunnel segments with the aim of supporting either national or international standards for structural design. The International Tunnelling Association (ITA) and the American Concrete Institute (ACI) published guidelines for FRC segmental linings (ACI 544.7R-16, 2016; ITA report n. 16, 2016). Similarly, *fib* TG 1.4.1 has prepared design guidelines for FRC precast segments (*fib* bulletin No. 83, 2017). It is worth noting that ITA report n. 16 (2016) and *fib* bulletin No. 83 (2017) refer to the performance-based design approach suggested by *fib* Model Code 2010 (2013), hereafter MC2010.

The post-cracking residual strength provided by FRC is a fundamental parameter to be included in the design approach of segmental lining. In this regard, it is worth noting that, for analytical calculations or numerical analyses of FRC precast tunnel segments, designers have to assume a FRC performance class. As an example, according to MC2010, a post-cracking class equal to '4c' means that the characteristic value of f_{R1k} (i.e. nominal residual strength at CMOD (crack mouth opening displacement) equal to 0.5 mm according to EN 14651, 2005) must range between 4 and 5 MPa, while the ratio f_{R3k}/f_{R1k} should range between 0.9 and 1.1 (according to EN 14651, 2005, f_{R3k} is the nominal residual strength at CMOD equal to 2.5 mm). Alternatively, the designers can directly specify the minimum values of f_{R1k} and f_{R3k}. According to MC2010, FRC residual strengths must fulfil minimum performance requirements.

A design procedure taking properly into account the FRC residual tensile strength is better adopted for FRC precast segments. In particular, the crack development in a FRC element can be accurately considered by using non-linear numerical analyses based on non-linear fracture mechanics (NLFM; Hillerborg et al., 1976), as is described in detail in *fib* bulletin 83 (2017). However, simplified analytical methods are introduced in the current MC2010 and specifically contextualized to precast tunnel segments in *fib* bulletin 83 (2017).

Design choices and assumptions regarding FRC tunnel segments based on analytical calculations or non-linear numerical simulations could be eventually validated by full-scale tests on precast tunnel segments (Caratelli et al., 2011; Conforti et al., 2019).

Based on experiences collected in the last 20 years (ITA report n. 16, 2016), two possible solutions based on fibre reinforcement can be used in precast tunnel segments:

- FRC: segments reinforced by fibre only (without bars); this solution is preferable for speeding up the production process, but depending on the lining slenderness and/or ground/load conditions, it may not optimize the reinforcement costs.
- Hybrid solution: segment reinforced by a combination of fibre (FRC) and an optimized amount of traditional reinforcement (RCO); this solution is also referred to as RCO + FRC.

A possible hybrid solution proposed by Plizzari and Tiberti (2009) is reported in Figure 10.24a. The main design principles of this solution can be listed as follows:

- A minimum amount of longitudinal curved rebars (e.g. 0.2%) is provided in order to guarantee the necessary flexural lining capacity at ULS. It is obvious that this amount should be verified with respect to the final ground/load conditions.
- The conventional longitudinal reinforcement is concentrated around the segment borders; in particular, two chords of steel rebars extending along the two longer side of the segment are able to guarantee a better behaviour with respect to spalling cracks arising in the lining between the TBM thrust shoes.
- The conventional longitudinal reinforcement proposed is particularly adequate to withstand the local tensile stresses arising during the thrust jack phase in case of irregularities such as an outward eccentricity of TBM shoes.
- A minimum amount of stirrups may be adopted for practical reasons (construction purposes) and to further contribute with respect to local splitting radial stresses. It is worth noting that, when referring to typical shear forces arising in the lining

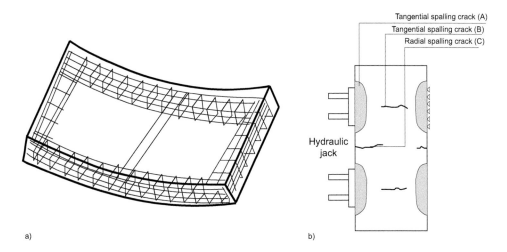

a)   b)

Figure 10.24 Possible hybrid reinforcement solution for use in precast segments (Plizzari & Tiberti, 2009) (a); typical crack pattern exhibited by segment during TBM thrust phase (Tiberti et al., 2017) (b).

in case of the embedded ground condition and grouting process, the stirrups can be generally entirely replaced by FRC.

- With regard to the splitting stresses (bursting stresses) arising under local loads, it should be underlined that FRC allows a progressive stable redistribution of stresses under the loading areas. Hence, fibre reinforcement is effective in withstanding local tangential and radial splitting tensile stresses due to the force exerted by TBM thrust shoes.

In case of segments reinforced by fibre reinforcement only (FRC segment), all the above mentioned tensile stresses mainly related to splitting, spalling and flexural phenomenon are withstood by fibre reinforcement, making necessary the adoption of a FRC exhibiting a proper post-cracking tensile performance.

The effectiveness of the two previously mentioned typologies of FRC segments was proven by means of a broad parametric study based on non-linear numerical analyses (Tiberti et al., 2017) and, in case of hybrid solution, confirmed by experimental full-scale test results (Conforti et al., 2019). In both cases, a crucial temporary phase was studied, namely the TBM thrust phase occurring during the excavation of the tunnel, when high concentrated forces are exerted by TBM hydraulic jacks. This particular stage was investigated by comparing the solutions based on fibre reinforcement with that based on traditional steel rebars only (RC segment).

In Figure 10.24b, the typical crack pattern appearing in precast tunnel segments during TBM thrust phase is reported. Based on the numerical parametric study, Tiberti et al. (2017) demonstrated that a segment made by steel fibre-reinforced concrete (exhibiting a post-cracking class '2b', according to MC2010) in combination with an amount of about $45\,\mathrm{kg/m^3}$ of steel rebars is capable of guaranteeing the same performance of the reference RC segment in terms of both local and global behaviour during this stage. On the other hand, it was also demonstrated that a segment reinforced by steel fibres only (SFRC classified as '6c', according to MC2010) is able to guarantee a similar or even better structural response, making clear that different reinforcement solutions based on fibre reinforcement can be used according to the post-cracking performance exhibited by FRCs. Regarding this point, based on full-scale experimental test results obtained by simulating high point loads transferred by TBM in case of a typical metro lining configuration, Conforti et al. (2019) confirmed that a combination of PFRC, polypropylene fibre-reinforced concrete (class '2.5e' according to MC 2010) and a low amount of conventional bars (steel content of about $40\,\mathrm{kg/m^3}$) exhibited a structural response in terms of bearing capacity, stiffness and crack width control similar to RC segments. Referring to FRC segments without rebars, a typical FRC performance, often adopted in practice, is the '4c'.

10.2.3   Design of waterproofing

The waterproofing of tunnels and underground structures is a key aspect in order to guarantee quality and effectiveness of the infrastructures and to reduce the potential impact on the surrounding areas. The attention on this aspect has intensified during the last decades as a consequence of the increase in requirements on the internal operational quality of the tunnel (e.g. dry surface) and of the design lifespan of the infrastructures (sometimes bigger than 150 years).

Inflows of water have been recognized as essential actors inducing damage in underground structures (CETU, 2015; Howard, 1991; ITA WG 6, 2001; Richards, 1998; Sandrone & Labiouse, 2011). This damage implies high costs for maintenance and refurbishment and can lead to expensive disruptions of service.

10.2.3.1 Design bases

The design of waterproofing solutions and technology is based on three main aspects:

- Groundwater level and properties: The design level of the groundwater must be evaluated taking into account all possible variations during the lifespan of the work (e.g. seasonal oscillations, tidal effects and anthropic impact). The pressure of the groundwater must be measured, and the presence of pressurized aquifer must be considered. Moreover, chemical and physical properties of the water have to be considered: the presence of solutes, salts or pollutants and temperature can impact the technical solution, avoiding the use of some materials or reducing the long-term effectiveness of drainages.
- Surrounding constraints due to structures or environmental aspects, which can impose the need of different solutions (e.g. to avoid any water drawdown).
- Internal quality requirements: The admissible water inflow in the tunnel is usually defined as a function of operation effectiveness and project design life. The required degree of tightness is to be correlated to the function of the tunnel: higher for structures where people are present or subjected to ice and lower for sewage tunnel. For traffic tunnels, the requirement is usually of dry or almost dry internal surface in the long term.

Once these three aspects have been satisfactorily analysed, the possible approaches and technology that fulfil all the requirements must be evaluated through a deep analysis combining the needs for the best technical solution for the long term and a suitable economical solution. It is noteworthy that the economic impact of a waterproofing system can not only be based on the initial investment in terms of costs and installation, but must take into account the costs for maintenance of the system during the life of the work and the possible refurbishment works needed in case of ineffectiveness. With this aim, a risk analysis-based approach can be suitable (Luciani & Peila, 2019).

10.2.3.2 Waterproofing approaches

Generally speaking, two different approaches to water management for underground structures exist: drained and undrained. The former permits a controlled water inflow in the tunnel through the drainage system (drainage layer and drainage pipes) and induces water table drawdown and hydraulic load reduction. On the contrary, the latter avoids any water inflow, eliminating environmental impact on the groundwater level, but applying higher loads on the linings. A further difference in the waterproofing approach exists between full-round waterproofing, usually applied in tunnels under the water table, and waterproofing only in the crown, also known as 'umbrella approach'. While the former is usually an undrained system (although it could be drained if radial drains are implemented), the latter is necessarily a drained approach. The choice of the approach to be applied in a project depends on the geological and hydrogeological conditions and on the excavation technologies. At the same time, the approach chosen influences the structural design and the maintenance plan; for example, the lining of an undrained tunnel should withstand a higher load than those of a drained one, while a drained tunnel will require more maintenance to ensure the effectiveness of the drainage system.

In shallow tunnels below the water table, the main issue is to avoid drainage in order to prevent possible effects on the surface due to groundwater level drawdown. Therefore, the tunnel has to be sealed with full-round undrained waterproofing. Secondary lining has to be designed to bear the hydraulic load. A grouted annulus around the tunnel can be created to reduce the filtration through the soil and to bear part of the load.

In deep tunnels in rocks, the main problem is usually the high hydrostatic load that is often impossible to bear for watertight linings. In these cases, both drained and undrained full-round approaches are possible. The undrained approach has to be limited to those situations where important environmental constraints limit the drainage of the water table and where it is technically and economically possible to design a lining that is able to bear the hydrostatic load. An undrained full-round waterproofing can be achieved by a membrane or by sealing of the rock by systematic injection of pre- and post-grouting for conventional tunnelling and by gaskets for TBM tunnels. In most of the cases, the environmental constraints are not so strong and therefore a drained approach is usually preferred. This can be achieved by a lining permitting a controlled water inflow, keeping a defined height of ground water level around the tunnel by means of horizontal drainage pipes installed at the base of the waterproofing membrane.

In tunnels excavated above the water table level, the prevention of water leakage through the lining is the key waterproofing aim, in order to avoid damage to tunnel structures and infrastructures and reduction in the efficiency of the tunnel. A drained waterproofing only in the crown is possible: the water, stopped by the waterproofing membrane, flows through the drainage layer to the base of the tunnel, where it is collected in the drainage pipes and taken outside the tunnel.

In tunnels excavated with segmental lining, the waterproofing is guaranteed by waterproof concrete of the segments and by the use of gaskets between one segment and the other. This is usually a full-round undrained solution, even if drains can be drilled in the lining to have a controlled drained approach, if needed.

10.2.3.3   Waterproofing technologies

In the long term, to waterproof a tunnel, the following solutions are usually adopted in conventional tunnelling:

- Waterproofing membranes at the extrados of the final lining: Waterproofing membranes are the most diffused technology for the waterproofing of linings cast in situ. Different technologies, materials and installation details exist for the use of this kind of solution. The general principle is to insert between the primary and final lining a watertight layer that prevents water from incoming. The most common materials used are geomembranes, usually made of plasticized polyvinyl chloride (PVC) or thermoplastic polyolefin (TPO). Less common is the use of geomembranes with clay layers. The presence of geomembranes makes the first-phase lining and the final lining mechanically independent of each other, since the shear stress transmission is prevented.

- Sprayed waterproofing membranes: Waterproofing sprayed membranes are some-times used in tunnelling by applying a thin layer of sprayed polymeric membrane on the shotcrete. The most relevant advantage of this technology is that it can be easily applied in situations with an irregular or complicated geometry. The application can be done both manually and by robots with the same technologies used for shotcrete (Makhlouf & Holter, 2008).

   Intrados waterproofing, i.e. installation of waterproofing or drainage systems at the intrados of the final lining. These solutions do not resolve the issues related to lining degradation due to water inflows and require quite high maintenance. This can be considered as a suitable solution only in existing tunnels.

For tunnels with segmental lining, gaskets are used to guarantee the watertightness. Gaskets are elements made of elastomeric elements applied on the lateral side of seg-ments in such a way that their contact will allow a watertight joint between segments. They have a specifically designed shape that is able to maximize their ability to con-trol the water pressure when compressed. Gaskets may be glued to a groove in the segment or can be bonded to the segment during casting. These elements provide effective water control if correctly installed and designed and can tolerate high water pressures.

   The mainly required inputs for a correct design of the waterproofing with gaskets are as follows:

- Design water pressure.
- Lifespan of the tunnel.
- Installation geometry and tolerances on the position of the segments: the distance between concrete of opposite joints (gap) and the relative displacement between the segments (offset).
- Interaction with other elements connecting the segments (connectors or bolts).

The value of resisting pressure must be evaluated taking into account the gap and offset. While the design offset value is defined based on construction tolerances of the project, the design gap is a function of segment geometry and geometrical tolerances and of the interaction between gasket and the connection elements. Indeed, when the segment is installed the minimum gap value is obtained due to the compression forces action on the segments. However, during the life of the tunnel, the compression can reduce and the gap can increase. This effect is limited by the connectors that avoid the detachment of segments. If steel bolts are used, the detachment is completely avoided and the gap value does not change. In this case, the design gap value is the minimum one (taking into account the geometry of the segment face and tolerances). Other-wise, when connectors are used, a certain amount of movement between segments is possible, before the complete effect of connection, and the gap increases. In this case, the design value of gap can be evaluated from the interaction diagram of the load–displacement curve of the connector with the load–displacement curve of the gasket.

The water pressure that a gasket can withstand can be evaluated through laboratory pressure tests as proposed by STUVA (2005).

It is important to remember that gasket design is strictly connected to the design of other elements of the segmental lining (connectors, segment lateral surface geometry, gasket groove geometry and position, and concrete strength) and any variation in one of these can influence the others.

The complete description of waterproofing technologies and the design choices can be found in Chapter 3 of Volume 2 of the book.

10.2.4   Interferences

10.2.4.1   Displacement field induced by tunnel excavation

As already mentioned in the introduction, the excavation of tunnel represents a hydro-mechanical disturbance for the environment, whose consequences may affect the already existing infrastructure. In this frame, a relevant role is played by the subsidence induced by tunnel excavation. In relation to this aspect, the most widely used empirically based method is that originally proposed by Peck (1969), according to which the greenfield transverse settlement trough is described by a normal Gaussian function:

$$w(x) = w_0 \exp\left(-\frac{x^2}{2i^2}\right) \tag{10.19}$$

where

$$w_0 = \frac{0.31V'D^2}{i} \tag{10.20}$$

with V' (volume loss) equal to the ratio of the volume enveloped by the settlement trough V_{ST} to the excavated volume ($V_T = \pi D^2/4$) per unit length at the tunnel face, where V_{ST} is obtained by integrating 10.21:

$$V_{ST} = \sqrt{2\pi} i w_0 \tag{10.21}$$

The assumption that V_{ST} equals the extra-volume excavated around the cavity is common, although this assumption is realistic only for clay under undrained conditions.

Many authors have shown that V' depends on the type of ground and on the excavation method. (Mair & Taylor, 1999): in open face excavations in stiff clay, V' ranges between 1% and 2% (even lower if a sprayed concrete lining is adopted); in excavation with TBM, V' can be lower than 0.5% in sand and between 1% and 1.5% in soft clay.

In Equation 10.19, the parameter i is the horizontal distance of the inflection point of the settlement trough from the tunnel axis. O'Reilly and New (1982) assumed that this parameter increases linearly with the depth of the tunnel axis, z_0. On this basis, common is the assumption that:

$$i = Kz_0 \tag{10.22}$$

According to Mair and Taylor (1999), K generally varies between 0.4 and 0.6 for clayey soils, while for sandy soils, K ranges between 0.25 and 0.35.

For mixed soil profiles, some suggestions are given by Chiriotti and Grasso (2001), taking into account both the prevailing conditions at the face and in the overburden.

At a given depth z, settlement profiles are still predictable with a Gaussian function, where:

$$i = K(z_0 - z) \tag{10.23}$$

with K being a linear function of depth:

$$K = \frac{0.175 + 0.325(1 - z/z_0)}{(1 - z/z_0)} \tag{10.24}$$

An alternative formulation has been proposed by Moh et al. (1996) for granular soils:

$$i(z) = i_1 D \left(\frac{z_0 - z}{D} \right)^m \tag{10.25}$$

where D is the tunnel diameter and b and m are two parameters. For granular soils $i_2 = 0.4$, while $i_2 = 0.8$ for fine-grained soils. The value of i_1 can be determined accordingly, assuming that $i = K z_0$ when $z = 0$.

Horizontal displacements can be calculated (O'Reilly & New, 1982) as:

$$u(x,z) = -\frac{x \cdot w(x,z)}{\alpha_u(z_0 - z)} \tag{10.26}$$

where $\alpha_u = 1$ when assuming that displacement vectors point towards the centre of the tunnel. Horizontal displacement can be used to derive horizontal strains ε_h as is illustrated in the next section.

Attewell and Woodman (1982) suggested fitting the profile of the maximum settlement along the tunnel axis by a complementary error function (*erfc*). Accordingly, the settlement can be expressed as:

$$\frac{w}{w_0} = \frac{1}{2} \text{erfc} \left(-\frac{y}{\sqrt{2}i} \right) \tag{10.27}$$

where w_0 is the maximum settlement under plane strain conditions calculated as reported above.

The measured value of settlements at the face may be lower than $0.5\,w_0$ (as follows from the above equation), depending on the stiffness of the face support. For instance, when using pressurized TBM, the settlement at the tunnel face can be very small and even null.

The empirical method is used also with the superposition effects for not so closely spaced tunnels. Generally, the superposition works well when the distance is larger than 1.5 times the tunnel diameter (Hansmire & Cording, 1985; Mair & Taylor, 1999).

It is also worth noting that the calculation of long-term settlements in fine-grained soils requires the knowledge of the lining permeability that affects the dissipation of the induced excess pore pressures around the excavation (Samarasekera & Eisenstein, 1992; Wongsaroj et al., 2007).

10.2.4.2   Simplified assessment of the effect on buildings of tunnelling-induced displacements

Existing buildings may be affected by ground movements induced by tunnelling. Nevertheless, since a tunnel is a linear infrastructure typically involving a large number of buildings, at preliminary stages, the study of soil–structure interaction problems tends to be neglected or largely simplified. In other words, to predict building damages the presence of buildings is often simply neglected and an uncoupled approach with green-field settlement profile is adopted; a slightly more accurate approach is to consider the building as an equivalent elastic beam.

A complete procedure for building damage assessment, appropriate for various steps of the design process, was established by Burland et al. (2004). The procedure is summarized in Figure 10.25.

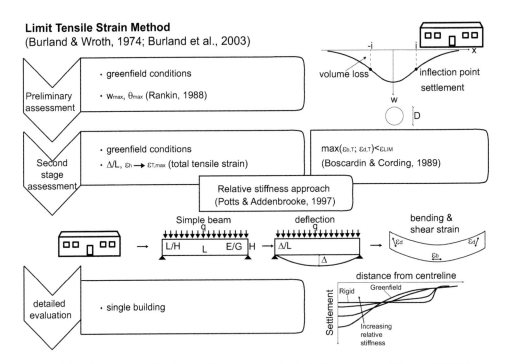

Figure 10.25 Overview of limit tensile strain method. (Adapted from Bilotta (2017).)

A preliminary screening of buildings is based on the estimated maximum green-field settlement w_{max} and slope θ_{max}. According to Rankin (1988), buildings for which $w_{max} \leq 10$ mm and $\theta_{max} \leq 1/500$ can be excluded from further assessments.

At a second stage of analysis, a maximum deflection ratio, Δ/L, and an average horizontal strain, ε_h, at the foundation level are used to calculate a resultant tensile strain, $\varepsilon_{t, TOT}$, in the building, schematized as a simply supported elastic beam. The maximum value between those resulting as maximum resultant bending strain, $\varepsilon_{b, TOT}$, or diagonal strain, $\varepsilon_{d, TOT}$, is called total tensile strain and is compared with the limit tensile strain ε_{LIM} defined by Boscardin and Cording (1989) for different damage levels. For those buildings falling within an unacceptable class of damage, the same procedure can be repeated by including the effect of building stiffness on the induced settlement trough (Potts & Addenbrooke, 1997). Hence, two relative stiffness parameters are defined and modification factors can be computed, based on relative stiffness, to reduce the greenfield deflection ratio, Δ/L, and horizontal strain, ε_h, and thus the assessed class of damage.

If the level of damage is still unacceptable, the problem must be analysed in a third-stage detailed analysis, using numerical methods.

10.3 APPLICATION OF OBSERVATIONAL METHOD IN TUNNELLING

The so-called 'observational method' is a design procedure aimed at optimizing the management of the unavoidable geological and geotechnical uncertainties affecting both tunnel design and construction. During the design phase, any method requires the full characterization of those threshold situations for which the adopted design solution can approach a limit state, and the detailed design of a safer solution (or even a more favourable solution), to be substituted to the former one on the basis of the response observed during construction. Besides reducing geotechnical and structural hazards, the application of the observational method limits costs and forces the execution times within the predicted terms, reducing the occurrence of legal disputes between the parties. For all these reasons, the observational method, including the monitoring plan, is recommended by Eurocode 7.

In case of tunnelling, the phases characterizing the implementation of the observational method are described here below.

1. Design phase:

   a. Planning and execution of the proper soil investigations considered to be sufficient for the definition of both effective geological and geotechnical models of the tunnel alignment.

   b. Definition of both geological and geotechnical models. Identification of the geotechnically homogeneous alignment stretches and relevant physico-mechanical characterization. For each of these stretches, the designer should identify the most likely geotechnical situation as well as the relevant (less favourable) deviation reasonably predictable and assess the stability of the tunnel face and of the excavation, without improvement/reinforcement works and supports.

c. Design of the excavation procedure, temporary and permanent supports and improvement/reinforcement techniques for each of the above geotechnical situations.

d. Assessment of the design solutions, concerning tunnel excavation (stability of cavity and structures, water inflow and excavation-induced damage) and prediction of the pertinent fields (displacements, stresses, pore pressures, water flows, etc.). Such calculations also make possible the identification of the measurable quantities best representing the response to excavation.

e. Definition of the admissible limits (the so-called 'threshold values') of the measurable quantities, representative of tunnel response. These thresholds are obtained from further calculations based on the less favourable geotechnical conditions, as identified by means of the characterization at point (1b).

f. Definition and assessment of alternative design solutions, to be adopted if the aforementioned admissible limits are exceeded or are far to be reached.

g. Design of a monitoring plan for the control of the quantities defined at points 1d and 1e. Frequency and timing of acquisition must be accurately set, along with the proper procedures of monitoring data processing, to ensure the timeliness of safety interventions.

2. Construction phase:

a. In situ assessment of the reliability of geological and geotechnical models of the tunnel alignment during the excavation face advancement. Identification of the geotechnical situation at face and application of the pertinent excavation, support and improvement techniques defined in the project (point 1c).

b. Recording, processing and interpretation of monitoring data, according to the requirements at point 1g, and comparison with the threshold values defined at point 1e.

c. If the admissible limits are exceeded or are far to be reached: modification of excavation, support and improvement/reinforcement techniques, by promptly adopting the already designed alternative solution (point 1f).

It can be remarked that the crucial phase 2b may require the back-analysis of the monitoring data, which can be developed by means of simplified analytical or empirical methods, or by employing the more advanced tools provided by computational mechanics (see Chapter 9).

10.4 LONG-TERM TUNNEL PERFORMANCES

After the construction, during its lifetime, the tunnel may experience a non-negligible evolution of stresses and strains due to either hydro-thermo-chemo-mechanical coupled processes (Chapter 9, Section 9.4.2) or accidental loads. In this section, the authors have decided to approach exclusively the accidental loads in terms of interaction with landslides (soil movements induced by external environmental actions), seismic actions and fire. As already mentioned (Chapter 9 Section 9.4.2), the evolution of stresses and strains with time may also be induced by hydro-thermo-chemo-mechanical processes (consolidation, squeezing, swelling and in general viscous effects), but these are strictly related to the excavation process and they have been partially tackled in Section 10.2 and Chapter 9.

10.4.1 Interferences between landslides and tunnels

Tunnel excavation in a sloping ground induces stress and strain changes in the surrounding medium significantly different from those occurring when the ground surface is horizontal, resulting in a modified subsidence trough and loading pattern on the lining. In particular:

- For a tunnel perpendicular to the slope dip direction, settlements and horizontal displacements in a transversal cross section are no longer symmetric with respect to the tunnel axis. For the subsidence assessment, a graphical approach named 'rays projection method', based on the projection of the classical semi-empirical curve valid for a horizontal ground surface to an inclined plane, was proposed by Whittaker and Reddish (1989) and Shu and Bhatacharya (1992). The results reveal that, for increasing inclination angles, the subsidence trough expands upstream and shrinks downstream, the maximum settlement increases and moves downstream, and the maximum horizontal displacement and the maximum tensile strain increase downstream and reduce upstream. Elastoplastic finite element analyses, carried out by D'Effremo (2015), confirmed these observations and showed that, for increasing slope inclinations, displacement vectors at the ground surface tend to be directed downstream, also for those points located upstream of the tunnel (Figure 10.26).
- The stability of the slope, investigated by the limit equilibrium method, decreases for increasing inclination angles (Picarelli et al., 2002). Elastoplastic finite element calculations (D'Effremo, 2015) show that plasticized zones around the tunnel become widespread for larger slope inclinations and that downstream horizontal

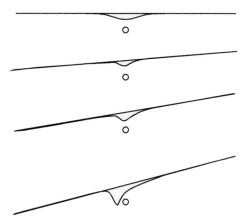

Figure 10.26 Tunnelling-induced displacements at the ground surface of an elastic–plastic medium for different slope inclinations. (Modified from D'Effremo 2015.)

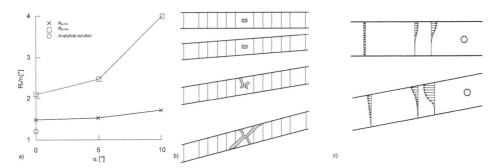

Figure 10.27 Influence of slope inclination on the magnitude (a) and shape (b) of the plasticized zones around the tunnel and (c) on the horizontal displacement profiles along verticals downstream of the tunnel. (Modified from D'Effremo 2015.)

displacements may be similar to those occurring due to a landslide movement, but are rather due to local plasticization phenomena (Figure 10.27). In fact, if the slope is stable before the tunnel excavation, only local possible failures are to be expected. All these phenomena lose their importance as tunnel depth increases and for earth coefficient at rest k_0 close to 1.

The intersection of a dormant landslide body by a tunnel can mobilize the shear strength along the failure surface and consequently reactivate the movements (D'Effremo et al., 2016). The affected portion of failure surface increases with slope inclination and decreases with tunnel cover and with the distance between the tunnel and the pre-existing slide surface.

Tunnel lining is significantly loaded at the cross sections intersecting a landslide body, as also suggested by pipeline studies (O'Rourke & Lane, 1989; Rajani et al., 1995; Spinazzè et al., 1999; Casamichele et al., 2004). In fact, the landslide failure surface intersects progressively the tunnel cross section at different elevations (Figure 10.28) and observed damage patterns (Wang, 2010) as well as convergence measurements (Boldini et al., 2004; Bandini et al., 2015) can provide information about the landslide characteristics in terms of direction and extension, respectively. Huang et al. (2010), disregarding the rigid motion components (i.e. translations and rotations), suggest that the observed deformation of the cross section can clearly indicate the location at which the landslide surface is intersected by the tunnel.

Limit analysis calculations performed by Scandale (2017) highlight the different possible failure mechanisms occurring in relation to the percentage of lining located inside the moving landslide body (Figure 10.29). The maximum loads in the lining at failure significantly increase with the tunnel depth, i.e. with the thickness of the landslide mass.

Soil–structure interaction analyses allow assessing the relationship between landslide movements and strains (Figure 10.30) and loads induced in the tunnel lining. The amount of landslide displacement needed to reach limit conditions is strongly dependent on the stiffness of the lining and the soil.

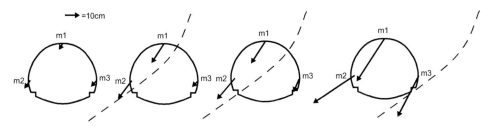

Figure 10.28 Convergence measurements in the entrance area of a landslide body. (Modified from Koronakis 2004.)

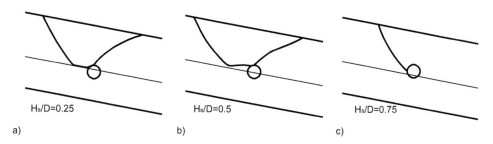

Figure 10.29 Results of limit analysis calculations showing the failure mechanisms. (Modified from Scandale 2017.)

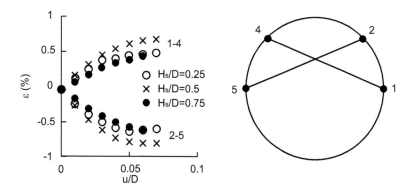

Figure 10.30 Relationship between strains of the tunnel cross section and normalized displacement of the landslide body (*u/D*) for different values of (*Hs/D*) along diagonal chords. (Modified from Scandale 2017.)

10.4.2 Seismic loads

Although tunnels have traditionally been considered less susceptible to earthquake perturbations than above-ground structures, several cases of seismic-induced damage to tunnels are reported in the literature (e.g. Dowding & Rozen, 1978; Oven & Sholl,

1981; Sharma & Judd, 1991; Power et al., 1998). Typical damages include lining cracks, portal failures, concrete lining spalling, groundwater ingress, exposed and buckled reinforcements, rockfalls (in unlined sections), collapses due to slope failure, lining shear-off (by displaced fault) and dislocations. In general, the level of damage is higher at the portals and for shallow tunnels.

A number of seismic events that occurred in the last decades can be examined (Tsinidis et al., 2020): among them, those that have produced significant damage to tunnels are the following: the 1995 Hyogoken-Nanbu earthquake in Kobe (Japan), the 1999 Kocaeli earthquake in Turkey, the 1999 Chi-Chi earthquake in Taiwan, the 2004 Mid Niigata Prefecture earthquake and the 2007 Niigata Prefecture Chuetsu offshore earthquake in Japan, and the 2008 Wenchuan earthquake in China. Earthquake magnitude, source depth and epicentral distance, tunnel depth, geometrical characteristics of the lining and abrupt changes in tunnel size or ground conditions are the main factors affecting the behaviour of mountain tunnels (Chen et al., 2012). In most cases, the damage occurring to tunnels during earthquakes is due to the lack of consideration of appropriate seismic loads in design and during construction.

Differently from above-ground structures, the seismic behaviour of tunnels is determined mostly by the loads induced by the surrounding ground, rather than by the inertial response of the tunnel itself (Wang, 1993; Hashash et al., 2001; Pitilakis & Tsinidis, 2014). Therefore, in seismic areas tunnels should be designed to respond to ground shaking (Section 10.4.2.1) and ground failure (Section 10.4.2.2). Ground shaking is produced by seismic wave propagation in the ground, and it affects underground structures immersed in such a kinematic field. Ground failure, on the other hand, may be due to soil liquefaction, slope failure or fault movements, and it generally produces large permanent deformation or dislocation of tunnels. In the following, the two aspects are considered separately.

10.4.2.1   Ground Shaking

Ground shaking induces deformation of the longitudinal axis of the tunnel, either as bending or as axial deformation and ovalization (circular tunnels) or racking (box-shaped tunnel) of the tunnel transverse section (Owen & Scholl, 1981). The tunnel response to ground shaking is a typical soil–structure interaction (SSI) problem. It depends on the ground motion characteristics and on the tunnel depth and geometry, and it is mainly regulated by the soil-to-tunnel relative stiffness and the characteristics of the soil–lining interface. Since the effects of shaking in the longitudinal and transverse directions are different, their analysis is generally carried out separately.

Some authors have proposed analytical solutions for a pseudo-static calculation of the increments of the internal forces in the transverse section of the tunnel lining due to ground shearing, considering the relative stiffness between the soil and the structure. The maximum shear strain, γ_{max}, due to the propagating shear waves, can be calculated in free-field conditions, from the seismic demand (e.g. PGA or PGV).

Earlier solutions for external loading on underground structural cylinders (Burns & Richards, 1964; Hoeg, 1968; Peck et al., 1972) were adapted by Wang (1993) to solve the case of seismic loading on a circular tunnel lining.

Compressive and flexural relative stiffness parameters, C and F, were defined as:

$$C = \frac{E_g\left(1-v_s^2\right)r_0}{E_s t_r\left(1+v_g\right)\left(1-2v_g\right)} \qquad (10.28)$$

$$F = \frac{E_g\left(1-v_s^2\right)r_0^3}{6E_s I\left(1+v_g\right)} \qquad (10.29)$$

where t_r is the lining thickness and E and v are the elastic parameters (subscript s is for 'structure' and g is for 'ground').

If full-slip conditions occur at the tunnel/ground interface, Wang (1993) proposed the following equation to calculate the diameter change due to ovalization:

$$\frac{\Delta D}{D} = \pm \frac{1}{3} K_1 F \gamma_{max} \qquad (10.30)$$

where K_1 is a coefficient that depends on the flexural relative stiffness parameter F and the Poisson's ratio of the ground:

$$K_1 = \frac{12\left(1-v_g\right)}{2G+5-6v_g} \qquad (10.31)$$

Hence, the maximum changes in hoop force and bending moment in the tunnel transverse section can be calculated as (Wang, 1993):

$$N_{max} = \pm \frac{1}{6} K_1 \frac{E_g}{\left(1+v_g\right)} R\gamma_{max} \qquad (10.32)$$

$$M_{max} = \pm \frac{1}{6} K_1 \frac{E_g}{\left(1+v_g\right)} R^2 \gamma_{max} \qquad (10.33)$$

In no-slip condition at the tunnel/ground interface, Wang (1993) proposed the following equation for the maximum hoop force:

$$N_{max} = \pm K_2 \frac{E_g}{\left(1+v_g\right)} R\gamma_{max} \qquad (10.34)$$

where

$$K_2 = 1 + \frac{F\left[\left(1-2v_g\right)-\left(1-2v_g\right)C\right]-\frac{1}{2}\left(1-v_g\right)^2+2}{F\left[\left(3-2v_g\right)+\left(1-2v_g\right)C\right]+C\left[\frac{5}{2}-8v_g+6v_g^2\right]+6-8v_g} \qquad (10.35)$$

The maximum bending moment is assumed equal to that in full-slip conditions (Equation 10.33).

Similar analytical solutions have been proposed by other authors under slightly different assumptions (Table 10.2).

It is worth noting that all the analytical solutions listed in Table 10.2 assume elastic behaviour for both the lining and the ground. Therefore, it is suggested that they are used by equivalent linear properties and the corresponding compatible strain level, as an approximate way of accounting for non-linear soil response. Although an equivalent linear approach is generally adopted in practice, the two main factors influencing the seismic behaviour of tunnels, i.e. the soil–lining relative stiffness and the behaviour of the tunnel/ground interface, may change during shaking due to the non-linear and irreversible behaviour of the soil and of the lining system (Pitilakis & Tsinidis, 2014). Hence, the use of analytical solutions should be limited to preliminary design stages. For a more accurate calculations of the seismic loads in the tunnel lining, a full dynamic numerical analysis should be carried out, modelling incrementally the non-linear behaviour of the materials and its effects on dynamic SSI.

Similar analytical solutions have been developed to analyse the tunnel deformation under propagating seismic waves along the tunnel longitudinal direction. In the simplest approach, the structure deformation is computed by using equations that estimate the elastic ground strains in free-field conditions. Hence, the 'seismic load' is assumed as an equivalent static ground deformation. For instance, St. John and Zahrah (1987) proposed the following equations to compute the seismic internal forces in the tunnel lining in longitudinal direction:

$$N = \left(\frac{2\pi}{\lambda}\right)\sin\phi\cos\phi E_s AD^* \cos\left(\frac{2\pi x}{\lambda/\cos\phi^*}\right) \tag{10.36}$$

Table 10.2 Summary of assumptions of analytical solutions for the analysis of tunnels under ground shaking in transverse section (modified after Tsinidis et al. 2020)

Solution	Saturation conditions		Soil–tunnel interface			Cross section	
	Dry	Saturated	No-slip	Frictional-slip	Full-slip	Circular	Rectangular
St. John and Zahrah (1984)	✓	✗	✓	✗	✓	✓	✗
Wang (1993)	✓	✗	✓	✗	✓	✓	✗
Penzien and Wu (1998)	✓	✗	✓	✗	✓	✓	✗
Penzien (2000)	✓	✗	✓	✗	✓	✓	✗
Bobet (2003)	✓	✓	✓	✗	✓	✓	✗
Huo et al. (2006)	✓	✗	✓	✗	✓	✗	✓
Park et al. (2009)	✓	✗	✓	✓	✓	✓	✗
Bobet (2010)	✓	✗	✓	✗	✓	✓	✗
Kouretzis et al. (2011)	✓	✗	✗	✗	✓	✓	✗
Kouretzis et al. (2014)	✓	✗	✗	✗	✓	✓	✗

✗, the method does not cover it; ✓, the method covers it.

$$V = \left(\frac{2\pi}{\lambda}\right)^3 \cos^4\phi E_s I D^* \cos\left(\frac{2\pi x}{\lambda/\cos\phi^*}\right)$$ (10.37)

$$M = \left(\frac{2\pi}{\lambda}\right)^2 \cos^3\phi E_s I D^* \sin\left(\frac{2\pi x}{\lambda/\cos\phi^*}\right)$$ (10.38)

where E_s is Young's modulus of the structure, A the area of the transverse section, I the second moment of area, D^* the motion amplitude, λ the wavelength, and $\phi*$ the angle of incidence between the wave and the tunnel. Equations 10.36–10.38 were obtained without considering the dynamic soil–structure interaction. More accurate predictions can be achieved by using subgrade reaction methods, where the tunnel is modelled as a beam on elastic springs (ISO 23469, 2005). The equivalent static ground deformation is applied to the springs. It may account for the spatial variation of the ground motion. When adopting a subgrade reaction method, an adequate estimation of characteristics of the spring is crucial. Once that spring coefficients in transversal and axial directions (K_t and K_a) are determined, reduction factors can be calculated to be applied to bending moment (R_1) and axial force (R_2) calculated in Equations 10.38 and 10.40:

$$R_1 = \frac{1}{1 + \dfrac{E_s I}{K_t}\left(\dfrac{2\pi}{\lambda}\right)^4 \cos^4\phi^*}$$ (10.39)

$$R_2 = \frac{1}{1 + \dfrac{E_s A}{K_a}\left(\dfrac{2\pi}{\lambda}\right)^2 \cos^2\phi^*}$$ (10.40)

10.4.2.2 Ground failure

As mentioned above, large permanent deformations of tunnels may be caused by ground liquefaction, slope instability and fault movements. In this case, design should be aimed at either mitigating ground failure, with stabilization or ground improvement techniques, or allowing the lining structure to accommodate large displacements.

Liquefaction mainly affects immersed tunnels, and it can induce non-negligible buoyancy or reconsolidation loads and uplift. Usually, to prevent the effect of liquefaction, mitigation measures can be undertaken, such as anchoring or increasing the dead load on the tunnel or replacing the soil around to enhance drainage (Power et al., 1998).

Tunnels are very vulnerable to seismically induced slope instability. However, it is unlikely that a tunnel can be designed to resist or accommodate these movements unless the amount of movement is small (Pitilakis & Tsinidis, 2014): the only feasible way to mitigate the effects of slope instability or lateral spreading movements is to stabilize the ground slope.

An estimation of the maximum expected fault displacement, D_{max}, can be performed using simplified relations such as that proposed by Wells and Coppersmith (1994):

$$\log D_{max}(m) = -7.03 + 1.03 M_w \qquad (10.41)$$

where M_w is the moment magnitude.

Such a displacement may be applied as a static dislocation. If the extension of the fault may be determined a priori, a possible countermeasure is to design a local enlargement of the tunnel section in the fault zone, thus allowing an easier restoration after dislocation. On the other hand, in case of small and distributed displacements the tunnel lining can be designed with transverse joints providing sufficient ductility to accommodate the deformation in the longitudinal direction.

10.4.3 Design of lining against fire

The fire analysis of concrete tunnels requires the comprehension of several problems and their solution requires the expertise in different engineering fields. In particular, beyond all the aspects related to the material degradation during the fire exposure and the approaches to be adopted in a fire analysis of a typical reinforced concrete structure, in tunnel the interaction between the lining and the surrounding soil or rock is mandatory to be considered. For these reasons, also the geotechnical aspects have to be correctly considered. In fact, the main loads on reinforced concrete linings, present when the fire is expected to occur, are due to soil or rock actions. Furthermore, the soil acts as a constraint on the tunnel lining, inhibiting its dilatation when temperature increases. The information herein presented can be used either for newly designed or for existing tunnels.

10.4.3.1 Structural requirements

The definition of failure for tunnels involves several aspects, and it is not always possible to assign a unique criterion. From the point of view of fire analysis, it is necessary to avoid the tunnel failure for all the duration of the fire scenario. Failure is generally defined as the condition at which the structure fails to perform as a tunnel. The failure is related to the structure collapse and not to the collapse of single component. Since the tunnel is considered to consist of primary and secondary load-bearing components, damage to secondary load-bearing components usually does not cause failure of the tunnel structure.

Some damage cannot be avoided after fire exposure, and after the fire scenario, a repair intervention is expected and accepted. Even when the surrounding ground is stable and the lining is not strictly needed, the collapse of the lining can jeopardize the human safety (e.g. for fire brigades). It is also evident that local damage or collapses of the tunnel may cause severe economic losses due to operational interruptions. Furthermore, possible settlements at the ground level associated with the fire scenario can jeopardize the safety of the building or structures located above the tunnel area. For defining the structural performance, in many cases, considering also the cooling phase could be necessary. All the critical situations before the collapse have to be addressed in the design procedure.

10.4.3.2   Material behaviour relevant for fire design

In thermal analysis of reinforced concrete tunnels, the thermal and physical properties to be defined are related to the concrete only (EN 1992-1-2). The effect of thermal and physical properties of the steel can be usually neglected (e.g. in the thermal analysis the presence of the steel reinforcement can be disregarded). These properties depend on the adopted aggregate used in the concrete. This information is not often available during the design phase. From the design point of view, the adoption of the properties of siliceous aggregate is on the safe side. For this reason, if the designer cannot know the type of aggregate adopted, in the fire analysis, the assumption of using siliceous aggregates is suggested.

The hot properties of the concrete are usually defined in the national codes. The main parameters to be defined are the strength and stiffness at different temperatures, under both compression and tension. The mechanical properties of concrete in each point can then be related to the maximum temperature reached locally by means of the temperature-dependent stress–strain curves. In EN 1992-1-2, stress–strain relationships are proposed to simulate concrete behaviour under compression. As far as concrete behaviour at high temperature is concerned, an irreversible loss of stiffness and strength takes place, these in the literature are referred to as thermal damage (or softening) and thermal decohesion, respectively. These effects are taken into account in the stress–strain relationships proposed by EN 1992-1-2.

The evaluation of concrete and steel residual properties is important, when a fire analysis with fire scenario considering the cooling phase is performed. The residual properties related to the cooling phase, immediately after the fire exposure, can remarkably differ from what measured several hours after heating. In the present version of EN 1992-1-2, this information is not given. Some information on the residual properties of the concrete is present in EN 1994-1-2. Regarding the residual properties of steel, some data are available in the literature (see Felicetti et al., 2009).

10.4.3.3   Spalling

Explosive spalling is a phenomenon that can occur in concrete structures, especially when medium- to high-strength concrete with dense cement matrix is used. Explosive spalling is usually related to the steam pressure linked to the water present in the concrete matrix. When this phenomenon occurs, concrete pieces separate from the exposed surface in a sudden and explosive way.

Several factors can contribute to the development of spalling, such as concrete porosity, concrete moisture, type of cement, type of aggregates and stress in the concrete. It is usually difficult to predict if spalling can occur, and specific tests have to be conducted. In these tests, it is important to simulate the compressive stresses in the lining before and during the occurrence of the fire scenario, since these can facilitate the spalling.

There is relative consensus about the beneficial effects of inclusion of polypropylene microfibres in fresh concrete on reducing the occurrence of spalling. However, this is not an effective solution for every concrete type or fire scenario.

The available standards for the protection of structures against spalling are at the moment inadequate for simulating, predicting or avoiding spalling, and usually, specific tests are needed. These tests are aimed at predicting the spalling depth and velocity, and this information can be used in the structural model of the tunnel.

AUTHORSHIP CONTRIBUTION STATEMENT

The chapter was developed as follows. di Prisco and Callari: chapter coordination; di Prisco and Flessati: §10.1, 10.2, 10.3, 10.4; Barbero and Graziani: §10.2; Bilotta: §10.2, 10.4; Boldini and Desideri: §10.4; Callari: §10.1, 10.2, 10.3; Luciani: §10.2.3; Meda: §10.4; Pescara: §10.2; Plizzari and Tiberti: §10.2 Russo: §10.2; Sciotti: §10.4. All the authors contributed to chapter review. The editing was managed by di Prisco and Flessati.

REFERENCES

ACI Committee 544 (1996) Report on fibre reinforced concrete ACI Report 544.1R-96. Amer Concr Inst., pp. 1–64.

ACI Committee 544 (2016) Report on design and construction of fiber reinforced precast concrete tunnel segments ACI 544.7R-16. Amer Concr Inst., pp. 1–36.

Anagnostou, G. & Kovári, K. (1994) The Face stability in Slurry-shield-driven Tunnels. *Tunnelling and Underground Space Technology*, 9(2), 165–174.

Anagnostou, G. & Kovári, K. (1996) Face stability conditions with earth-pressure-balanced shields. *Tunnelling and Underground Space Technology* (2), 165–173.

Anagnostou, G. & Perazzelli, P. (2015) Analysis method and design charts for bolt reinforcement of the tunnel face in cohesive-frictional soils. *Tunnelling and Underground Space Technology*, 47, 162–181.

Anagnostou, G., Schuerch, R., Perazzelli, P., Vrakas, A., Maspoli, P. & Poggiati, R. (2016) Tunnel face stability and tunnelling induced settlements under transient conditions. Eidgenössisches Departement für UVEK, Bundesamt für Strassen, 1592.

Andersland, O.B. & Ladanyi, B. (2003). *Frozen Ground Engineering*, 2nd Edition, Hoboken, NJ: Wiley.

Attewell, P.B. & Woodman, J.P. (1982) Predicting the dynamics of ground settlement and its derivatives bytunnelling in soil. *Ground Engineering*, 15, 13–22.

Bandini, A., Berry, P. & Boldini, D. (2015) Tunnelling-induced landslides: the Val di Sambro tunnel case study. *Engineering Geology*, 196, 71–87.

Barla, G. (1995) Squeezing rocks in tunnels. *ISRM News Journal*, 3(4), 44–49.

Barton, N., Grimstad, E. & Grimstad, G. E. (2014). Q-system-an illustrated guide following forty years in tunnelling. Technical report.

Barton, N., Lien, R. & Lunde, J. (1974) Engineering classification of rock masses for the design of tunnel support. *Rock Mechanics*, 6(4), 189–236.

Bell, F.G. (1993) *Engineering Treatment of Soils*, Boca Raton, FL: Taylor & Francis.

Bezuijen, A., Pruiksma, J.P. & van Meerten, H.H. (2006) Pore pressures in front of tunnel, measurements, calculations and consequences for stability of tunnel face. Tunnelling. A decade of progress. GeoDelft 1995–2005, pp. 35–41.

Bieniawski, Z.T. (1984) *Rock Mechanics Design in Mining and Tunneling*, Balkema: Rotterdam.

Bieniawski, Z.T. (1989) *Engineering Rock Mass Classifications*, Wiley: New York.

Bilotta, E. (2017) Soil-structure interaction in tunnel construction in soft ground. *Rivista Italiana di Geotecnica*, 51(2), 5–30.

Bobet, A. (2003) Effect of pore water pressure on tunnel support during static and seismic loading. *Tunnelling and Underground Space Technology*, 18(4), 377–393.

Bobet, A. (2010) Drained and undrained response of deep tunnels subjected to far-field shear loading. *Tunnelling and Underground Space Technology*, 25(1), 21–31.

Boldini, D., Graziani, A. & Ribacchi, R. (2004) Raticosa tunnel, Italy: characterization of tectonized clay-shale and analysis of monitoring data and face stability. *Soils and Foundations*, 44(1), 59–71.

Boscardin, M. D. & Cording, E. J. (1989) Building response to excavation-induced settlement. *Journal of Geotechnical Engineering*, 115(1), 1–21.

Broere, W. (2001) Tunnel Face Stability & New CPT Applications, Ph.D. Thesis. Delft University of Technology.

Broere, W. & Van Tol, A.F. (2000) Influence of infiltration and groundwater flow on tunnel face stability. Geotechnical aspects of underground construction in soft ground, 1, 339–344.

Burland, J. B., Mair, R. J., & Standing, J. R. (2004). Ground performance and building response due to tunnelling. In *Advances in geotechnical engineering: The Skempton conference: Proceedings of a three day conference on advances in geotechnical engineering, organised by the Institution of Civil Engineers and held at the Royal Geographical Society*, London, 29–31 March, pp. 291–342. Thomas Telford Publishing.

Burns, J.Q. & Richard, R.M. (1964). Attenuation of stresses for buried cylinders. *Proceedings of Symposium on Soil-Structure Interaction*, Tucson, AZ, pp. 378–392.

Callari, C. (2009) Three-dimensional effects of ground desaturation due to tunneling. In: *Computational Methods in Tunneling-EURO:TUN 2009*, G. Meschke, G. Beer, J. Eberhardsteiner, D. Hartmann & M. Thewes (Eds.), Aedificatio Publishers Freiburg, pp. 517–524.

Callari, C. (2015) Numerical assessment of tunnel face stability below the water table. In: *Computer Methods and Recent Advances in Geomechanics*, Oka, Murakami, Uzuoka & Kimoto (Eds.), Taylor & Francis Group: London, pp. 2007–2010.

Callari, C. (2021) Assessment of tunnel stability: safety factors and numerical techniques. In: *Challenges and Innovations in Geomechanics*. IACMAG 2021. Lecture Notes in Civil Engineering, M. Barla, A. Di Donna & D. Sterpi (Eds.), vol 126, Springer: Cham.

Callari, C., Alsahly, A. & Meschke, G. (2017) Assessment of stand-up time and advancement rate effects for tunnel faces below the water table. *4th ECCOMAS Conference on Computational Methods in Tunnelling and Subsurface Engineering (EURO:TUN 2017)*, Innsbruck, April 2017.

Callari, C. & Casini, S. (2005) Tunnels in saturated elasto-plastic soils: three-dimensional validation of a plane simulation procedure. In: *Mechanical Modelling and Computational Issues in Civil Engineering*, M. Frémond & F. Maceri (Eds.), Springer: Cham, pp. 143–164.

Callari, C. & Casini, S. (2006) Three-dimensional analysis of shallow tunnels in saturated soft ground. *Geotechnical Aspects of Underground Construction in Soft Ground – Fifth International Symposium*. TC28, 2006, Balkema, pp. 495–501.

Callari, C. & Pelizza, S. (2020) The role of 19th-century tunnelling engineering in binding Italy to Europe and bridging the gap between the north and south of a new state, 135–136, pp. 19–33, Gallerie e Grandi Opere Sotterranee.

Canetta, G., Cavagna, B. & Nova, R. (1996) Experimental and numerical tests on the excavation of a railway tunnel in grouted soil in Milan. Geotechnical aspects of underground construction in soft ground, pp. 479–484.

Caratelli, A., Meda, A., Rinaldi, Z. & Romualdi, P. (2011) Structural behaviour of precast tunnel segments in fibre reinforced concrete. *Tunnelling and Underground Space Technology*, 26, 284–291, doi: 10.1016/j.tust.2010.10.003.

Carigi, A., Luciani, A., Todaro, C., Martinelli, D. & Peila, D. (2020a) Influence of conditioning on the behaviour of alluvial soils with cobbles. *Tunnelling and Underground Space Technology*, 96, Article number 103225.

Carigi, A., Todaro, C., Martinelli, D., Amoroso, C. & Peila, D. (2020b) Evaluation of the geo-mechanical properties property recovery in time of conditioned soil for EPB-TBM tunneling. *Geosciences* (Switzerland)Open Access, 10(11), 1–10, Article number 438.

Carranza Torres, C. & Diederichs, M. (2009) Mechanical analysis of circular liners with particular reference to composite supports. For example, liners consisting of shotcrete and steel sets. *Tunnelling and Under-ground Space Technology*, 24, 506–532.

Casamichele, P., Maugeri, M. & Motta, E. (2004) Non-linear analysis of soil-pipeline interaction in unstable slopes. *XIII World Conference on Earthquake Engineering*, Vancouver, Canada, August 1–6, 2004, Paper No. 3161.

C.E.N. (1997) Eurocode 7: Geotechnical design. European Standard EN 2004.

Centre d'étude de tunnels (2015) Guide de l'inspection du génie civile des tunnels routiers – Livre 1: du désordre á l'analyse, de l'analyse á la cotation.

Chen, Z., Shi, C., Li, T. & Yuan, Y. (2012). Damage characteristics and influence factors of mountain tunnels under strong earthquakes. *Natural Hazards*, 61(2), 387–401.

Chiriotti, E. & Grasso, P. (2001) Porto Light Metro System, Lines C, S and J. Compendium to the Methodology Report on Building Risk Assessment Related to Tunnel Construction. Normetro – Transmetro, Internal technical report (in English and Portuguese).

Conforti, A., Trabucchi, I., Tiberti, G., Plizzari, G.A., Caratelli A. & Meda, A. (2019) Precast tunnel segments for metro tunnel lining: a hybrid reinforcement solution using macrosynthetic fibers. *Engineering Structures*, 199, doi: 10.1016/j.engstruct.2019.109628.

Contini, A., Cividini, A. & Gioda, G. (2007) Numerical evaluation of the surface displacements due to soil grouting and to tunnel excavation. *International Journal of Geomechanics*, 7(3), 217–226.

Croce, P., Flora, A. & Modoni, G. (2014) *Jet Grouting*, CMC Press: Boca Raton.

D'Effremo, M.E. (2015) Meccanismi di riattivazione di movimenti di versante indotti dallo scavo di una galleria. Tesi di dottorato. Dipartimento di Ingegneria Strutturale e Geotecnica – Sapienza Università di Roma.

D'Effremo, M.E., Desideri, A., García Martínez, M.F. & Simonini, P. (2016) Analysis and monitoring of a tunnelling-induced deep landslide reactivation. Landslides and Engineered Slopes. *Experience, Theory and Practice*, 2, 735–742.

di Prisco, C., Flessati, L., Cassani, G. & Perlo, R. (2019) Influence of the fibreglass reinforcement stiffness on the mechanical response of deep tunnel fronts in cohesive soils under undrained conditions. *Tunnels and Underground Cities: Engineering and Innovation meet Archaeology, Architecture and Art - Proceedings of the WTC 2019 ITA-AITES World Tunnel Congress*, pp. 1323–1331.

di Prisco, C., Flessati, L., Frigerio, G., Castellanza, R., Caruso, M., Galli, A. & Lunardi, P. (2018a) Experimental investigation of the time-dependent response of unreinforced and reinforced tunnel faces in cohesive soils. *Acta Geotechnica*, 13(3), 651–670, doi: 10.1007/s11440-017-0573-x.

di Prisco, C., Flessati, L., Frigerio, G. & Lunardi, P. (2018b) A numerical exercise for the definition under undrained conditions of the deep tunnel front characteristic curve. *Acta Geotechnica*, 13(3), 635–649, doi: 10.1007/s11440-017-0564-y.

di Prisco, C., Flessati, L. & Porta, D. (2020) Deep tunnel fronts in cohesive soils under undrained conditions: a displacement-based approach for the design of fibreglass reinforcements. *Acta Geotechnica*, 15(4), 1013–1030, doi: 10.1007/s11440-019-00840-8.

Dowding, C.H. and Rozen, A. (1978) Damage to rock tunnels from earthquake shaking. *Journal of the Geotechnical Engineering Division*, 104, 175–191.

EN 12715 (2000) Execution of special geotechnical work – grouting. European Committee for Standardisation, rue de Stassart 36, B-1050 Brussels.

EN 14651 (2005) Test method for metallic fibre concrete – measuring the flexural tensile strength (limit of proportionally (LOP), residual). European Committee for Standardization, p. 18.

EN 1992-1-2 (2004). Eurocode 2: design of concrete structures – part 1–2: general rules – structural fire design. CEN. Brussels.

EN 1994-1-2 (2005) Eurocode 4 – design of composite steel and concrete structures – part 1–2: general rules – structural fire design. CEN. Brussels.

Falk, E. & Schweiger, H.F. (1998) Shallow tunnelling in urban environment – different ways of controlling settlements. *Felsbau*, 16(4), 215–223.

Felicetti, R., Gambarova, P.G. & Meda, A. (2009) Residual behavior of steel rebars and R/C sections after a fire. *Construction and Building Materials*, 23, 3546–3555.

fib bulletin No. 83 (2017) Precast tunnel segments in fibre-reinforced concrete. fib WP 1.4.1, ISBN: 978-2-88394-123-6, pp. 1–168.

fib Model Code for Concrete Structures 2010 (2013) Ernst & Sohn, ISBN 978-3-433-03061-5, pp. 1–434.

Flessati, L. & di Prisco, C. (2020) Deep tunnel faces in cohesive soils under undrained conditions: application of a new design approach. *European Journal of Civil and Environmental Engineering*, 2020, doi: 10.1080/19648189.2020.1785332.

Flessati, L., di Prisco, C. & Lunardi, P. (2020) A displacement based approach for reinforced tunnel faces under undrained conditions. *Gallerie e Grandi Opere Sotterranee*, 135–136, 22–31.

Galli, G., Grimaldi, A. & Leonardi, A. (2004) Three-dimensional modelling of tunnel excavation and lining. *Computers and Geotechnics*, 31(3), 171–183.

Galli, M. & Thewes, M. (2014) Investigations for the application of EPB shields in difficult grounds / Un-tersuchungen für den Einsatz von Erddruckschilden in schwierigem Baugrund. *Geomechanics and Tunnelling*, 7(1), 31–44.

Gehring, K.H. (1996) Design criteria for TBMs with respect to real rock pressure. *Proceedings of the "International Lecture Series TBM Tunnelling Trends – Tunnel Boring Machines, Trends in Design & Construction of Mechanized Tunnelling"*, Hagenberg December 1995, Austria: A.A. Balkema Rotterdam Brookfield, pp. 43–53.

Grasso, P., Mahtab, A. & Pelizza, S. (1989) Reinforcing a rock zone for stabilizing a tunnel in complex formations. *Proceedings of International Congress on Progress innovation in Tunnelling*, Toronto, 2, 671–678.

Grimstad, E. (2007) The Norwegian method of tunnelling – a challenge for support design. XIV European Conference on Soil Mechanics and Geotechnical Engineering. Madrid.

Guglielmetti, V., Grasso, P., Mahtab, A. & Xu, S. (Eds.). (2008) *Mechanized Tunnelling in Urban Areas: Design Methodology and Construction Control*. London: Taylor & Francis.

Hansmire, W. H. & Cording, E.J. (1985) Soil tunnel test section: Case history summary. *Journal of Geotechnical Engineering*, 111(11), 1301–1320.

Hashash, Y. M., Hook, J. J., Schmidt, B., John, I. & Yao, C. (2001) Seismic design and analysis of underground structures. *Tunnelling and Underground Space Technology*, 16(4), 247–293.

Hillerborg, A., Modker, M. & P.E. Petersson (1976) Analysis of crack formation & crack growth in concrete by means of fracture mechanics and finite elements. *Cement and Concrete Research*, 6, 773–782.

Hoeg, K. (1968) Stresses against underground structural cylinders. *Journal of Soil Mechanics and Foundations Division*, ASCE, 94(4), 833–858.

Horn, N. (1961) Horizontaler erddruck auf senkrechte abschlussflächen von tunnelröhren. Landeskonferenz der ungarischen tiefbauindustrie, pp. 7–16.

Howard, A. (1991) Report on the damaging effects of water on tunnels during their working life. *Tunnelling and Underground Space Technology*, 6(1), 11–76.

Huang, K.P., Wang, T.T., Huang, T.H. & Jeng, F.S. (2010) Profile deformation of a circular tunnel induced by ambient stress changes. *Tunnelling and Underground Space Technology*, 25(3), 266–278.

Huo, H., Bobet, A., Fernández, G. & Ramírez, J. (2006) Analytical solution for deep rectangular structures subjected to far-field shear stresses. *Tunnelling and Underground Space Technology*, 21(6), 613–625.

International Standardization Organisation (2005) ISO 23469:2005 Bases for design of structures-Seismic actions for designing geotechnical works

ITAtech (2014) ITAtech guidelines on best practices for segment backfilling, ITAtech Activity Group Excavation, ITAtech report N°4- MAY 2014.

ITA report n. 16 (2016) Twenty years of FRC tunnel segments practice: lessons learnt and proposed design principles. ISBN 978-2-970-1013-5-2, pp. 1–71.

ITA Working Group 6 (2001) Study of methods for repair of tunnel linings. International Tunnelling Association – Working group 6.

ITA WG 19 (2009) General report on conventional tunneling. [Online] Available from: https://about.ita-aites..org/publications/wg-publications/content/24/working-group-19-conventional-tunneling.

Jessberger & Hans, L. (1980) Theory and application of ground freezing in civil engineering. *Cold Regions Science and Technology*, 3(1), 3–27.

Koronakis, N., Kontothanassis, P., Kazilis, N. & Gikas, N. (2004). Stabilization measures for shallow tunnels with ongoing translational movements due to slope instability. Tunnelling and Underground *Space Technology*, 19(4–5), 495.

Kouretzis, G.P., Bouckovalas, G.D. & Karamitros, D.K. (2011). Seismic verification of long cylindrical underground structures considering Rayleigh wave effects. *Tunnelling and Underground Space Technology*, 26(6), 789–794.

Kouretzis, G.P., Andrianopoulos, K.I., Sloan, S.W. & Carter, J.P. (2014). Analysis of circular tunnels due to seismic P-wave propagation, with emphasis on unreinforced concrete liners. *Computers and Geotechnics*, 55, 187–194.

Kummerer, C. (2003) Numerical modelling of displacement grouting and application to case histories, PhD dissertation, Institut für Bodenmechanik und Grundbau, Technische Universität Graz.

Leonards, G.A. (1962) *Foundation Engineering*, McGraw-Hill Book Company: New York.

Littlejohn, S. (2003b) The development of practice in permeation and compensation grouting: a historical review (1802–2002): part 2 compensation grouting. In: Reston, V.A. & American Society of Civil Engineers (eds.) *Third International Conference on Grouting and Ground Treatment: Proceedings of Third International Conference on Grouting and Ground Treatment*, 10–12 February 2003, New Orleans, LO.

Luciani, A. & Peila, D. (2019) Tunnel waterproofing: available technologies and evaluation through risk analysis. *International Journal of Civil Engineering*, 17(1), 45–59.

Lunardi, P. (2008) *Design and Construction of Tunnels*, Springer-Verlag Publishing: Berlin.

Mair, R.J. (1994) Report on session 4: displacement grouting. In: *Proceedings on International Conference on Grouting in the Ground*, Bell (Ed.), Th. Telford: London, pp. 375–384.

Mair, R.J. & Taylor, R.N. (1999) Bored tunnelling in the urban environments. *Fourteenth International Conference on Soil Mechanics and Foundation Engineering. Proceedings International Society for Soil Mechanics and Foundation Engineering*, vol. 4.

Makhlouf, R. & Holter, K. (2008) Rehabilitation of concrete lined tunnels using a composite sprayed liner with sprayed concrete and sprayable waterproofing membrane; the Chekka road tunnel, Lebanon. *Proceedings of World Tunnel Congress 2008*, pp. 1175–1182.

Martinelli, D., Todaro, C., Luciani, A. & Peila, D. (2019) Use of a large triaxial cell for testing conditioned soil for EPBS tunnelling. *Tunnelling and Underground Space Technology*, 94, Article number 103126.

Moh, Z. C., Ju, D. H., & Hwang, R. N. (1996). Ground movements around tunnels in soft ground. *Proceedings of Symposium on Geotechnical Aspects of Underground Construction in Soft Ground*, London, pp. 725–730.

Muir Wood, A.M. (1975) The circular tunnel in elastic ground. *Géotechnique*, 25(1), 115–127.

Müller-Kirchenbauer, H. (1977) Stability of slurry trenches in inhomogeneous subsoil. *Proceedings of 9th International Conference on Soil Mechanics and Foundation Engineering*, vol. 2, Tokyo.

Nicholson, D.P., Gammage, C. & Chapman, T. (1994) The use of finite element methods to model compensation grouting. Grouting in the Ground, November 1994, The Institution of Civil Engineers, Thomas Tel-ford: London, pp. 297–312.

Nicolini, E. & Nova, R. (2000) Modelling of a tunnel excavation in a non-cohesive soil improved with cement mix injections. *Computers and Geotechnics*, 27(4), 249–272.

O'Reilly, M.P. & New, B.M. (1982). Settlements above tunnels in the United Kingdom - their magnitude and prediction. *3rd International Symposium, Institute of Mining and Metallurgy*, London, pp. 173–181.

O'Rourke, T.D. & Lane, P.A. (1989) Liquefaction hazard and their effects on buried pipelines. Technical Report NCEER-89-0007, National Center for Earthquake Engineering Research, State University of New York at Buffalo.

Osgoui, R.R. & Ünal, E. (2009) An empirical method for design of grouted bolts in rock tunnels based on the Geological Strength Index (GSI). *Engineering Geology*, 107, 154–166.

Owen & Scholl. (1981) Earthquake engineering of large underground structures. Report no. FHWA/RD-80/195. Federal Highway Administration and National Science Foundation.

Panet, M. (1995) Calcul des Tunnels par la Me'thode de Convergence–Confinement. *Presses de l'Ecole Nationale des Ponts et Chausse'es*, Paris, p. 178.

Park, K.H., Tantayopin, K., Tontavanich, B. & Owatsiriwong, A. (2009) Analytical solution for seismic-induced ovaling of circular tunnel lining under no-slip interface conditions: A revisit. *Tunnelling and Underground Space Technology*, 24(2), 231–235.

Peck, R.B. (1969) Deep excavation and tunneling in soft ground. *State-of-the-Art, Proceeding of the 7th International Conference on Soil Mechanics and Foundation Engineering*, Mexico City, pp. 225–290.

Peck, R.B., Hendron, A.J. & Mohraz, B. (1972) State of the art in soft ground tunneling, Rapid Excavation and Tunneling Conference. *The Proceedings of the Rapid Excavation and Tunneling Conference*. American Institute of Mining, Metallurgical, and Petroleum Engineers, New York, pp. 259–286.

Peila, D., Martinelli, D., Todaro, C. & Luciani, A. (2019) Soil conditioning in EPB shield tunnelling – an overview of laboratory tests. *Geomechanik und Tunnelbau*, 12(5), 491–498.

Penzien, J. & Wu, C.L. (1998) Stresses in linings of bored tunnels. *Earthquake Engineering and Structural Dynamics,* 27, 283–300.

Penzien, J. (2000) Seismically induced racking of tunnel linings. *Earthquake Engineering & Structural Dynamics*, 29(5), 683–691.

Perazzelli, P. & Anagnostou, G. (2017) Analysis method and design charts for bolt reinforcement of the tunnel face in purely cohesive soils. *Journal of Geotechnical and Geoenvironmental Engineering*, 143(9), 0401704.

Picarelli, L., Petrazzuoli, S. & Warren, C.D. (2002) Interazione tra gallerie e versanti. Le opere in sotterraneo in rapporto con l'ambiente. IX ciclo di Conf di Meccanica e Ingegneria delle Rocce MIR 2002, Patron Editore, pp. 219–248.

Pimentel, E., Sres, A. & Anagnostou, G. (2012) Large-scale laboratory tests on artificial ground freezing under seepage-flow conditions. *Geotechnique*, 62(3), 227–241.

Pitilakis, K. & Tsinidis, G. (2014) Performance and seismic design of underground structures. In: *Earthquake Geotechnical Engineering Design, Geotechnical Geological and Earthquake Engineering*, M. Maugeri & C. Soccodato (Eds.), Springer International Publishing: Switzerland, vol. 28, pp. 279–340.

Plizzari, G. & Tiberti, G. (2009) Tunnel linings made by precast concrete segments. In: *Construction Methodologies and Structural Performance of Tunnel Linings*, G.A. Plizzari (Ed.), pp. 131–136, Vol. unico, ISBN/ISSN: 978-88-96225-31-8, Starrylink, Brescia, Italy, 226 p.

Potts, D. M. and Addenbrooke, T. I. (1997): A structure's influence on tunnelling-induced ground movements. *Proceedings of Institution of Civil Engineers, Geotechnical Engineering.,* 125, 109–125.

Power, M., Rosidi, D., Kaneshiro, J., 1998. Seismic vulnerability of tunnels-revisited. In: *Proceedings of the North American Tunneling Conference,* L. Ozedimir, (Ed.), Elsevier: Long Beach, CA.

Rajani, B., Robertson, P.K. & Morgenstern, N.R. (1995) Simplified design methods for pipelines subjected to transverse and longitudinal soil movements. *Canadian Geotechnical Journal,* 32(2), 309–323.

Ramoni, M. & Anagnostou, G. (2006) On the feasibility of TBM drives in squeezing ground. *Tunnelling and Underground Space Technology,* 21(3–4), 262, Elsevier Ltd.: Oxford.

Ramoni, M. & Anagnostou, G. (2007a) Numerical analysis of the development of squeezing pressure during TBM standstills. *11th ISRM Congress. International Society for Rock Mechanics and Rock Engineering.*

Ramoni, M. & Anagnostou, G. (2007b) The effect of advance rate on shield loading in squeezing ground. *Underground Space–The 4th Dimension of Metropolises, ITA World Tunnel Congress,* pp. 673–677.

Ramoni, M. & Anagnostou, G. (2010) Thrust force requirements for TBMs in squeezing ground. *Tunnelling and Underground Space Technology,* 25(4), 433–455.

Rankin, W. (1988) Ground movements resulting from urban tunnelling: Predictions and effects. Engineering geology of underground movements. *Proceedings of the 23rd Annual Conference of the Engineering Group of the Geological Society,* Nottingham University, 13–17 September, 1987. Publication of: Geological Society: England.

Richards, J. (1998) Inspection, maintenance and repair of tunnels: international lessons and practice. *Tunnelling and Underground Space Technology,* 13(4), 369–375.

Riedmuller, G. & Schubert, W. (1999) Critical comments on quantitative rock mass classifications. *Felsbau,* 17(3), 164–167.

RILEM TC 162-TDF (2003) Test and design methods for steel fibre reinforced concrete. Design with σ-ε method. Brite-Euram BRPR-CT98–0813-RILEM TC 162-TDF Workshop, Bochum, Germany, pp. 31–46.

Sandrone, F. & Labiouse, V. (2011). Identification and analysis of Swiss National Road tunnels pathologies. *Tunnelling and Underground Space Technology,* 26(2), 374–390.

Scandale, S. (2017) Effetti dei movimenti di versante sui rivestimenti di una galleria. Tesi di dottorato. Dipartimento di Ingegneria Strutturale e Geotecnica – Sapienza Università di Roma.

Schweiger, H.F. & Falk, E. (1998) Reduction of settlements by compensation grouting - numerical studies and experience from Lisbon underground. *Proceedings of the World Tunnel Congress'98 on Tunnels and Metropolises,* Sao Paulo, Brazil, 25–30 April, 1998, Balkema: Rotterdam, pp. 1047–1052.

Sharma, S., & Judd, W. R. (1991). Underground opening damage from earthquakes. *Engineering Geology,* 30(3–4), 263–276.

Shi, Y., Fu, J., Yang, J., Xu, C. & Geng, D. (2017) Performance evaluation of long pipe roof for tunneling below existing highway based on field tests and numerical analysis: Case study. *International Journal of Geomechanics,* 17(9), 04017054.

Shu, D.M. & Bhattacharyya, A.K. (1992) Modification of subsidence parameters for sloping ground surfaces by the rays projection method. *Geotechnical and Geological Engineering,* 10, 223–248.

Shuster, J.A. (1972) Controlled freezing for temporary ground support. *Proceedings of North America Rapid Excavation and Tunneling Conference,* vol. 2, Chicago, June 5–7.

Spinazzè, M., Bruschi, R. & Giusti, G. (1999) Interazione tubazione-terreno in pendii instabili. *Rivista Italiana di Geotecnica,* 1, 65–72.

St-John, C.M. & Zahrah, T.F. (1987) Asiesmic design of underground sructures. *Tunneling and Underground Space Technology,* 2, 165–197.

Sterpi, D. & Moro, G. (1995) Finite element analyses of a shallow tunnelling process. *Gallerie e grandi opere sotterranee*, 46, 58–67.

STUVA (2005) Recommendations for Testing and Application of sealing Gaskets in segmental Linings. *Tunnel-Gutersloh*, 8, 8–21.

Su, J. & Bloodworth, A. (2016) Interface parameters of composite sprayed concrete linings in soft ground with spray-applied waterproofing. *Tunnelling and Underground Space Technology*, 59, 170–182.

Terzaghi, K. (1943) *Theoretical Soil Mechanics*, John Wiley & Sons: New York.

Thewes, M. (2007) TBM tunneling challenges redefining the state of the art. Keynote lecture ITAAITES WTC, PRAGUE.

Thewes, M. & Budach, C. (2010) Soil conditioning with foam during EPB tunnelling. Konditionierung von Lockergesteinen bei Erddruckschilden. *Geomechanics and Tunnelling*, 3(3), 256–267.

Tiberti, G., Trabucchi, I. & Plizzari, G.A. (2017) Numerical study on the effects of TBM high-concentrated loads applied to precast tunnel segments. *Proceedings of the IV International Conference on Computational Methods in Tunneling and Subsurface Engineering (EUROTUN 2017)*, Innsbruck (Austria), pp. 285–293, ISBN: 978-3-903030-35-0.

Todaro C, Carigi A., Peila L., Martinelli D., Peila D. (2021) Soil conditioning tests of clay for EPB tunnelling, Underground Space, in press, avaible on line, https://doi.org/10.1016/j.undsp.2021.11.002.

Tsinidis, G., de Silva, F., Anastasopoulos, I., Bilotta, E., Bobet, A., Hashash, Y. M. & Fuentes, R. (2020) Seismic behaviour of tunnels: From experiments to analysis. *Tunnelling and Underground Space Technology*, 99, 103334.

UFC (2004) Dewatering and groundwater control, UFC-3-220-05, Unified Facilities Criteria, 16 January 2004.

Ünal, E. (1983) Design guidelines and roof control standards for coal mine roof, PhD Thesis, Pennsylvania State University, University Park, 355 p.

Volkmann, G.M. & Schubert, W. (2007) Geotechnical model for pipe roof supports in tunneling. In: *International Conference on Underground Space–the 4th Dimension of Metropolises*, Barták, Hrdina, Romancov & Zlámal (Eds.), Taylor & Francis Group: London.

Vu, M.N., Broere, W. & Bosch, J. (2016) Volume loss in shallow tunnelling. *Tunnelling and Underground Space Technology*, 59, 77–90.

Wang, J.-N. (1993) Seismic Design of Tunnels: A State-of-the-Art Approach, Monograph, monograph 7. Parsons, Brinckerhoff, Quade and Douglas Inc., New York.

Wang, T.T. (2010) Characterizing crack patterns on tunnel linings associated with shear deformation induced by instability of neighboring slopes. *Engineering Geology*, 115, 80–95.

Wells, D. L. & Coppersmith, K. J. (1994) New empirical relationships among magnitude, rupture length, rupture width, rupture area, and surface displacement. *Bulletin of the Seismological Society of America*, 84(4), 974–1002.

Whittaker, B.N. & Reddish, D.J. (1989) *Subsidence Occurrence Prediction and Control*, Elsevier: Amsterdam.

Wisser, C., Augarde, C.E. & Burd, H.J. (2005) Numerical modelling of compensation grouting above shallow tunnels. *International Journal for Numerical and Analytical Methods in Geomechanics*, 29(5), 443–471.

Wong, H., Subrin, D. & Dias, D. (2000) Extrusion movements of a tunnel head reinforced by finite length bolts—a closed-form solution using homogenization approach. *International Journal for Numerical and Analytical Methods in Geomechanics*, 24(6), 533–565.

Wong, H., Trompille, V. & Dias, D. (2004) Extrusion analysis of a bolt-reinforced tunnel face with finite ground-bolt bond strength. *Canadian Geotechnical Journal*, 41(2), 326–341.

Wongsaroj, J., Soga, K. & Mair, R. J. (2007). Modelling of long-term ground response to tunnelling under St James's Park, London. *Geotechnique*, 57(1), 75–90.

Xanthakos, P.P., Abramson, L.W. & Bruce, D.A. (1994) *Ground Control and Improvement*, John Wiley & Sons: New York

Zhang, C., Feng, X., Zhou, H., Qiu, S. & Wu, W. (2012) A top pilot tunnel preconditioning method for the prevention of extremely intense rockbursts in deep tunnels excavated by TBMs. *Rock Mechanics and Rock Engineering*, 45(3), 289–309.

Zienkiewicz, O.C., Humpheson, C. & Lewis, R.W. (1975) Associated and non-associated visco-plasticity in soil mechanics. *Geotechnique*, 25 (4), 671–689.

Zizka, Z., Kube, S., Schößer, B. & Thewes, M. (2020) Influence of stagnation gradient for face support calculation in Slurry Shield Tunnelling. *Geomechanics and Tunnelling*, 13, 372–381.

Zizka, Z. & Thewes, M. (2016) Recommendations for face support pressure calculations for shield tunnelling in soft ground, DAUB German Tunnelling Committee (ITA-AITES).

Chapter 11

Monitoring during construction

S. Miliziano
Sapienza University of Rome

G. Russo
University of Naples Federico II

A. de Lillis
Sapienza University of Rome

D. Sebastiani
Geotechnical and Environmental Engineering Group

CONTENTS

11.1 INTRODUCTION

During tunnel construction, as for many other civil engineering works, data are collected, processed and interpreted with the aim of observing the performance of the system under construction. The whole of all these activities is named monitoring, and it is commonly included in a risk control and management strategy. Monitoring should ensure the timely detection of any warning signs of instability in the response. Monitoring should also allow the optimization of the design during construction and the assessment of the technical-economic efficiency of the employed excavation procedures in terms of production parameters (e.g. advancement speed, power consumption and tool wear). Furthermore, monitoring data constitute a useful documentation for the resolution of any legal dispute between the parties (contracting entity, contractor and third parties). Another crucial goal of monitoring plans should be the real-time assessment of the environmental impact of tunnel excavation.

    Three main types of monitoring can be identified: for geotechnical and structural purposes, for the workers' health and safety and for avoiding machines damages. A common factor for all of them is the minimization of the typical risks of underground

DOI: 10.1201/9781003256175-11

construction works, already handled at the design stage. This chapter focuses on the first category; only brief hints are given about the other two.

The monitoring plan is an important part of the project. It consists in the identification of the physical quantities needed to understand the overall behaviour of the system, to select the appropriate measuring instruments (accuracy, precision and resolution), to choose their number, location and acquisition frequency and, finally, to define all the activities to register, process and share the recorded information correctly. In the early stages of the monitoring system design, it is also important to establish a significant volume where the measured physical quantities are expected to change, to support the decision of both the number and locations of the instruments. The aim of this chapter is to illustrate the relevant physical quantities typically measured, the measuring instruments usually employed and, overall, the criteria that should be followed in order to correctly develop a monitoring plan and to properly use the collected data. General criteria are illustrated here, while details about instrumentation and measurement techniques are reported in Book 2. The calculation procedures that can be followed to obtain quantitative predictions of the relevant physical quantities to be compared with the measurements are reported in Chapters 8 and 10.

Monitoring activities *lato sensu* should include the geological and geotechnical surveys carried out during the construction; these pieces of information, in fact, together with the monitoring data, is essential to make informed decisions during works. Last but not least, the monitoring activities should imply a careful record of the construction process with all the significant details. The availability of such information is of utmost importance for the successful interpretation of the monitoring data.

Ideally, in order to optimize the use of all the information acquired on site, analytical and/or numerical tools able to carry out real-time back-analysis and predictions should be developed to support the decision-making process with quantitative evaluations. Examples of monitoring systems applied to tunnelling and underground constructions are reported in Book 3, which gathers selected case histories of underground works.

Herein a distinction is made based on where the measurements are taken: from inside the tunnel (internal/tunnel monitoring) and from the ground surface (external/surface monitoring). The former aims to gather information about the response of the soil or rock immediately surrounding the tunnel and the soil–lining interaction. It is also tasked with investigating the geomechanical conditions along the advancement. The latter is aimed at recording the excavation-induced effects at the ground surface and also below, where underground pipes or structures are present. While the former is always a fundamental part of the project, the latter is mainly relevant for relatively shallow tunnels in urban environment.

A single section is devoted to the external monitoring, aimed at controlling tunnelling-induced effects on the surface and on pre-existing structures and infrastructures, regardless of the adopted excavation method. Vice versa, for internal monitoring – due to very different construction methods – separate sections are devoted to conventional and mechanized tunnelling.

In this chapter, only monitoring during construction is discussed. Surveys and monitoring activities performed during operational stages and refurbishment works are specifically illustrated in Chapter 12.

11.2 RELEVANT PHYSICAL QUANTITIES AND MEASUREMENT TECHNIQUES

Monitoring systems are tasked with periodically measuring the response of the soil or rock surrounding the excavation and of the tunnel structures. The most relevant physical quantities that should be measured during the excavation (and later when delayed effects are expected due to consolidation or creep phenomena) are those enabling to draw a clear picture of the system response and of the effects induced on the surrounding environment. These quantities, act as a matter of fact, as benchmark. Tunnel monitoring should start at the beginning of the works, and for some quantities even before, in order to clearly distinguish tunnelling-induced effects from others, such as those due to seasonal variations or traffic. In fact, it is extremely important to master the monitoring activities, fine-tune the decision-making process and gain insight into the ground response and soil–structure interaction, right as the excavation starts – which should only happen once the monitoring system is fully operational and full awareness of the natural environmental effects on the monitoring systems is achieved.

11.2.1 Tunnel monitoring

The information that should be acquired from inside the tunnel to get a clear, quantitative understanding of the soil–structure interaction relates to both the lining and the surrounding soil (Burland et al., 1996). The state of the lining can be observed by monitoring its displacements and strains. Sometimes internal forces in the lining or between segments of the lining are also directly monitored via appropriate instruments. The normal pressure exerted by the surrounding soil on the lining is sometimes measured. Lining displacements can be measured using a variety of instruments, while the state of strain in the lining is usually measured directly using strain gauges. The state of stress in such cases is derived assuming an appropriate constitutive relationship for the lining material. As the monitoring changes based on the type of lining (temporary/ final or precast segments) which is a function of the excavation procedure, this aspect will be covered in the following sections.

The effects of the excavation on the soil surrounding and close to the tunnel lining can be quantified installing instruments able to measure displacements, total stresses and pore pressures from inside the tunnel. The soil displacements can be measured using extensometers, the pore pressure, using piezometers or piezometric cells, and the stresses exerted by the soil on the lining, using pressure cells installed on the extrados of the lining.

A fundamental part of the tunnel monitoring concerns the observation of the excavation face. Although not classified as a monitoring activity in the more traditional sense, investigating the geological and geomechanical conditions along the alignment is of great importance as it allows to be aware of possible sudden variation of the soil/ rock and to react promptly to unexpected risks associated, for instance, with high inflow of water, fractured zones and the presence of gas. The investigation of the excavation face also enables to understand the mechanical response of the ground beyond (extrusion). As well as lining monitoring, face monitoring is closely related to the excavation method and thus separate subsections will deal with it in the following.

Very important are the monitoring activities that must be carried out inside the tunnel related to the workers' health and safety. For instance, many environmental factors shall be controlled to guarantee a safe construction environment. Among them, the most important are air quality, qualified via the presence of gas, the presence of asbestos, dust levels, heat and air flows, temperature and humidity. Since this chapter is concerned with monitoring activities for geotechnical and structural purposes, this aspect, albeit of fundamental importance, will not be discussed.

11.2.1.1 Conventional tunnelling

In conventional tunnelling, monitoring is primarily concerned with the stress and strain in the lining and with the geomechanical survey at the excavation face. The stresses and strains in the lining (both temporary and final) can be monitored by installing strain gauges on the steel ribs and load cells at the base of the ribs. Very often, the steel ribs are monitored with vibrating wire gauges that are considered as strain gauges, but are indeed a sort of stress gauge. A typical example of vibrating wire gauges bolted on brackets welded on the steel ribs is reported in Figure 11.1 (Bilotta & Russo, 2016).

Measurements of convergence (i.e. reduction in minimum distance between two points on the tunnel walls) are very important in conventional tunnelling, and for this reason, absolute and relative displacements of the temporary and of the final linings are usually monitored in a number of selected cross sections. Traditionally, convergence measurements were carried with the use of tape extensometer (made of steel or invar) connected by hooks to eyebolts positioned at different locations in the lining

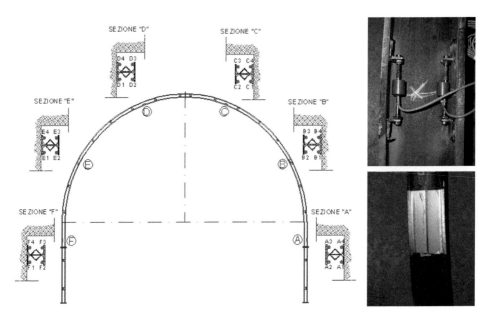

Figure 11.1 Typical distribution of vibrating wire gauges bolted to the rib. (Modified after Bilotta and Russo 2016.)

cross section. Displacements are now frequently measured by a total station that detects the position of 5–7 targets located on the ribs or on the final lining immediately after their installations. During the subsequent excavation, the targets are read many times, following the monitoring plan. One or more benchmarks located very far from the face and nominally fixed (targets on the final lining are generally considered fixed) are read at the same time in order to obtain absolute values of the displacements. Any observed asymmetric response permits to identify the portions of the tunnel characterized by strain concentration. Millimetric accuracy and precision can easily be achieved.

An emerging technique used to monitor displacements is laser scanning. The laser scanner is nowadays employed for the localization of the preliminary support immediately before casting the final lining, to control its actual thickness at various locations along the section. The instrument is positioned close to the measuring area, and it is able to detect in few minutes the absolute position of the exposed surface (the position of the instrument needs to be geo-referenced). If repeated many times during the excavation, the evolution of the field of displacements can be obtained by comparison with the first measurement (Figure 11.2). The interpretation of the displacement field requires mapping algorithms. The technique requires good lighting conditions. In the best overall conditions, the accuracy may be of the order of few millimetres.

As per face monitoring, two main types of activities are usually carried out: geomechanical surveys of the face and extrusion measurements (longitudinal displacement measurements into the nucleus beyond the face as the excavation proceeds). The main goal of these surveys is the evaluation of the geological characters of the encountered rock mass. Surveys can be performed at different levels of detail and allow identifying the geological and geomechanical characteristics of the face, aiming to detect and control the variability along the advancement of the excavation and also checking the compliance with the geological model defined at the design stage. Surveys can be carried out beyond the excavation face by means of exploratory adits or sub-horizontal

Figure 11.2 Laser scanner monitoring. (Courtesy of Anas S.p.A.)

boreholes. These solutions can be motivated by the need to ascertain in proper advance the possible presence of soil and rock masses with poor mechanical properties (e.g. fault zones) or the risk of water inrushes. Samples can be taken from the excavation face for laboratory tests, and also fast in-situ tests can be carried out, such as point load or uniaxial compression tests.

Extrusion measurements, usually recorded using incremental sliding micrometers, provide useful information regarding the soil response to the excavation. A plastic pipe equipped with magnetic elements is grouted to the ground inside a borehole drilled into the nucleus approximately along the tunnel axis. The comparison of the relative distance between the magnetic elements, measured inserting a sliding micrometer inside the pipe, after each excavation step, allows obtaining the extrusion curve (an example is reported in Figure 11.3). Differently from displacement measurements, which are a useful tool to evaluate the effects of soil–support interaction once the temporary supports are installed and the face moves further away, extrusion measurements enable to observe the response of the soil/rock mass of the nucleus before its excavation.

The extrusometer is usually installed in a borehole drilled into the nucleus destroying the core. In this case, DAC test technology should be used during drilling to indirectly acquire information about the mechanical properties of the rock massif, by recording perforation parameters such as velocity, torque and energy (Boldini et al., 2018). In the case of continuous core drilling, further information can be obtained from the samples extracted. Furthermore, before the installation of the column for extrusion measurements, video inspections of the hole can be carried out and further information about the number, nature and orientation of discontinuities of the rocks can be obtained. This information is essential to better identify the expected scenario as the excavation restarts (this is particularly important in the presence of faulted zones, water at high pressure and gas) and enables to take informed decision about how to excavate and the auxiliary works that should be implemented, if any.

11.2.1.2 Mechanized tunnelling

When preliminary support and final cast-in-place linings are employed, as for conventional tunnelling, criteria and instruments for monitoring the state of the lining are similar to those described in the previous section. For precast reinforced concrete segmental lining, the segments should be preliminarily equipped with strain gauges,

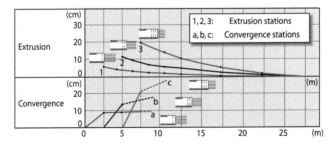

Figure 11.3 Example of extrusion measurements by means of a sliding micrometer combined with convergence measurements (Lunardi, 2008).

attached to the steel cage, in order to measure local deformations from which the state of stress can be inferred (Figure 11.4a). In these applications, vibrating wire gauges are often used, and in the last years, optical fibres have gained a significant slice of the overall market. Interpretations of these measurement are not straightforward, and internal forces may be derived only with accurate recording of the overall installation procedure of the segmental lining (Bilotta & Russo, 2013). Using pressure cells installed on the extrados of the segmental lining (Figure 11.4b), it is possible to measure the contact stresses arising from the soil–structure interaction. This type of monitoring, as all the contact measurements are carried out at the interface between different materials, is not always successful due to the stiffness of the instruments – i.e. the pressure cells – that may affect the quantity to be measured. Figure 11.5 shows a typical evolution of internal stresses in the segmental lining after the installation due to consolidation processes in clayey soils. Note that 1 year after the installation, the process is still ongoing.

Although the convergences are usually very small, and their measurement is less important than in conventional excavation (especially for temporary support), displacements are systematically measured. Absolute displacements are rarely relevant. As an example, the absolute displacement can be relevant when a tunnel portion is directly involved in movements caused by external factors, for instance a landslide. Due to the presence of the backup, in mechanized tunnelling it is quite difficult to

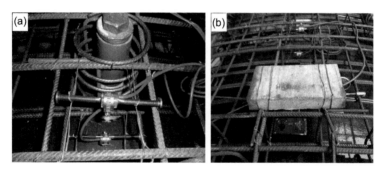

Figure 11.4 Precast lining segments monitoring instruments: (a) strain gauges attached to the steel cage; (b) load cell for contact pressure measurements. (Courtesy of Geotechnical Design Group s.r.l.)

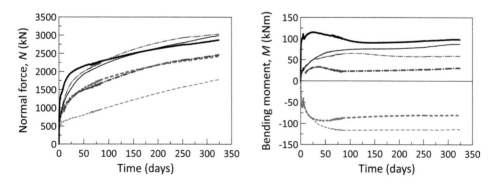

Figure 11.5 Stress evolution in the precast lining due to consolidation processes (Genzano hydraulic tunnel in central Italy; modified after De Gori et al. 2019.)

timely install the monitoring system to measure displacements immediately after the installation of the lining.

Tunnel boring machines require advanced managements and controls procedures (Maidl et al., 2013) during the construction phase, usually included in an overall risk management approach. Therefore, a long list of operational parameters is monitored in real time by modern TBMs (AFTES, 2005; ITA/AITES, 2011). These parameters are monitored both for avoiding damages of the TBM, which will be explicitly discussed in the following, and for the possible effects at the surface on existing structures (see Section 11.2.2). For TBMs operating in closed mode, both slurry shields (SS) and earth pressure balance (EPB), the face pressure, evolving over time and during different excavation phases, is the prominent parameter and it is monitored through pressure cells specifically installed in the excavation chamber (ITA-AITES, 2016). Thrust, torque, penetration, flow and properties of the slurry for SS technology and soil conditioning parameters for EPB technology, volume and pressure of grout injection are also important and monitored. The weight of the excavated muck for each advancement is another relevant measured parameter. Figure 11.6 reports an example of earth pressures in the excavation chamber measured during drilling and the subsequent standstill phase for the lining installation; the charts show the purposeful management of earth pressures aimed at minimizing tunnelling-induced settlements as the excavation nears the Carducci School in Rome (Miliziano & de Lillis, 2019). In terms of safety, it is important to outline the role of some measurements as, for example, the pressure in the excavation chamber. Sudden changes in these parameters should be carefully monitored because of the potential adverse effects on the surrounding soil and on the above structures and buildings.

For TBMs operating in open mode (as rock TBMs), since a frequent access to the tunnel face could be considered, the direct cutterhead consumption level and the geological examination of the excavation tunnel face could be included in the monitoring protocols (Zou, 2017).

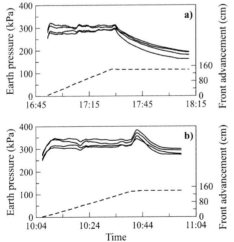

Figure 11.6 Earth pressure measurements inside the excavation chamber (a) far away from and (b) near the Carducci School (Miliziano & de Lillis, 2019); (c) pressure sensors' position in an EPB TBM. (Courtesy of Roma Metropolitane s.r.l.)

During closed-mode tunnelling, measurement at the tunnel face cannot be obtained with the same type of procedures described for conventional tunnelling. That is because of course the tunnel boring machines forbid frequent access to the excavation face. In fact, this can only be accessed at the cost of a significant slowing of the advancement; thus, such operations are carried out only in particular situations. However, modern TBMs are equipped with drilling systems that allow forward prospecting at the face. The use of techniques of geophysical prospection beyond the face is also increasingly frequent.

As stated before, several important TBM operational parameters are constantly monitored during the construction process just to prevent and avoid possible damages to the machine. As a matter of fact, one of the priorities of the TBM monitoring system is to ensure the correct functioning of the TBM itself, preventing electrical faults, electronic malfunctions or mechanical failures in order to avoid temporarily or permanently compromising the tunnel construction. As an example, the breaking of the bearing is particularly feared because its replacement is a very complex task, expensive and time-consuming. Monitoring of power systems, working pressure, thrust cylinders, engines and temperature of all the mechanic components belongs to this category.

One of the further aims of monitoring TBM operational parameters is to evaluate in real time the excavation performance and possibly to improve it by modifying some of them. To this aim, TBMs are equipped with a complex multilevel monitoring framework connected in real time with a server allowing everyone involved in the project to have full access anytime and anywhere to all relevant information (clients, contractor, designer, suppliers, etc.). Within the framework are integrated:

- the positioning and tracking system which includes a navigation system, based on a total station and a target unit installed within the TBM shield, required to determine the current advance position;
- the control system of the operating TBM parameters, able to monitor in real time the following quantities: all the main soil conditioning parameters, penetration/advancement speed, cutterhead and screw conveyor rotation speeds, pressures at the face in the chamber and pressures in the screw conveyor, cutters wear/consumption, weight of the material extracted from the excavation chamber, backfilling injection parameters (volumes and pressures), start and end times of the excavation and assembly phases of each rings, and clockwise and anticlockwise cutterhead rotation time. The analysis of all these parameters can be very useful to undertake actions to improve/optimize the excavation performance.

The monitoring framework allows to set an early warning system able to provide several alarms (through automatic acoustical signal, mail, text message and phone call) in case preset warning thresholds are exceeded.

11.2.2   Surface monitoring

Monitoring activities carried out on and near the ground surface aim at observing the ground response and the response of pre-existing structures and infrastructures to the excavation. Clearly, surface effects grow larger as the depth of the tunnel decreases or its diameter increases; thus, the monitoring plan has to be designed accordingly.

The ground response can be observed referring primarily to displacements. Tunnelling-induced movements can be measured at surface using a variety of techniques: topographic, interferometric and high-resolution imaging.

- Topographic measurements: when settlements (i.e. vertical displacement) are the focus, precision optical or digital surveys may be selected as the best choice. These instruments are hand-operated and work on a network of invar steel brackets installed on the buildings and the structures to be monitored. The system when appropriately used is capable of high resolution and accuracy (i.e. submillimetric). When displacements in all directions are needed, topographic measurements are made by using a total station working on a number of benchmarks or optical targets. The precision of this traditional technique is usually millimetric and depends on the system's geometry (relative distances and angles between benchmarks and total station; Figure 11.7a) and on the overall installation layout.
- Satellite interferometry: the technique is based on the phase modification of waves emitted by a satellite, reflected on the ground and detected back by the satellite. The precision of the measurements is millimetric for displacements along the target–satellite alignment, while it is centimetric for the perpendicular displacement components. Hence, the technique is most suitable for the monitoring of settlements. Sometimes, the use of properly built and oriented corner reflectors is convenient, as it can significantly improve the precision of the technique. It should also be noted that when monitoring building displacements, this is seldom necessary as multiple pre-existing persistent scatterers can usually be identified. The main disadvantages are that satellite interferometry requires a complex numerical algorithm for the interpretation of the data and that the frequency of the measurement is based on the passage of the satellite above the monitoring area (usually weekly). An example of the results obtained from different satellite sensors is reported in Figure 11.7b.

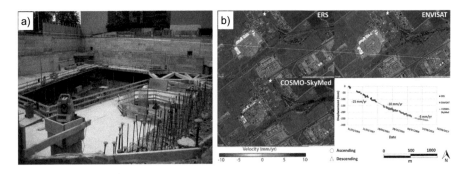

Figure 11.7 (a) Topographic measurement system installed to monitor displacements during an underground excavation in an urban area. (Courtesy of Geotechnical Design Group s.r.l.) (b) Interferometry results obtained from different satellite sensors (Mazzanti et al., 2020).

Figure 11.8 Terrestrial interferometer installed (Mazzanti et al., 2020).

- Terrestrial interferometry: based on the same principles of the satellite interferometry, in this case the waves are generated by an antenna located on the ground, which is also capable of receiving the reflected waves. The precision is significantly higher (up to millimetric), and it is possible to measure accurately both vertical and horizontal displacements by properly placing the instruments (Figure 11.8). It has the clear advantage of allowing free setting of the measurement frequency.
- High-resolution imaging: this technique is relatively new and very promising; it is based on the comparison of high-resolution images taken at different times using advanced cameras. The cameras can also be installed on drones to perform aerial mapping. The accuracy depends strongly on a number of details that rarely can be controlled in large works with an expected long duration. Its application at the moment is still limited to small areas and on a relatively short-term basis.

Sub-surface movements can be measured installing extensometers and inclinometers into boreholes above and laterally to the tunnel. Single-point or multi-point extensometers (Mikkelsen, 1996) can be used to measure absolute and relative settlements along the tunnel axis and other verticals of interest, while inclinometers measure relative horizontal displacements along vertical alignments. In Figure 11.9, an example of a monitoring system tasked with measuring displacements from the surface is reported.

When variations of pore pressures are expected, the monitoring system should also include piezometric measurements; especially in the case of shallow tunnels, the instruments can be installed from the surface.

Buildings are typically monitored by observing, beyond the displacements, also the opening or width evolution of cracks (crack gauges), local tilting (clinometers) and relative displacements of selected points (hydraulic settlement gauges with precise pressure transducer). Hydraulic settlement gauges allow real-time monitoring of the

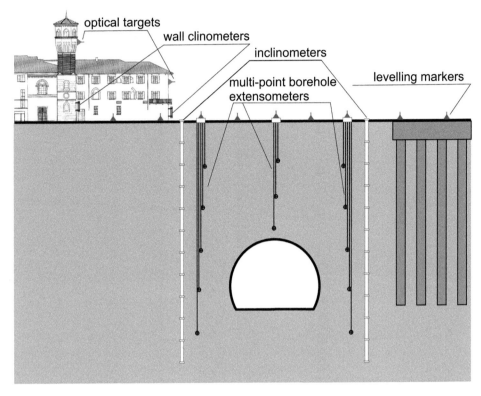

Figure 11.9 Schematic representation of a surface and sub-surface movements monitoring system. (Courtesy of C. Callari.)

relative settlements of multiple points. They are systematically used when grouting compensation is carried out to minimize the induced settlements on existing buildings. Their accuracy is very high (millimetric). An example of a rich monitoring system is shown in Figure 11.10.

Figure 11.10 Monitoring instruments installed on the Colosseum during the excavation on the new metro line C. (Modified after Romani et al. 2020.)

Another aspect of surface monitoring is that concerned with the tunnel entrance. In this case, monitoring activities are typically the same of those performed when dealing with open excavations. The main physical quantities that should be measured are the displacement of the excavation walls (in some instances, especially in urban areas, movements occurring above and behind the excavation walls are measured too) and the strains and stresses developing in the support elements (i.e. struts and anchors).

The same physical quantities are monitored in both conventional and mechanized tunnelling. The acquisition frequency is generally higher when TBMs are employed, due to much greater excavation rates.

In all cases, nowadays the tendency is towards automatic readings and real-time transmission of the monitoring data.

11.3   CALCULATIONS FOR MONITORING PURPOSES – WARNING THRESHOLDS

Ultimate limit state (ULS) calculations are carried out for safety evaluations following the approaches proposed by the Eurocode and other international codes. These approaches are usually based on the use of partial safety factors. In order to ensure adequate safety margins against collapse, minimizing the related risks, USL calculations are performed employing design values of soil mechanical parameters, obtained by reducing the characteristic values, and design load values, obtained by amplifying the characteristic ones (see Chapter 10). Thus, stresses and displacements associated with ULS calculations are conceptually and purposely overestimated if compared to the live conditions that will be monitored.

Based on the above, specific calculations should be carried out for monitoring purposes, assuming soil parameters and external loads in a range of variation close to the average most probable values (neither characteristic values, nor design ones).

Moreover, calculations should be aimed to follow the evolution of the relevant physical quantities (see Section 11.2) during the construction process. This allows to compare the expected values (predictions) with the measured values acquired during works, long before the most critical configurations are reached. A typical example, provided in order to clarify this concept, is the prediction of tunnelling-induced settlements on pre-existing building resting at the ground level. The calculation model should be able to predict not only the final values of settlements and distortions induced by tunnelling, but also how they evolve as the excavation face approaches the building, passes under it and proceeds further. When consolidation phenomena are expected in clayey soils, the prediction should of course take into account the time effects, evaluating the evolution of deformations and stresses until final drained conditions are reached. Creep effects on both soils and rocks require time effects to be taken into account, too.

In addition to the most probable expected range of values for each scenario and for each phase of the construction process, two warning thresholds for each measured quantity should be defined: attention and alarm thresholds. The alarm threshold should be determined by the designer, taking into consideration the needs of all the stakeholders involved (client and third parties) and the level of residual risk accepted by the owners/clients. The attention threshold is usually assumed as a percentage of the alarm one.

The alarm threshold is usually associated with a well-defined limit state. As an example, when tunnelling in urban areas, an alarm threshold is a precise level of damage induced on pre-existing structures. Frequently, this threshold is associated with a serviceability state. For example, the formation of few and small cracks that do not compromise serviceability or functionality, and do not require evacuation, can be accepted. Obviously, the alarm threshold is a function of the vulnerability of the structure to be protected and of its historical, monumental, or also strategic importance. Based on the acceptable level of the damage selected, maximum values of settlements and distortions at the foundation level are calculated and assumed as alarm thresholds. Other examples are the following:

- the limit state of deformation of temporary supports, for which the alarm threshold is the value of the maximum convergence admissible to avoid excessive reductions and consequent under-thickness of the final lining;
- the limiting values of the stresses in the ribs employed for temporary support, for which the alarm threshold is usually determined as a percentage of the yielding stress.

The availability of monitoring data since the very beginning of the construction process allows re-calibrating the models adopted in the design stage within the typical process of the observational method. When the predictions obtained after the model re-calibration based on monitoring data (as further discussed below) are close to the attention threshold, works can proceed as planned, but greater attention should be paid (the measurements frequency can be increased, the prediction model can be fine-tuned, and some construction activities can be slightly modified). When the predictions overcome the alarm threshold, more relevant changes should be implemented during the construction (such as boosting the soil improvement, increasing the pressure applied to the face and closing the temporary support at the invert), in order to steer the values of the monitored quantities below the target maximum one (below the assumed alarm threshold). If the variation rates of the measured representative quantities do not cancel out in a reasonable time, the excavation operations must be stopped and the temporary countermeasures must be promptly applied as planned, to counteract the observed trend towards a potentially unstable tunnel behaviour. All the countermeasures deemed necessary for the protection of people inside and outside the tunnel will be implemented.

Thus, the values of the monitored quantities to be compared with the thresholds are usually re-calculated during the construction phase – taking explicitly into account the physical quantities recorded by the monitoring system and the new information acquired through surveys at the excavation face. They do not necessarily coincide with design predictions. The need and the opportunity for re-calibration of the models adopted at the design stage are clarified by the following example. Typically, the constitutive models adopted in practice for soils and rocks are simple and robust – i.e. linear elastic–perfectly plastic (with Hoek–Brown/Mohr–Coulomb or Tresca strength criterion) – and therefore, soil non-linearity, time effects (creep/viscous behaviour) and softening phenomena are often neglected. Hence, deformations/displacements and stresses that arise and their evolution are poorly predicted. Also, operational values of strength parameters and particularly of the stiffness of soil and rock masses are very difficult to assess at the design stage. Moreover, the behaviour of the system is very much affected by rock non-homogeneity, anisotropy of mechanical properties and anisotropy of initial lithostatic state of stress. All of them are rather commonly encountered, very difficult to investigate and not susceptible of easy quantification and, finally, very difficult to account for in the calculation models usually adopted in practice. For all these reasons, the ranges of expected values should be re-calculated during works accounting for all the recorded data, including geomechanical surveys, and changing the operational values of the strength and of the stiffness parameters. Usually, for a geotechnical unit, the re-calibrated model can provide good prediction as the cover changes.

The improved accuracy of the updated predictions is useful to control the performance of the system in the deterministic design approach and crucial for a successful application of the observational approach.

11.4   CONCLUDING REMARKS

This chapter described the main aspects of the monitoring of tunnels during construction, focusing mainly on monitoring activities aimed at geotechnical and structural purposes. The monitoring plan, from the identification of the relevant physical quantities, to the definition of the suitable instruments and measurements technique, up to the interpretation of the recorded data, is a fundamental part of the project, ultimately aimed at further reducing residual risks. With the goal of providing the basic information useful for the development of the design and the monitoring plan, only a few information regarding the main monitoring techniques was given herein, briefly discussing their accuracy, applicability and limitations, and referring the reader to Book 2 for a more in-depth description.

It was remarked that specific calculations should be carried out in order to obtain predictions that are comparable with the measurements. To this aim, mechanical parameters and load values should not be reduced or increased through partial safety factors, but instead should reflect the most probable scenario. In addition, two warning thresholds – attention and alarm – should be defined for each relevant physical quantity being monitored and used as framework for the identification of the level of risk and its acceptability. These calculations should be performed for different stages of the construction, to compare predictions and measured values progressively acquired. If needed, as often is the case given the intrinsic difficulty of the task at hand, the model may be re-calibrated by back-analysing the monitoring data with the aim of updating and improving the predictions while the work is in progress.

Design, monitoring, back-analysis and re-prediction constitute an efficient engineering approach that enables to improve the understating of these complicated systems and constantly learn from experience.

AUTHORSHIP CONTRIBUTION STATEMENT

The chapter was developed as follows. Miliziano and Russo: chapter coordination. The chapter was joined developed by the authors. All the authors contributed to chapter review. The editing was managed by de Lillis and Sebastiani.

REFERENCES

AFTES (2005) Guidelines on monitoring methods for underground structures. Working Group GT 19.

AITES/ITA WG2-Research (2011) Monitoring and control in tunnel construction. ITA report no. 009/November 2011. Available from: https://about.ita-aites.org/publications/wg-publications/download/75_b0364940d7fbafd5a1b8564052aebb8b [accessed 15th January 2021].

Bilotta, E. & Russo, G. (2013) Internal forces arising in the segmental lining of an earth pressure balance-bored tunnel. *Journal of Geotechnical and Geoenvironmental Engineering*, 139(10), 1765–1780.

Bilotta, E. & Russo, G. (2016) Lining structural monitoring in the new underground service of Naples (Italy). *Tunnelling and Underground Space Technology*, 51, 152–163.

Boldini, D., Bruno, R., Egger, H., Stafisso, D. & Voza, A. (2018) Statistical and geostatistical analysis of drilling parameters in the Brenner Base Tunnel. *Rock Mechanics and Rock Engineering*, 51, 1955–1963.

Burland, J., Standing, J.R., Mair, R.J., Linney, L.F. & Jardine, F.M. (1996) A collaborative research programme on subsidence damage to buildings: prediction, protection and repair. *Proceedings of the International Symposium on Geotechnical Aspects of Underground Construction in Soft Ground*, April 15–17, London, UK.

De Gori, V., de Lillis, A. & Miliziano, S. (2019) Lining stresses in a TBM-driven tunnel: A comparison between numerical results and monitoring data. In: Peila, D., Viggiani, G. and Celestino, T. (eds.) *Tunnels and Underground Cities. Engineering and Innovation Meet Archaeology, Architecture and Art. Proceedings of the WTC 2019 ITA-AITES World Tunnel Congress (WTC 2019)*, May 3–9, 2019, Naples, Italy. CRC Press, London.

ITA-AITES, German Tunnelling Committee (2016) Recommendations for face support pressure calculations for shield tunnelling in soft ground. Available from: https://www.daub-ita.de/fileadmin/documents/daub/gtcrec1/gtcrec10.pdf [accessed 1st March 2021].

Lunardi, P. (2008) *Design and Construction of Tunnels: Analysis of Controlled Deformations in Rock and Soils (ADECO-RS)*. Springer-Verlag, Berlin Heidelberg.

Maidl, B., Herrenknecht, M., Maidl, U. & Wehrmeyer, G. (2013) *Mechanised Shield Tunnelling*. John Wiley & Sons, New York.

Mazzanti, P., Moretto, S. & Romeo, S. (2020) From the sky to underground: New Remote Sensing solutions in tunnelling and excavation works. *Gallerie e Grandi Opere Sotterranee*, 135.

Mikkelsen, P.E. (1996) Landslides: Investigation and mitigation. *Chapter 11- Field instrumentation*. Transportation Research Board Special Report, 247.

Miliziano, S. & de Lillis, A. (2019) Predicted and observed settlements induced by the mechanized tunnel excavation of metro line C near S. Giovanni station in Rome. *Tunnelling and Underground Space Technology*, 86, 236–246.

Romani, E., D'Angelo, M., Foti, V. & Sidera, L. (2020) Metro Line C in Rome: Two different case histories for the comparison between the design predictions and the experimental data acquired by the monitoring system during the TBM tunnels construction. *Gallerie e Grandi Opere Sotterranee*, 135.

Zou, D. (2017) Monitoring and instrumentation for underground excavation. In: *Theory and Technology of Rock Excavation for Civil Engineering*. Springer, Singapore. https://doi.org/10.1007/978-981-10-1989-0_22.

Chapter 12

Maintenance and refurbishment of existing tunnels

S. Miliziano and A. de Lillis
Sapienza University of Rome

S. Santarelli
Consultant

CONTENTS

12.1 INTRODUCTION

The extensive construction of tunnels started at the end of the 19th century as a consequence of the railway development and after the diffusion of hydroelectric power plants. Roughly at the same time, the construction of urban tunnels for subways and road tunnels for the urban development started. In the second half of the 20th century, the development of motorways prompted the construction of a large number of road tunnels, including very long ones across the Alps.

During this long period of time, the construction methods evolved in terms of both excavation techniques and lining types and materials. The stone elements initially employed for the construction of the linings were progressively replaced by bricks, concrete, reinforced concrete and reinforced concrete precast elements.

With time, several degradation phenomena started to appear – induced by the action of natural agents, the ageing of the materials and the use of the tunnels and

DOI: 10.1201/9781003256175-12

accelerated by construction defects – and required the implementation of systematic maintenance activities and reparation works.

Maintenance is a very important activity. A good maintenance enables good operational performances and increases the life of the tunnel. Thus, a day after the beginning of tunnel operations, the maintenance programme should start. Since the evolution of the state of the lining, as in general that of physical and mechanical properties of all materials employed, depends on a number of factors, this is quite difficult to predict accurately at the design stage. Therefore, the lining's state evolution should be kept under observation by means of systematic inspections and monitoring activities (ITA, 1995). Thus, the maintenance programme developed at the design stage should be constantly updated. During the tunnel operational life, a surveillance system should be implemented, with the aim of controlling the evolution of the state of the lining.

Maintenance, repair and refurbishment works can only be properly planned if detailed information about the state of the lining and the evolution of defects and degradation phenomena are available. The main goals of the maintenance and repair works are to set the conditions for a safe use, to extend the tunnel operational life and to safeguard the external environment, while keeping the interference with its use at a minimum. However, after a long time, systematic maintenance interventions and repair works can become so frequent and important that it may be better, as a result of a cost–benefit analysis, to consider a complete refurbishment of the tunnel to properly restart its operation life. It is also worth noting that the experience acquired carrying out maintenance activities can improve the design criteria for tunnels, reducing future maintenance and increasing the operational life.

After a brief description of the main damages and causes of degradation generally encountered in tunnels, this chapter illustrates the activities – inspections and monitoring – necessary to keep tabs on the evolution of the state of the tunnel. Maintenance, minor repairs and protection measures works are then briefly mentioned before a detailed description of refurbishment works. The more frequently employed refurbishment methods are illustrated, together with the design approach followed in refurbishment projects. Readers looking for further details regarding materials and technologies are referred to Book 2.

Widening of existing tunnels (Lunardi, 2003), re-profiling works and maintenance and refurbishment of electrical or ventilation plants, or other secondary systems, are out of the scope of this chapter and therefore are not addressed herein. Interested readers can find further information on these topics in Books 2 and 3.

12.2 DEGRADATION PHENOMENA AND DAMAGES

The structural capacity of the lining, due to constructive defects – which can be minimized, but not entirely avoided – is always inferior to the design one. This is particularly true for old tunnels, constructed adopting underdeveloped technologies and lacking adequate quality controls. Furthermore, the lining strength is bound to decrease in time. The durability of the lining and, in general, of all the materials employed is affected by several degradation factors, among which the most prominent are humidity, temperature (both average values and variations in time), use and intensity of use, nature of the surrounding soils/rocks, water table (both height and oscillation), chemical composition of water, frost, etc. Finally, the lining can be damaged by natural phenomena not duly considered in the project (consolidation processes, creep behaviour, etc.) or rare events such as earthquakes, landslides and fires. In this section, a brief description of the most common constructive defects and of the main causes of structural degradation is reported.

Constructive defects: constructive defects are related to imperfections in the construction of the tunnel. Regarding cast-in-place concrete linings, the main defects are honeycombing, formation of cavities or discontinuities due to casting interruption, and under-thickness. The development of cracks due to shrinkage or to thermal strains in the absence of sufficient expansion joints can be ascribed to constructive defects as well. Regarding reinforced concrete, other defects can be due to the wrong placement of the steel bars, inducing excessive or insufficient concrete covers.

With respect to precast lining segments, defects in the concrete or in the placement of the bars are quite rare due to accurate quality controls implemented at the production facilities. In this case, the main defects are associated with the stripping from the formworks, storage, transportation and assembling (assembling mistakes or jacks thrust) of the segments. Once the lining is put in place, further damages can be induced by stress states different from those assumed in the design or by defects in the grouting of the tail void. Finally, other defects are connected with the waterproofing; the circulation of water through joints or cracks, in fact, contributes to the acceleration in the degradation processes.

Material degradation: the degradation of the concrete is mainly a consequence of carbonation reactions, in which the carbon dioxide in the environment reacts with the calcium hydroxide in the cement. Carbonation induces a progressive worsening of the mechanical characteristics and increase in the permeability. Carbonation is widespread in road and rail tunnels – due to greater air circulation and exposition to exhaust fumes – while it is almost absent in hydraulic tunnels, especially if pressurized. In hydraulic tunnels, degradation phenomena are mainly associated with the scouring or the erosion due to the water flow.

The corrosion of the reinforcements is another relevant degradation phenomenon that can be worsened by natural or anthropic stray currents, sometimes even induced by the activities in the tunnel as for rail tunnels. The corrosion induces a progressive reduction in the bar thickness and a loss of adherence and also favours concrete detachments. These processes are part of the natural ageing of the tunnels; they can be minimized with regular maintenance, but cannot be entirely eliminated.

Lining degradation phenomena, both carbonation and steel corrosion, which generally proceed very slowly, are accelerated by the infiltration of water (ITA, 1991). Water infiltration can also facilitate the intrusion of chemical and biological contaminants that can further contribute to degrading concrete and corroding steel bars. The presence of construction defects accelerates the degenerative processes, too. Cracks and discontinuities, in fact, allow the steel bars to get in contact with the air, and eventually with the circulating water, and increase the exposed concrete surface.

Damages: beyond the slow degradation due to the above-mentioned physical and chemical phenomena, tunnels can suffer significant damages – ranging from cracks to detachments to potentially catastrophic instabilities – due to the occurrence of other events. The following is a brief synopsis of the more widespread causes of the damages.

Earthquakes: on the ground surface, seismic actions are much more severe than they are at depth due to the amplification effects. Nonetheless, seismic-induced loads can cause significant damages to the tunnel linings, especially those realized in masonry or without reinforcements. The damages are usually much more pronounced near the entrances of the tunnel (low covers and, often, poor soils), near zones characterized by large differences in stiffness (soil/rock contacts or fault zones) and near active faults crossing the tunnel, when permanent relative displacements develop due to seismic events.

Landslides: structural damages that in extreme cases can even lead to the loss of the tunnel may occur when the tunnel is involved in a landslide (or when the excavation of the tunnel itself reactivate or accelerate a landslide as in the case of the Val di Sambro Tunnel; Bandini et al., 2015). These phenomena are usually more significant for low-cover tunnels; thus, the entrance zones are generally more susceptible.

Changes in the loads acting on the lining: when tunnels are excavated in soils characterized by poor mechanical properties, significant swelling, creep behaviour or delayed response, the effects of the construction evolve with time, changing the short-time equilibrium of the soil–structure interaction and the loads acting on the lining. If the change in the lining loads is not correctly evaluated at the design stage, the stresses can exceed the lining strength inducing unacceptable deformations or the formation of cracks. In extreme cases, the stability of the system may be in jeopardy and onerous structural works may be needed.

Nearby constructions: the evolution of the stress and strain state induced by new constructions near the tunnel (excavation, embankments, twin tunnels, etc.) can induce significant damages to the lining. Especially in poor ground conditions, the generally acceptable three-diameter distance may not be sufficient and the induced damage can be considerable and even potentially dangerous for the users.

12.3   INSPECTIONS AND MONITORING

The evolution of the state of every tunnel should be kept under control during its entire operational life. Thus, it is crucial to check the state of the lining, the formation of cracks and the evolution of their length and width, the changes in the strength of the materials following concrete carbonation/degradation or reinforcements' corrosion, etc. Since one of the goals of maintenance is to minimize the risks to the users, detachments and falling of structural and non-structural elements (ventilation systems, lighting systems, etc.), platform deformations and water inflows (both quantity and quality) should also be monitored. To these aims, the implementation of inspections and monitoring systems – whose combination forms a complete surveillance system – can provide the information required to guarantee serviceability and safety.

All the information gathered from the inspections and the monitoring instruments should be archived and handled using a tunnel management system (TMS) based on BIM or GIS. The TMS should also contain other relevant information regarding the tunnel (as-built, maintenance and repair activities, etc.). Beyond making promptly available all these pieces of information, the TMS should be a tool for scheduling maintenance activities and inspections, installing monitoring instruments, carrying out refurbishment works, etc. More in general, the TMS has a twofold function: that of a dedicated database and that of a useful instrument to guide the decision-making and the overall management of the tunnel.

12.3.1   Inspections

Inspections should be carried out periodically, as part of the maintenance programme. The inspection frequency can vary depending on the specific state of the tunnel. Inspections should also be carried out immediately after every relevant natural event, such as landslides, floods or earthquakes, occurring near the tunnel. On this subject, several recommendations and codes have been drafted by specialized bodies and national authorities (CETU, 2012, 2015a, b; MIT, 2020; Sétra, 2010). Defects and damages detected during inspections can be classified according to one of several classification systems available. For instance, the IQOA (Image Qualité des Ouvrage d'Art; CETU, 2015a) proposes a classification accompanied by indications about possible actions to be implemented (waiting/no action, maintenance and minor repair works, service interruption and major refurbishment, etc.). Figure 12.1 reports some typical tunnel damages. In the following, a distinction is made between visual and instrumental inspections.

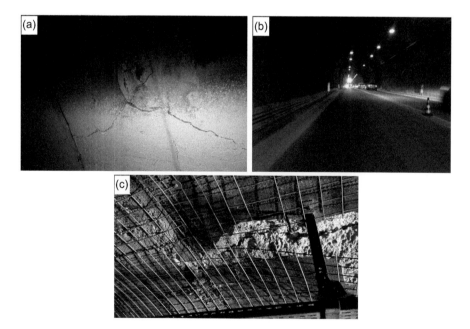

Figure 12.1 Examples of typical damages: (a) network of cracks; (b) road platform uplift and deformations; (c) spalling, concrete detachment. (Courtesy of Geotechnical Design Group srl.)

Visual inspections should be carried out by experienced engineers, whose expertise comprises tunnelling, lining behaviour and the use of the tunnel. Expertise is necessary because the possible effects of leakage and structural degradation of the lining are quite difficult to predict and because non-visible damages are not easily detected: for instance, extensive damages/corrosion of the reinforcements, induced by small water leaks through thin cracks, can develop over time without appreciable clues on the visible side of the lining. The inspections aim to detect damages and properly assess their severity. To enhance the effectiveness of the inspections, it is convenient to periodically clean the lining beforehand using pressurized water (the exercise should be temporarily limited or suspended). Following inspections, further investigation can be prompted if more information is needed, or operational decisions can be taken for safety reasons.

The survey of the network of cracks is an important tool for the identification of potential collapse mechanisms.

The evaluation of the current state of drainage pipes, drainage channels, fans and lighting devices should also be carried out to guarantee the safety of the users.

Every point of water inflow and all humidity areas should be detected and mapped. The comparison with the previously detected areas can provide precious information regarding the evolution of the state of the lining.

The state of the concrete can be quickly investigated using the hammer: sound differences are in fact useful to identify degraded areas and local discontinuities, which could induce concrete detachments. Video inspections, inexpensive and very useful, can also be carried out using micro-cameras inserted inside small-diameter drilled holes, in order to detect discontinuities, delamination surfaces, cavity, etc.

Visual inspections usually entail meaningful interference with the ordinary use of the tunnel. In the case of road tunnels, the inspections are usually carried out in night-time, temporarily suspending the use of the tunnel; in railway tunnels, night inspections can generally be carried out with little to no interference; in hydraulic tunnels, it is obviously necessary to temporarily put the tunnel out of service.

Instrumental inspections can be carried out using laser scanning, thermal scanning and high-resolution imaging techniques. Laser scanning and high-resolution imaging allow acquiring detailed and georeferenced images of all visible elements inside the tunnel and provide both qualitative and quantitative information regarding the state of the tunnel, the networks of cracks, detachments, deterioration, etc. Thermal scanning techniques are primarily concerned with the presence of water and allow detecting areas with different levels of humidity. An example of images resulting from these techniques is reported in Figure 12.2.

Modern instruments acquire data at a rapid pace and can be easily mounted on road or rail transportation moving at 3–30 km/h, depending on the requested accuracy (Figure 12.3). Thanks to these characteristics, in road and railway tunnels, the system can operate with just minor interference to the traffic (in hydraulic tunnels, instead, it is necessary to temporarily put the tunnel out of service).

Recently, several software applications have been developed to allow the generation of 3D virtual reality environments based on scanning and imaging data. Tools such as these can be highly effective to remotely carry out very realistic virtual inspections.

Figure 12.2 Images obtained with laser scanning, thermal scanning and high-resolution imaging technologies. (Courtesy of ETS srl.)

Figure 12.3 Laser scanning, thermal scanning and high-resolution imaging instruments mounted on rail transportation (Foria et al., 2019).

Irrespective of the type of inspection, attention should always be paid to the road/rail platform; uplifts and deformations, in fact, can be precursors of instability problems and are very dangerous for the users.

12.3.2 Monitoring

The monitoring of tunnels can be performed by periodically and systematically acquiring relevant data. The comparison between the results of subsequent inspections

Figure 12.4 Topographic monitoring via automatic total station. (Courtesy of Leica Geosystems.)

is, in fact, a useful way to keep tabs on the state of the tunnel. The comparisons allow observing the evolution of the networks of cracks (both opening and width), detachments, humidity areas, etc. If laser scanning technologies are employed, the comparison with previous measurements can also be used to quantify the field of displacements and its evolution over time. The comparison of the results of periodic in situ and laboratory tests – such as pull-out tests, concrete carbonation tests, flat jack tests, uniaxial compression tests on concrete specimens or tensile strength tests for the steel bars – can be used, too, to assess the evolution of the state of the lining.

Depending on the specific situation, or if the inspections detect something that deserves continuous attention, instrumental measurements can also be implemented to carry out periodic readings. The main physical quantities usually measured in tunnels during operations are crack width (using crack gauges), state of stress of the lining (using strain gauges and load cells), displacements or convergence (using automatic total stations, reported in Figure 12.4, or optical fibres), pore pressures in the ground close to the tunnel (using piezometers) and the amount of water collected by the drainage system.

In some cases, real-time monitoring can be prescribed. This is, for example, the case of shallow tunnels involved in slow-moving landslides. Typically, from inside the tunnel, displacements, cracks opening and local stresses are monitored, while, outside the tunnel, surface and sub-surface displacements and pore pressure are recorded. The measurements can be carried out with high frequency and transmitted in real time for analyses. In these cases, early warning thresholds are usually preset and associated with well-defined actions (visual inspection, exercise suspension, etc.).

For more detailed information about monitoring technologies, instruments, limitations, accuracy and precision, the reader is referred to Volume 2.

12.4   MAINTENANCE ACTIVITIES, REPAIR WORKS AND PROTECTIVE MEASURES

Ordinary maintenance activities generally consist in light works such as painting steel to protect from corrosion, replacing support elements of the ventilation and lighting

Figure 12.5 Installation of two steel nets of different grid sizes bolted to the lining to retain detached elements.

systems and cleaning of the drains and drainage channels (CETU, 2012; ANAS, 2018). Ordinary maintenance can include minor repair works which are generally local and address the safety of the users: removing concrete pieces of the lining in precarious stability conditions, treating the steel bars with anti-corrosion products, replacing the concrete cover with special protective mortar and sealing of concrete cracks by grout injections (cementitious mixtures or chemicals). Repair works are usually very fast and not expensive and can be carried out with minimum or no interference with the use of the tunnel (ITA, 2001).

When the safety cannot be guaranteed with acceptable residual risks by carrying out minor repair works, temporary passive protection systems should be implemented until major refurbishment works can be carried out. Among these are the installation of steel nets at the crown of the tunnel, adequately bolted to the lining, to minimize spalling risk by retaining detached elements (Figure 12.5), or the installation of drainage channels to avoid water flowing into the tunnel. However, the use of drainage channels should be limited to localized leakage areas (joints) and to short periods of time – until waterproofing works can be realized or draining pipes are installed – to avoid concealing the lining, which should always be visible to allow direct observation.

Irrespective of the kind of works, the available data should be sufficiently detailed to allow the identification of the causes of the damages, thus enabling to select the more appropriate repair method to directly address the root causes.

12.5 REFURBISHMENT WORKS

More extensive than ordinary maintenance activities can be classified as refurbishment works. At the end of its operational life, or in any case when the functionality or the operational safety is no longer fully guaranteed (due, for instance, to unacceptable risks and maintenance cost or excessive water leakage for hydraulic tunnels), the tunnel must be put out of service or extraordinary refurbishment works must be carried out. Refurbishment works generally involve the whole tunnel and aim to improve the tunnel structure (static reasons due to material degradation, requirements of new technical rules, etc.) or geometry (functional needs and operational aspects) and to organize and control the water flow (waterproofing and drainage systems), creating the conditions for a safe use, safeguarding the external environment and prolonging the life of the tunnel. Refurbishment works are very costly and time-consuming and usually require the suspension of tunnel operation.

The design of refurbishment work requires a thorough definition of the problem and the correct identification of the root causes of the damages. Remedial actions, whenever possible, should be addressed towards the elimination of the causes of the damages and not just towards temporary repairment. To identify the causes, all available data should be carefully analysed and specific surveys and investigations should be carried out.

Refurbishment works also offer the opportunity to implement further safety measures (possibly required by updated national codes) such as bypasses or escape routes. These topics, together with the complete demolition of the existing lining for re-profiling works, go beyond the scope of this chapter and therefore are not discussed.

12.5.1   Methods

In this section, the methods usually adopted to carry out refurbishment works are described. A distinction is made between water control works and structural works. Since the refurbishment methods are quite similar, independently of the lining's nature (precast reinforced concrete segments and both plain and reinforced cast-in-place concrete), no distinction is made between mechanized and conventional tunnelling.

12.5.1.1   Water control

As discussed above, the main cause of lining degradation is the infiltration of water. In fact, degradation phenomena will generally proceed very slowly without water's damaging effects. Therefore, the elimination of all sources of leakage is paramount for the safeguard of the tunnel.

Several methods can be employed to eliminate water leakages or their dangerous effects. Suitability and effectiveness of these methods can also depend upon the lining's nature.

A first method, suitable where controlled drainage is acceptable, consists in collecting all the water inflow using drainage pipes. The collected water is then conveyed towards a water sump for disposal through the main drainage system running through the invert for the whole length of the tunnel. The drainage pipes should be able to capture the water at the extrados of the lining in order to avoid it coming in contact with the concrete. The pipes should also be protected using geotextiles, to avoid the transportation of fine solid particles. A separate drainage system should be used to capture the platform water, which needs to be treated before disposal. Finally, periodic maintenance activities should be carried out or, if not possible, the pipes should be monitored and replaced when necessary. All these activities can be performed from inside the tunnel.

Drainage channels address the same need of water control. However, although inexpensive and rapidly installable, drainage channels should only be employed as temporary measures, as it is far preferable to capture water before it leaks through the lining, progressively contributing to the material's degradation.

An opposite strategy is that of waterproofing the lining through grout or chemical injections. The injections should be mainly concentrated along the surfaces of structural and constructive joints, where leakage usually occurs. If present, cracks should be treated to make the lining impervious (injecting the cracks also constitutes a structural repairment which increase strength and durability of the concrete). The injections must be done in more than one session, starting from the leaking points and then addressing the new leakages that form after the closing of the first ones. In such cases, it is preferable to make the lining impervious injecting at its extrados to avoid that filtrating water comes in contact with the internal concrete.

Very similar techniques and materials can be used to inject the joints between the precast lining segments. Furthermore, in this case, the injection of the tail void during mechanized construction (carried out to completely fill the void annulus) can greatly contribute to achieving the requested level of hydraulic tightness. Different products can be used: cementitious mixtures, and chemical products such as resins of different physical and mechanical characteristics and costs (refer to Book 2 for further details).

It is worth specifying that in the case of deep tunnels, sealing works should be realized in conjunction with draining systems, while for shallow urban tunnels, waterproofing is usually mandatory. In some cases, due to environmental prescriptions (mainly to preserve aquifers at risk), it can be necessary to resort only to waterproofing even for relatively deep tunnels.

Another frequently adopted method consists in creating an internal re-lining, which can become part of the original lining. This method can be successfully applied to seal leakages characterized by very small water inflows. The re-lining is usually carried out by removing a few centimetres of the lining, collecting water with small draining pipes and then spraying a thin layer of shotcrete, possibly reinforced with fibres or with electro-welded meshes. Alternatively, it is also possible to use PVC membranes nailed to the existing lining (special precautions must be taken to guarantee watertightness where the nails pass through the membranes) or even sprayed waterproofing membranes. The secondary lining, if properly reinforced, can also play an important role from a structural point of view (see the following). Finally, another option is that of carrying out the re-lining by using prefabricated shells, separated from the existing lining, covered by a PVC membrane. This last method, however, definitively conceals the original structural lining which cannot be maintained or inspected any longer, and therefore, it is not advised.

All the previous methods involve works carried out from inside the tunnel. Another possible course of action that only involves operations from outside the tunnel consists in removing the source by lowering the water table permanently. This can be accomplished using gravity drainage systems such as sub-horizontal draining pipes or draining wells. In this case, it is crucial to set up a monitoring system, possibly automated, able to signal malfunctioning in real time.

The above methods can be employed in conjunction and together with works aimed at structurally improving the lining.

12.5.1.2 Structural improvement

There are several viable methods for the structural improvement in the tunnel. The choice of the most appropriate method depends on many factors related to the current state and function of the tunnel (e.g. severity of damages, type of lining, tunnel use and environmental settings) and to the requested characteristics (e.g. strength and durability of the materials and works duration). Irrespective of the adopted technique, before proceeding with structural improvements, it is always advisable to carry out the water control works selecting one of the methods discussed in the previous section.

The most effective structural improvement technique consists in constructing an internal re-lining of reinforced concrete. When a reduction in the internal profile of the tunnel is admissible, the new lining can be made in contact with the existing one by using cast-in-place concrete and properly designed formworks. If the water table is absent, specific waterproofing systems can be avoided and a good concrete mix design, cast with great care, will result in an impervious enough lining. In any case, structural joints between segments must always be prescribed in order to avoid the formation of cracks due to concrete shrinkage during curing and thermal deformations. Nonetheless, the use of an impervious membrane attached to the existing lining is always advisable; alternatively, a sprayed waterproofing membrane can be employed. These measures protect the concrete and the reinforcements from percolating water leaking through the joints and the cracks of the existing lining. The reinforcement can be made of a classic steel cage, metallic fibres or both. Electro-welded wire meshes and lattice girders are frequently employed, too.

Shotcrete can be also employed instead of ordinary cast-in-place concrete for a very fast construction, using alkali-free accelerators for durability reasons. In this case, if impervious PVC membranes have been installed, it is necessary to use systems to allow the adhesion of the shotcrete, minimizing the rebound phenomena, such as wired meshes in contact with the membrane, and the shotcrete must be sprayed multiple times, thus creating overlapping thin layers. Further information on this topic can be found in several handbooks (e.g. Sika, 2020). Finally, it is also advisable to use steel support elements (lattice girders) to strengthen the lining and ensure the final tunnel profile.

When a reduction in the internal profile is not admissible, or a small increment is necessary, before the construction of new internal lining the existing one must be preliminarily demolished. Two different methods can be employed. Hydro-demolition can be adopted, which is an efficient, cost-effective and precise method that removes concrete with jets of high-pressure water. Alternatively, a portion of internal lining can be milled mechanically by means of specially designed roadheaders (Figure 12.6). In the presence of reinforced concrete, before milling the steel bars must be cut. To minimize the amount of dust produced, the machines are equipped with special protection caps and numerous water-spraying nozzles (ventilation and dust-processing systems must be prepared carefully, too).

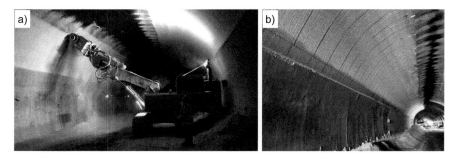

Figure 12.6 Mechanical milling of the internal lining concrete aimed at re-lining the tunnel. (Courtesy of CIPA SpA.)

Usually, few decimetres of new, good reinforced concrete lining are generally enough to guarantee the requested margins of safety against general collapse mechanism and local instabilities (collapse of pieces of concrete of the lining). After milling, the works can proceed as described above for the simple internal re-lining (without milling): elimination or control of water inflows, installation of a waterproofing system, installation of reinforcements and casting of concrete.

For both safety and construction site reasons, usually, the works are carried out slowly, operating segment after segment. This can be necessary, for instance, when the lining is realized with reinforced concrete and is heavily loaded and the thickness of the remaining portion of the lining after milling is relatively thin.

A complete structural connection between the new lining and the existing one is practically impossible to realize, especially when waterproofing membranes are used. Nonetheless, even if the two linings can only exchange normal forces/stresses, the composite system can work quite well, guaranteeing the required safety conditions.

It is important to note that, frequently, the new lining will be essentially unloaded. In most cases, during their operational life, the new linings are only stressed in the occasion of rare events such as earthquakes, vehicle impacts and fire. They are mainly tasked with a prevention/protection role, lasting as long as possible, guaranteeing a safe use of the tunnel and the safety of the users (road and railway tunnels).

For hydraulic tunnels, the internal surface of the new lining can be treated with special products in order to obtain a smooth surface (reducing the roughness) and, at the same time, minimizing scouring and erosion phenomena.

Much less frequent is the use of liner plates, realized using steel plates, calendered and welded in place. The plates are structurally connected to the existing lining through injections of high-resistance cementitious mixtures. This method is mainly employed in the refurbishment of hydraulic tunnels. Usually, few centimetres of steel are sufficient; thus, the re-modulation of the internal profile is minimal. It is a very costly solution.

Figure 12.7 reports some examples of the structural improvement works described in this paragraph.

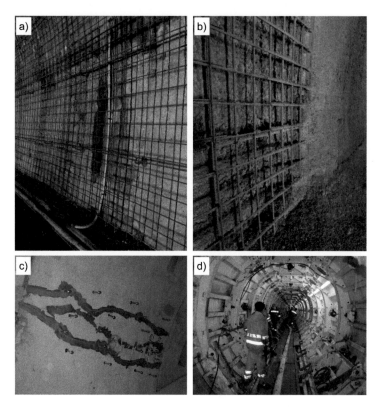

Figure 12.7 Examples of structural improvement works: (a) water collection before con-
creting; (b) shotcrete re-lining reinforced with steel nets; (c) cracks sealing
via resin injections on precast lining segments; (d) mobile formworks for the
re-lining of segmental linings. (Courtesy of Geotechnical Design Group srl.)

12.5.2   Design

The design of refurbishment works for old tunnels is a very peculiar topic. The correct
development of the design requires the knowledge of all geometrical elements and the
current mechanical properties of the lining (masonry, plain and reinforced concrete).
Furthermore, a geotechnical model (stratigraphy, physical and mechanical parameters, pore pressures, water table and seepage regime) must be set up.

When a complete and detailed documentation of the tunnel is available, investigations should be devoted essentially to verifying (sample checks) the actual geometry
and to evaluating the current mechanical characteristics – strength and stiffness – of
the lining. Vice versa, if the documentation is not available, as frequently happens when
dealing with old tunnels, surveys are more burdensome and must be extended to the
ground surrounding the tunnels, with special care to the entrances. Regarding ground
investigations, beyond the usual investigations carried out from the surface, tunnels
offer the very advantageous possibility – especially for deep tunnels – of operating
from inside, the only downside being the mandatory operational suspension (railway

or hydraulic tunnels) or limitations (roadway tunnels). In any case, the knowledge of the state of stress of the lining, due to the interaction between ground and structure, is extremely relevant because the design of refurbishment works depends very much on it, especially when the demolition of significant parts of the lining is necessary.

Beyond the quality and state of the materials, if the project involves the milling of the lining, it is crucial to obtain a good knowledge of the thickness of the lining; dedicated investigations – direct investigations, if possible – should be carried out along the tunnel to this aim. Furthermore, considering the importance of the lining stress state, numerous flat jack tests should be performed, also taking into consideration the type of soil/rock surrounding the tunnel, the cover and the general state of the lining.

In order to obtain a reliable model, parametric analyses should be carried out to overcome the uncertainties related to: (i) geotechnical operational values of mechanical parameters; (ii) initial state of stress in the ground; and (iii) stress release percentages associated with the excavation techniques. The parametric analyses allow establishing the best set of parameters, capable of reproducing the stresses acting on the lining as close to the measured ones as possible. This calibration of the structural and geotechnical model – which relies on trustworthy measurements of the lining stress state – is crucial to the development of the design of refurbishment works.

Even though complex 3D numerical models can be developed (Galli et al., 2004; de Lillis et al., 2018; Ochmański et al., 2018), in most cases, 2D models can be effectively used to design the refurbishment works. The stress release technique, either uniform (Karakus, 2007; Möller & Vermeer, 2008) or differential (Altamura et al., 2007), is usually suitable to simulate the excavation. Thanks to the calibration process described above, very simple constitutive laws (linear elastic–perfectly plastic with a tension cutoff for both the soil and the lining) are generally able to provide good enough results. The simulation of the lining using continuous elements (properly taking into account the structural joints) and appropriate meshing makes quite easy the simulation of the milling of the lining, if this activity is foreseen in the design.

Usually, since the uncertainties are relatively few, the design approach adopted for refurbishment projects is deterministic.

In general, it is useful to include in the design a monitoring system tasked with measuring the geo-structural response to refurbishment works through some key parameters. This allows either confirming the design predictions or, vice versa, timely detection of significant departure from the expected behaviour. For instance, when a relevant portion of the existing lining is to be demolished, the strains in the lining or the convergences can be usefully monitored or, if water control works (waterproofing, draining, etc.) are to be performed, measuring the pore pressure around the tunnel can provide meaningful information.

12.6 CONCLUDING REMARKS

The maintenance of tunnels is a very important activity because it allows ensuring good operational performances over time and increasing the life of the structure. The knowledge of the evolution of the state of a tunnel during its entire operational life enables not only to correctly schedule maintenance activities, execute repair works and put in place protective measures with the aim of minimizing the risks to users, but also to guide the decisions regarding potential use suspensions for safety reasons and, eventually, to carry out relevant refurbishment works.

Since the evolution of the state of the lining and, more in general, of the materials properties is quite difficult to predict as it depends on many natural and anthropic factors, the tunnel should be kept under observation by means of systematic inspections and monitoring activities. Even though modern technologies constantly provide more effective tools, such as laser and thermal scanners or high-resolution images, that also allow minimizing the interference with the tunnel use, visual inspections, carried out by experienced tunnel engineers, cannot be fully replaced by instrumental inspections. Engineering judgement is always essential, especially in tunnelling.

The correct identification of the causes of the damages should always precede remedial actions that should address the causes of the problems whenever possible, rather than simply repair the damages. It is also crucial that the best repair methods and the most suitable materials are selected taking into account the specificity of the site.

The works can be distinguished between ordinary maintenance activities, including minor repairs, and extraordinary refurbishment works. The former, associated with light works, are carried out periodically and are generally local, cheap and fast. Their main aims are to guarantee a safe use, prolong the tunnel life and safeguard the external environment. Protective measures such as steel nets or drainage channels are frequently installed in order to ensure the safety of the users while waiting for more impacting refurbishment works. Refurbishment works deal with the elimination of water inflow and its dangerous effects (water leakage is one of the main causes of concrete degradation) and the improvement in the tunnel lining. Refurbishment works generally involve the entire tunnel and are very costly and time-consuming.

The correct development of the design of refurbishment works requires the knowledge of all geometrical elements and current mechanical properties of the lining. A good geotechnical model is necessary, too. Thus, specific surveys are generally carried out for design purposes, especially in the case of old tunnels, for which information is usually scarcer. The refurbishment design requires models specially calibrated based on the measured state of stress acting on the lining; therefore, these measurements are mandatory. The calibration is crucial because it enables to overcome several uncertainties (geotechnical parameters, stress history and events occurred during the construction) and the shortcomings of the simple constitutive models usually adopted: it is one of the keys of the project success. As for the construction of new tunnels, also for refurbishment works, especially when a relevant portion of the existing lining is foreseen, a monitoring system plays an important role and it is a fundamental part of the design.

AUTHORSHIP CONTRIBUTION STATEMENT

The chapter was developed as follows. Miliziano: chapter coordination. The chapter was joined developed by the authors. All the authors contributed to chapter review. The editing was managed by de Lillis.

REFERENCES

Altamura, G., Burghignoli, A. & Miliziano, S. (2007) Modelling of surface settlements induced by tunnel excavation using the differential stress release technique. *Rivista Italiana di Geotecnica*, 41(3), 33–47.

ANAS (2018) I quaderni tecnici per la salvaguardia delle infrastrutture. Volume IV. Available from: https://www.stradeanas.it/sites/default/files/pdf/1.3.3/Quaderni_tecnici_Volume_4.pdf [accessed 1st March 2021].

Bandini, A., Berry, P. & Boldini, D. (2015) Tunnelling-induced landslides: the Val di Sambro tunnel case study. *Engineering Geology*, 196, 71–87.

CETU, Centre d'Études des Tunnels (2012) Guide d'application de l'instruction technique pour la surveillance et l'entretien des ouvrages d'art. Fascicule 40: Tunnels. Available from: http://www.cetu.developpement-durable.gouv.fr/IMG/pdf/Fascicule40-2_cle05a564.pdf [accessed 30th January 2021].

CETU, Centre d'Études des Tunnels (2015a) Road tunnel civil engineering inspection guide. Book 1: from disorder to analysis, from analysis to rating. Available from: http://www.cetu.developpement-durable.gouv.fr/IMG/pdf/guide_inspection-book1_hv.pdf [accessed 5th February 2021].

CETU, Centre d'Études des Tunnels (2015b) Road tunnel civil engineering inspection guide. Book 2: catalogue of deteriorations. Available from: http://www.cetu.developpement-durable.gouv.fr/IMG/pdf/road_tunnel_inspection_guide_book2_hv.pdf [accessed 5th February 2021].

de Lillis, A., De Gori, V. & Miliziano, S. (2018) Numerical modelling strategy to accurately assess lining stresses in mechanized tunneling. In: Cardoso, S.A., Borges, J.L., Costa, A.P., Gomes, T.A., Marques, J.C., and Vieira, S.C. (eds.) Proceedings of the 9th European Conference on Numerical Methods in Geotechnical Engineering, NUMGE 2018, 25–27 June 2018, Porto, Portugal. London, CRC Press, pp. 1295–1302.

Foria, F., Avancini, G., Ferraro, R., Miceli, G. & Peticchia, E. (2019) ARCHITA: an innovative multidimensional mobile mapping system for tunnels and infrastructures. MATEC Web of Conferences, 295, 01005. EDP Sciences.

Galli, G., Grimaldi, A. & Leonardi, A. (2004) Three-dimensional modelling of tunnel excavation and lining. *Computers and Geotechnics*, 31(3), 171–183.

ITA. Working Group on Maintenance and Repair of Underground Structures (1991) Report on the damaging effects of water on tunnels during their working life. *Tunnelling and Underground Space Technology*, 6(1), 11–76.

ITA. Working Group on Maintenance and Repair of Underground Structures (1995) State-of-the-art of non-destructive testing methods for determining the state of a tunnel lining. *Tunnelling and Underground Space Technology*, 10(4), 413–431.

ITA. Working Group on Maintenance and Repair of Underground Structures (2001). Study of methods for repair of tunnel linings. Available from: https://about.ita-aites.org/publications/wg-publications/download/112_ec82e7e2c738b6b06cbc46dfdc409895 [accessed 1st March 2021].

Karakus, M. (2007) Appraising the methods accounting for 3D tunnelling effects in 2D plane strain FE analysis. *Tunnelling and Underground Space Technology*, 22(1), 47–56.

Lunardi, P. (2003) Widening the load at Nazzano. *Tunnels & Tunnelling International*, 35(7), 16–19.

MIT, Ministero delle Infrastrutture e dei Trasporti (2020) Manuale ispezione gallerie (25/05/2020).

Möller, S.C. & Vermeer, P.A. (2008) On numerical simulation of tunnel installation. *Tunnelling and Underground Space Technology*, 23(4), 461–475.

Ochmański, M., Modoni, G. & Bzówka, J. (2018) Automated numerical modelling for the control of EPB technology. *Tunnelling and Underground Space Technology*, 75, 117–128.

Sétra, Service d'études sur les transports, les routes et leurs aménagements (2010) Instruction technique pour la surveillance et l'entretien des ouvrages d'art. Available from: http://dtrf.cerema.fr/pdf/pj/Dtrf/0005/Dtrf-0005955/DT5955.pdf?openerPage=notice [accessed 30th January 2021].

Sika (2020) Sika sprayed concrete handbook. Available from: https://www.sika.com/dms/getdocument.get/664de18e-dbcb-4184-87b7-162e9dd260d5/glo-sprayed-concrete-handbook-2021.pdf [accessed 8th March 2021].

Chapter 13

Project acceptance strategy

G. Dati
Tunnel Euralpin Lyon Turin

CONTENTS

13.1 INTRODUCTION

Project acceptance strategies should be considered an integral part of the design of a major project such as tunnels. These projects must be conceived upstream, with a multidisciplinary team including engineers, architects, geologists (and other experts 'typically' integrated in the design phase) and also professionals who can support the client in managing the message perception and in the dialogue with the local areas, before, during and after the construction. According to international design experiences of different projects all over the world, these strategies are clearly needed (Chen et al., 2019). Mass communication theories and techniques can provide strategic support, especially to avoid the perception of a division between technicians and experts, on the one hand, who draw up the perfect project on paper, and a community, on the other, which has its own very clear concept of the common good (Lippmann, 1922).

The research of recent years has started from different approaches (from project management to social and communicative processes), comparing experiences of major projects around the world and evaluating their successes and failures (Boateng et al., 2012). In summary, the results that emerged offer a key concept: the importance of communication in project management.

Some more elements are as follows:

- The existence (and communication) of a strong strategic vision with regard to the project is the best starting point for the process of building consensus.
- Open and transparent communication must be part of upstream planning (Hertogh et al., 2008).
- Communication with local authorities, supervisory agencies and environmentalists is particularly delicate, as these are usually the stakeholders with the most negative approach (Di Maddaloni & Davis, 2018).

DOI: 10.1201/9781003256175-13

- There is a cause–effect relationship (Shenhar & Holzmann, 2017) between stake-holder involvement (satisfaction and management of interest groups) and the success of a project.
- The proactive approach in the involvement of all stakeholders (compared to the reactive approach commonly adopted) – and in particular with local communities – confers a greater added value to the project (Di Maddaloni & Davis, 2017).
- Stakeholders are not, by definition, an opposing party, but a complex and diversified body to be mapped and classified in order to adopt specific engagement techniques (Eskerod & Ang, 2017).
- Stakeholder engagement processes require communication strategies to be planned in a strategic key not only in small construction sites, but also in megaprojects (Mok et al., 2015).

In this chapter, you can find some useful advice about this important task.
Let's start with some questions:

- Who influences the creation of consensus/opposition with respect to major projects and why?
- How much do different interest groups affect the implementation of major projects and why?
- What mechanisms and procedures in public discourse can facilitate/complicate the construction of an infrastructure?
- Are there strategies to win over the willingness of a community to accept a major construction site and manage its impact itself?

For example:

> Public decision-maker XY is formalising the details of a tunnel project in the territory YZ. It is all in the hands of shareholders, ministries and the executor. The local community hears some vague news about the project; outsiders are noticed on the site; local citizens and authorities in the area learn from experience (and not through knowledge) that there will be a change and start to question and talk about it. A vicious circle begins: the 'rumours' develop uncontrolled and without verification, with pre-constituted and unfounded hypotheses and theses, which develop into fake news; local citizens call on their own experts; associations are set up to understand how to mobilise themselves. At the same time, the employees of the company to which the work contract has been awarded wonder what resources will be hired and how they will organise themselves.

The example above is a typical case. Here are two examples more: Notre-Dame-des-Landes airport (France) and the Lyon–Turin high-speed railway line (Italy–France): in these cases, infrastructure has become a symbol, and they offer useful cases for exploring these aspects. Communication and news reporting have led to a media exposure that has predominated over the technical aspects.

Example 1: In France, the NDDL airport (south of Nantes), following 50 years of controversy and violent clashes, will not be built: the Macron government decided to abandon the project in 2018, despite the establishment in 2016 of a local referendum which saw 55% in favour of the airport, more than 178 judicial decisions in favour of the construction, and the project's decree of 'public usefulness' in 2008. The story began in 1968, with the identification of the area for the construction of a new airport infrastructure and the creation of the 'Zad', which officially means 'zone d'aménagement différée' (deferred development area) and which the protest and occupation movement interpreted as '*zone à défendre*' (area to be defended). Years of occupation, violence and struggles have led to an abandonment with legal consequences for the termination of the contracts signed for the construction.

Example 2: The case of the Susa Valley in northern Italy on the border with France has been different: here, opposition movements have created a 'No TAV' brand, often more recognisable than the project itself, which has slowed down construction with physical and paper barriers, but the work is still in progress.

In order to analyse similar situations, it is necessary to look beyond: the realisation of a project has an impact on the territory and on the people living there, not only during the construction site phase, but also afterwards (Brookes & Locatelli (2015);. Mover, an underground work feeds mysteries and 'hidden' elements that, even more, should be revealed, narrated and made accessible.

The debate on the quality of impacts opens as soon as the first project hypothesis arises. Often, it focuses on the dichotomy between 'common good' at a broad level and community benefits at a local level, in tunnels, as well as in other infrastructure projects or production systems.

For infrastructures, arguments about sustainable modes of transport, faster passenger trains and greater exchanges between areas affected and positive economic benefits for an entire country are typical of the approach that looks to the 'common good'. Interventions and changes in the local landscape, disruption in everyday life – perhaps limited, but very concrete – brought about by the construction of the works and the construction site area causing tourists to stay away are instead the typical content of the negative approach of the local population.

13.2   THE SOCIAL AND COMMUNICATION CONTEXT OF THE 20TH AND 21ST CENTURIES

The 21st century sees some key steps in democratic countries:

- the expansion of civil conquests in a climate of opposition between activists and governments, with the growth of awareness of objectives that can be achieved through struggle in the streets (Gallino, 2007);
- disintermediation, where once there was a rigid mediation of different structures (parties, media and traditional communication structures, for example), and the arrival of social networks to represent and accentuate this shift;
- the widespread sharing of the value of transparency, which overcomes guarded fortresses of public offices and the closed doors of large companies, but which also leads to the revelation of political, economic and financial scandals and to public institutions suffering a crisis in trust (Short, 1984).

Talking about consensus regarding a project cannot overlook this scenario.

13.2.1   The stakeholders

Born as an 'economic constraint', meaning shareholders, suppliers and clients (Forester, 1999), the word stakeholder today comes to exemplify the wider public of subjects affected by a work (Evan & Freeman, 1993). In the case of a public developer, in fact, it has to take into account both its own shareholders and a vast array of parties affected and involved in various ways:

*   local, national, international institutions;
*   supervisory bodies;
*   research and training institutions;
*   local communities;
*   potential and future users;
*   civil associations;
*   politicians and political parties;
*   public promoter's consultants;
*   public promoter's suppliers;
*   individual citizens with a small interest, perhaps as owners of property, in the area;
*   public opinion at different levels.

In setting up a process of creating awareness (first) and consensus (subsequently), we need to start by mapping all the involved areas and stakeholders, in order to avoid the risk of the project itself being perceived as too self-referential and imposed from above (Boateng et al., 2015). It is no longer a question, therefore, of working only on a technically perfect project, but at the same time, on a plan to make it known even before arriving on site: it means taking on the communicative dimension as a key element of the whole upstream process.

A concrete case in point is that of the Los Angeles subway: the potential users, however, have never been informed of its potential, nor has this been conveyed. The result is an efficient project, with a confirmed budget and time frame, but with few users. Seen fully in the round, has it been a successful or unsuccessful project? At the other end of the spectrum, the World Trade Center project was conceived upstream as a large-scale project, and during both the design and implementation, the long-term strategic perspective for various stakeholders was highlighted. The academic community agrees that this is a successful case from every point of view.

Let us return to the initial example:

> While the rumour mill is churning, the public promoter is conspicuous by its absence: it has not officially shown itself in the territory and, as a consequence, has been perceived as an outsider. It exists, it operates at a central level with its structure of technicians and engineers; it plans from afar on the maps of a territory.

Since 'one cannot not communicate' (Watzlawick, 1967), the absence of the developer is itself a form of communication; at best, it is perceived by the population as a negative communication, a form of closure with respect to information to be given out to the territory; at worst, as a wish to hide something.

The typical situation can be compared with that of the new Lyon–Turin line, which had already been hypothesised in the 1990s to replace the historic line crossing the Alps via the Fréjus Tunnel, designed in 1857 and inaugurated in 1871. The public decision-makers (the Italian and French governments and Europe) carried the project forward within the framework of the new European infrastructure network and defined the public developer for the different phases (Alpetunnel first and then LTF – Lyon Turin Ferroviaire).

In France, an *enquête publique* was carried out: an independent commission gathered the comments and needs of the local area through public meetings and sessions with all the stakeholders. This phase ended in 2006 with the favourable opinion of the dedicated Commission, followed in 2007 by the declaration of the project being of public usefulness issued by the French Prime Minister.

In Italy, a traditional approach was adopted, that is used for all infrastructure projects up to that moment: it may be summarised as DAD (Decide, Announce, Defend). As a result of this approach, far from the local area itself, protests broke out in Piedmont: citizens organised themselves, constructed a narrative about the project and catalysed media attention, building on the fact that the lack of adequate communication activities concerned not only the local population, but also the local administrators.

A key and effective element was the overturning of the decision-maker's perspective: the European sustainable project, part of Corridor 5 (now Mediterranean), became a local project (Lyon–Turin, indeed) that destroys the environment and the future of people (families, children, elderly people on the march with plausible reasons and without counter-movement). Some key elements in this story are the following:

- There is only one narrative and only one side that speaks to the media, with the success of the 'agenda setting' by the new-born movement.
- The decision-makers and the developer are obliged to try and catch up and defend themselves, mainly bringing into the debate technicalities which are difficult for ordinary people to understand.
- The identity of the project – and also the common name of the infrastructure itself (TAV – 'high-speed train') – would, for a long time in the eyes of public opinion, be those 'dictated' by the opposition movement.
- The public political decision-maker addresses the conflict as a public order issue, which contributes to the radicalisation of the struggle.

After the protests, but in particular after 2005, came the establishment of the Lyon–Turin Railway Observatory, 'a technical forum for the confrontation of all interested parties, with the analysis of critical issues and the establishment of solutions for political and institutional decision-makers'. The new body was the first to define who was entitled to participate and, for the first time, set up a participatory process. Among the important results achieved, there was the analysis of ten alternative routes for the new line, with the convergence on the current route, accepting some requests that emerged from the local area. A part of the movement continued to exist and to fight on, but local administrations sat at the table and had the opportunity to participate.

What this experience highlights:

- The first element of cooperation is the definition and sharing of a method, which becomes the first content of communication.

- A correct approach to project acceptance with the identification of stakeholders, dialogue and listening to different sides can bring improvements to a pre-established project.
- The communication can emphasise the constructive aspects and the authoritativeness of third parties at the table.
- The 'relational' approach will lead to the birth of a 'rational' project that is scientifically and technically convincing.

In 2015, the Observatory was called upon by the Italian government to establish some guidelines to be successively written into the law on public debate in Italy passed in 2016. In Virano (2015), some key activities are identified as part of the path to planning a major work; to mention just two:

- presenting and living the moment in which the identity of a major project is traced out (purpose, layout, characteristics, costs, etc.), as a real 'constituent' phase of the work;
- associating the decision to undertake a major work of communication and participation of citizens, both in a preliminary phase and in progress, with some ad hoc tools (visitors' centre, creation of a role of a person in charge of relations with the territory, non-scientific language, etc.) (Figure 13.1).

Figure 13.1 Cover of the magazine *Politis*, reproducing a historical image of the No TAV protest. (Author's pic. – Lyon; February 2016.)

13.2.2   The path of creation of awareness and/or cooperation

The path of the creation of awareness and cooperation can be explained starting from the following sentence by Forester (1999):

> I once asked a very famous mediator what drives a person who enters a room where everyone is angry with each other to stay there instead of leaving. And the answer was: "When I enter a room and there is this energy of conflict, I think this is always better than apathy, and I start from there. If I am myself angry, I am so because I think something is not working and I want things to work better".

In France, the *Débat Public* procedure is based on information and discussion meetings, with the gathering of proposals and points of view from citizens, associations and institutions, at a stage in which there are still project alternatives. It provides for the mapping of concerns and needs and an exchange of positions expressed by different stakeholders.

In the English-speaking world, the creative confrontation (CC) or consensus building approach is widespread; it provides for a way of discussing and deciding that turns differences into resources and is based on listening, the multiplication of options and creative co-planning, with a criterion of qualitative representativeness in which the project is a process shared around the table.

While several legislators have proposed or structured consultation procedures in the last 30 years, a change of approach in the planning of works and infrastructures is evident. Protests against major works have a solid history in North America, while in Italy, the notion arrived a few decades later (Bobbio & Zeppetella, 1999), although replicating a similar script. The latest data from the Italian Nimby Forum (Blanchetti & Seminario, 2017) reveal 317 infrastructures and plants currently being disputed in Italy and identifies the typical types of opposition (Table 13.1).

Table 13.1 Comparison of point of view

Promoters	Opponents
Useful works	Useless works
Irrefutable scientific data and opinions in favour	Other irrefutable scientific data and opinions against
Opponents have prejudices and do not understand the economic and scientific scenarios	Promoters are self-referential (often even corrupt and motivated by economic concerns)
The common good is more acceptable than that of each local area	The local area is the only one being impacted, and it is therefore its welfare that needs to be protected
Great general economic spin-offs	Important spin-offs only for the benefit of private individuals
The project improves the development of the local	The project worsens the quality of life

Where to start then? An ideal example.

The political decision-maker is thinking about a tunnel in the area of YZ, with strong integration in an international infrastructural context. The technical project has yet to be defined. Actions planned:

- identify all stakeholders in different capacities – stakeholder mapping;
- assess the experience gained from similar cases – benchmarking;
- set up a round table to 'plan the planning' – communication (listening and stakeholder engagement).

This is an advocacy process that contemplates some of the fundamentals of project acceptance: communication, listening to outside points of view and reversing earlier perspectives (top-down) towards a strategy of transversal stakeholder engagement. The best possible project is being planned.

Two advantages of this approach:

- project logic (Baiguini, 2020) that focuses on a constructive and inclusive approach aiming at creating a solid consensus, already as to the method and then as to the contents of the project (Won et al., 2017). This approach defuses the 'friend–enemy' logic of possible counterparts, based on the assumption that 'we are the good guys and there is a bad guy we need to fight outside';
- creation of the favourable context (first) and the positive narrative (second), laying the foundations for an improved reconciliation and alignment of objectives, at the right time to prevent fake news and prejudice.

13.2.3   The strategies, the 'significant' contents and the communication tools

If stakeholder involvement and the process of building consensus regarding a project is a matter of communication, we need also to address at least two other elements: the content and tools of communication.

Strong contents serve to build the 'corpus' of stakeholders around some key elements and so avoid both a misalignment between the parties and a possible chasing after replies to opponents' theses, with consequent reputational crises.

There are some specific cases in which convergence on certain elements was at the centre of the evolution of a project: an Australian icon, the Sidney Opera House as an example. The architectural jewel, UNESCO heritage site and tourist destination also faced a moment of conflict due to strong disagreement between the architect Jørn Utzon, the Municipality and the city's political stakeholders, which led to increased costs. The failure here, therefore, was to have underestimated the level of complexity surrounding the work, not only in terms of construction, but also as a symbol of the community. A strong sharing of needs and 'tools' given by the architect to the client would have avoided the long disputes with different positions and criticisms that followed one another for years. Again, in architecture, an opposite example is the construction of the Guggenheim Museum in Bilbao. Here, American architect Frank Gehry, who signed the project, imposed his project management structure on the client, which included designers, planners, financial analysts, marketing experts and builders. This structure was to have complete control of the project, from concept to realisation, including all financial and managerial decisions. The vision of the project in the architect's hands was assured by total coordination of the entire chain, and the result was a successful and equally iconic project.

Research in recent years has shown that the vision is the most important content to give the correct framework to the theme, upstream of the project. This is true for many infrastructures: a project is realised because it creates added value; this is the key concept of the then president of the Lyon–Turin Observatory, Mario Virano, architect and public manager, and now General Director of TELT, this major work's public developer.

> A megaproject is done because it creates added value, and the worst unsustainability of an infrastructure is the decline of relations and the local area; therefore, in the design of megaprojects the challenge is to generate value and combat the decline of the local areas.

The strategic content around which to consolidate the table of stakeholders must be closely linked to the vision of the project and include:

- the method to tackle the project;
- the reasons behind a work (the requirement that has to be satisfied, the expected spin-offs at a broad level and those for stakeholder groups);
- the management of change (such as the steps that have an impact on the territory and on the other stakeholders) and the vision of the future based on 'a policy of small steps towards big objectives';
- the local area as the first inspirer of the basic planning approach itself.

During the realisation of the work and in the unfolding of the idea–project–site–construction process the contents also evolve: while the vision must remain a constant, there are several other elements that may be used to support it, among which are:

- impact assessment by the different bodies;
- the dialogue with employers' organisations and trade unions concerning the capacity to create jobs;
- the search for innovative tools to meet the need for transparency and respond to legitimate fears.

In the evolution of the project, it is necessary to identify a specific communication department which, in a professional, structured and organised manner, carries out this work, addressing different targets. The principles of openness of communication and the need to address as many different targets as there are stakeholders remain valid. Several tools can be provided, traditionally different elements within a pure communication structure, such as:

- press office and media/public relations;
- position papers;
- digital communication (social networks, websites, newsletters);

- construction site communication;
- info point;
- thematic brochures, information leaflets and videos;
- public meetings;
- construction site visits/open doors;
- fairs and expos;
- roadshows.

In general, external relations are an important tool for dialogue with the company's large number of interlocutors and for reaching institutions and public opinion, which are often directly and indirectly involved and mobilised by various players in civil society, both for and against a project. The aim is not only to raise awareness of the project, but also to provide service information (change management) at a local level and to stakeholder groups, for example:

Communication with the local areas
At a local level, the presence of the site produces changes that must be included in the relations with citizens, mediated or not by local administrations. Content regarding acquired data and monitoring processes concerning health, circulation of dust and CO_2 production must be periodically disseminated, with the aim of responding to any fears. Progress in the management of positive effects must also be communicated, and an integrated communication plan can be envisaged, with an ad hoc site, public meetings at various levels and relations with the press.

Communication with companies
With the planned start of works, information on tender procedures and access to calls for tenders are a core element of a public promoter's communication, both for the tender phases and for strengthening the knowledge of the project in general. With this objective, international roadshows and periodic events (including online) may be envisaged, with the involvement of what sociology calls 'network nodes' (associations) which are the equivalent of today's influencers, with the added point that they can coordinate the access of different players and strengthen the network, thanks to their own structure (Van Marrewijk et al., 2008).

Communication about sustainability
The design of megaprojects, by definition, includes aspects of sustainability, which should be highlighted in the communication, depending on the project characteristics: circular economy, recycling of materials, economic sustainability and all those elements universally shared in line with the United Nations guidelines (Oliomogbe & Smith (2012)". Dedication of an ad hoc corporate structure, publication of a sustainability report (where possible) and a CSR with particular inclusion of original content in this area are elements that help the positioning of a public developer and project acceptance.

Talking construction sites
The construction sites constitute the most immediate information tool regarding the work, the technologies applied, the safeguards adopted to protect the environment,

etc. All the work areas (offices, conveyor belts, machines, etc.) can be physically used to explain the work, its history and context, as well as current and future changes.

13.3   CONCLUSIONS

The planning of a megaproject is often very long and project acceptance strategies need to be adapted to the specific phase, but in general, the initial approach can determine the success or failure (with all the ambiguity that these words may contain) of the strategy itself.

An active approach by the decision-maker and developer serves to set the path in the most productive way to prevent rumours and misinformation.

While admitting that it is unlikely to succeed in demolishing networks of confirmatory relationships, particularly on the contesting side, one cannot fail to communicate actively; the whole strategy must therefore be established in the name of transparency and of actions aimed at gaining and confirming credibility and trust, while possibly succeeding in mobilising authoritative 'ambassadors' for the project, who can give the public opinion a third view.

The planning of the works can no longer see dialogue with the local areas and communication as a mere appendix, but must be an integral part of the process. In this, the vision and approach of the engineer plays a fundamental role, as he is understood as the technical manager at the helm of the arduous challenge of building an infrastructure from scratch. This leads to the development of the professional figure: from the design engineer to the engineer who is also a communicator.

The project's early round tables must speak and make themselves understood. Simplifying, clarifying complexity and using the right words that all stakeholders can understand are the basic actions to get off on the right foot. Every construction site has a multifaceted nature within it, which has repercussions on the outside world: traffic, noise, citizens' concerns, expectations, protests and curiosity for knowledge. It is not enough to put a sign on the outside declaring 'we are working for you' and stating some information about the work, before closing the doors and opening them again only for the ribbon cutting at the inauguration. The work starts much earlier, in parallel with the birth of the idea. Dialogue must accompany all phases and must not be skimped on, because errors in approach become indelible during the practical implementation of the work. Corrective actions are possible, but less effective than a shared planning with the territories.

In short, the identity of the project in a community depends on when and how it is proposed and on the stakeholders mobilised; even though it is a moving variable, the construction of this aspect must be effected as soon as possible, in order to be managed in a productive form for the project.

Finally, to answer the questions in the introduction:

- the creation of consensus/opposition with respect to megaprojects is influenced by all the players involved, which the public developer must identify and involve;
- the various stakeholders in the realisation of megaprojects carry weight in the project acceptance strategies to the extent that they can enhance a vision, participate and are credible and authoritative, thanks also to the support of strong data and content, produced at the right time and with a coherent and transparent 'narrative';

- strategies for gaining the willingness of a community to host a large construction site and to manage its impact itself cannot overlook the territory at the centre of it all, with particular emphasis on the paths of change (Tipaldo, 2011);
- the mechanisms that can facilitate/complicate the implementation of an infrastructure can be foreseen upstream and can be defined in paths of stakeholder engagement, advocacy, consultation and communication.

ACKNOWLEDGEMENTS

Viviana Corigliano, Silvia Cerrati and Valentina Canossi are gratefully acknowledged for the help and assistance in the preparation of this chapter.

REFERENCES

Baiguini, L. (2020) Creare consenso attorno ad un progetto (e non contro un nemico). [Online], Available from: //https://www.econopoly.ilsole24ore.com/2020/05/07/coraggio-creare-consenso/.

Blanchetti, E. & Seminario, S. (2017–2018) L'era del dissenso. Osservatorio Nimby Forum. [Online] 13° edizione, Available from: https://www.nimbyforum.it/wpcontent/uploads/2019/04/Nimby_forum_2018_doppia.pdf.

Boateng, P., Chen, Z. & Ogunlana, S.O. (2015) An Analytical Network Process model for risks prioritisation in megaprojects. *International Journal of Project Management*. [Online] 33(8), 1795–1811. Available from: https://www.sciencedirect.com/science/article/abs/pii/S0263786315001386?via%3Dihub.

Boateng, P., Chen, Z., Ogunlana, S. & Ikediashi, D. (2012) A system dynamics approach to risks description in megaprojects development. Organization, Technology & Management in Construction. *An International Journal*. [Online] 4(3), 593–603. Available from: http://www.grad.hr/otmcj/clanci/vol%204%20spec/OTMC_SI_3.pdf.

Bobbio, L. & Zeppetella, A. (1999) Perché proprio qui? Grandi opere e opposizioni locali. FrancoAngeli Edizioni.

Brookes, N. & Locatelli, G. (2015) A megaproject research framework: a guide for megaproject researchers. Report. University of Leeds, Leeds.

Chen, Z., Boateng, P. & Ogunlana, S.O. (2019) A dynamic system approach to risk analysis for megaproject delivery. Proceedings of Institution of Civil Engineers: Management, Procurement and Law. [Online] 172(6), 232–252. Available from: https://www.icevirtuallibrary.com/doi/pdf/10.1680/jmapl.18.00041.

Di Maddaloni, F. & Davis, K. (2017) The influence of local community stakeholders in megaprojects: rethinking their inclusiveness to improve project performance. International Journal of Project Management. [Online] 35(8), 1537–1556. Available from: https://www.apm.org.uk/media/31637/research-summary-akt12237-stakeholder-engagement-lr.pdf.

Di Maddaloni, F. & Davis, K. (2018) Project manager's perception of the local communities' stakeholder in megaprojects. An empirical investigation in the UK. International Journal of Project Management. [Online] 36(3), 542–565. Available from: https://www.researchgate.net/publication/321729186_Project_manager's_perception_of_the_local_communities'_stakeholder_in_megaprojects_An_empirical_investigation_in_the_UK.

Eskerod, P. & Ang, K. (2017) Stakeholder value constructs in megaprojects: a long-term assessment case study. *Project Management Journal*. [Online] 48(6), 60–75. Available from: https://journals.sagepub.com/doi/10.1177/875697281704800606.

Evan, W. & Freeman, R.E. (1993) A stakeholder theory of the modern corporation: Kantian Capitalism. In: Beauchamp, T. & Bowie, N.E. (eds) *Ethical Theory and Business*, 4th edition. Englewood Cliffs, NJ: Prentice Hall.

Forester, J.F. (1999) *The Deliberative Practitioner: Encouraging Participatory Planning Processes*. MIT Press, Massachussetts Institute of Technology, Cambrige, Massachussetts 02142.

Gallino, L. (2007) Tecnologia e democrazia. Conoscenze tecniche e scientifiche come beni pubblici. Biblioteca Einaudi.

Hertogh, M., Baker, S., Staal-Ong, P. & Westerveld, E. (2008) Managing large infrastructure projects: Research on best practices and lessons learnt in large infrastructure projects in Europe. [Online] AT Osborne BV. Available from: http://netlipse.eu/media/18750/netlipse%20 book.pdf.

Lippmann, W. (1922). *Public Opinion*. New York: Free Press.

Mok, K.Y., Shen, G.Q. & Yang, R.J. (2015) Stakeholder management studies in mega construction projects: A review and future directions. *International Journal of Project Management* 33(2), 446–457.

Oliomogbe, G.O. & Smith, N.J. (2012) Value in megaprojects. Organization, technology & management in construction. *An International Journal*. [Online] 4(3), 617–624. Available from: https://pdfs.semanticscholar.org/09d1/2143402a40aeda03f3e406c85cbfa23cfa2d.pdf.

Shenhar, A. & Holzmann, V. (2017) The three secrets of megaproject success: clear strategic vision, total alignment, and adapting to complexity. Project Management Journal. [Online] 48(6), 29–46. Available from: https://www.researchgate.net/publication/326140514_The_Three_Secrets_of_Megaproject_Success_Clear_Strategic_Vision_Total_Alignment_and_Adapting_to_Complexity.

Short, J.F. (1984) The social fabric at risk: toward the social transformation of risk analysis. *American Sociological Review* 49(6), 711–725.

Tipaldo, G. (2011) Né qui né altrove! Critica alle grandi opere: un problema di «cultura civica»? *Rassegna Italiana di Sociologia* 52(4), 607–638.

Van Marrewijk, A.H., Clegg, S.R., Pitsis, T.S. & Veenswijk, M.B. (2008) Managing public-private megaprojects: paradoxes, complexity, and project design. *International Journal of Project Management* 26(6), 591–600.

Virano, M. (2015) Contributi all'approccio decisionale alle grandi opera. Spunti di analisi e discussione in un'ottica operativa sulla base dell'esperienza dell'Osservatorio. I Quaderni dell'Osservatorio. [Online] Quaderno dell'Osservatorio n.9. Available from: http://presidenza.governo.it/osservatorio_torino_lione/quaderni/quaderno9.pdf.

Watzlawick, P. (1967) *Pragmatics of Human Communication*. W.W. Norton & Company, Inc. New York - London

Won, J.W., Jang, W., Jung, W., Han, S.H. & Kwak, Y.H. (2017) Social conflict management framework for project viability: case studies from Korean megaprojects. *International Journal of Project Management*. 35(8), 1683–1696.

Glossary

Abrasivity the property of a soil or rock to cause abrasion and wear of metallic elements.

Accuracy the closeness of the measurements/predictions to the real value.

Adhesion (of clay) the phenomenon that occurs when the steel surfaces of the machine are completely covered by the excavated material (see **stickiness**).

Advancement ratio (of a TBM) the excavated length divided by the operating time during a continuous boring phase.

Analytical solution a closed-form expression derived from the exact integration of the differential equation system governing a hydro-thermo-chemo-mechanical problem.

Anisotropy the characteristic of a mechanical or physical property to be different in different directions.

Aperture *see* **opening.**

Aquiclude a solid, impermeable area underlying or overlying an aquifer, the pressure of which could create a confined aquifer.

Aquifer an underground layer of water-bearing permeable rocks, rock fractures or soils.

Aquitard a bed of low permeability along an aquifer arching transfer of stress from a part of a soil/rock mass subject to excavation to adjoining parts of the mass.

Arching effect the stress redistribution occurring near the tunnel face from unsupported to supported zones.

Asbestos the silicate minerals with a fibrous morphology, belonging to the mineralogical group of serpentine and amphibole, whose health risk is linked to the inhalation of airborne fibres (**asbestos respirable fibre:** elongated asbestos with length $L \geq 5$ μm, diameter $\phi \leq 3$ μm and elongation ratio $L/\phi \geq 3$).

As-built model the report of the monitoring data and the notes recorded during the excavation and construction of a tunnel.

Associated flow rule the condition for which the plastic strain flow occurs in a direction normal to the yielding surface (i.e. the plastic potential corresponds to the yielding surface and the dilatancy angle is equal to the friction angle).

Annulus (annulus gap or void) the space, gap or void left around the body of a full-face Tunnel Boring Machine (TBM) by the slightly larger cut diameter of the TBM cutterhead and around the rings of the segmental lining, which are smaller again in outside diameter.

Anti-collision systems   a technological system designed to prevent or reduce the severity of a collision between vehicles or vehicle and pedestrian.

Atterberg limits (for clays)   the boundaries (determined by laboratory tests) of moisture content in a clayey soil between the liquid state and the plastic state (known as the liquid limit); between the plastic state and the semi-solid state (known as the plastic limit); and between the liquid limit and the plastic limit (known as the plasticity index).

Axial symmetry   the symmetry around an axis, often corresponding to the tunnel axis.

Axisymmetry   *see* **axial symmetry**.

Backfilling   the filling of the annulus gap, usually obtained with grout or mortar.

Beam element   a structural element suitable to model slender linear structures with axial, bending and torsional (not always) stiffness.

Bentonite slurry   a viscous mixture of bentonite and water.

Biodegradability   the measure of how much chemical substances are liable to be decomposed by microorganisms naturally present in the environment.

Block theory   the theory developed to identify the critical blocks among those created by the intersection of discontinuities in the rock mass with the boundary of an excavation.

By-product   a substance or object resulting from a production process, the primary aim of which is not the production of that item, but can be reused.

Bypass   connection drifts usually connecting the two tubes of the main tunnel usually used for safety reason in exercise.

Borehole   the hole created or enlarged by a drill or auger used for ground investigation.

Boundary conditions   the constraints imposed to the HTCM system of differential equations at the boundary of the domain.

Boundary value problem   a HTCM problem governed by a system of differential equations with constraints at the domain boundary.

Brittle behaviour   the behaviour of a material that shows an abrupt reduction in strength after a peak.

Brittleness   the property of a material manifested by failure without appreciable prior plastic deformation.

Brittleness value   the indirect measure of the ability to resist crushing by repeated impacts.

Bulkhead   the steel partition within the TBM shield that contains the positive operating pressure within the excavation chamber (or plenum) and allows the reminder of the tunnel to be at atmospheric pressure. Through the bulkhead must pass pressure-tight seals, the drive unit of the cutterhead, the outlet and intake of the excavated material (different depending on the type of TBMs) and the air locks that allow for man and materials entry into and exit from the pressurized chamber.

Carbonation   the chemical reaction between carbon dioxide in the air and calcium hydroxide and hydrated calcium silicate in the concrete.

Carbon footprint   the amount of carbon dioxide released into the atmosphere as a result of activities of a particular individual, organization or community.

c–φ reduction method   the simplified numerical procedure adopted to estimate the safety factor consisting in the progressive reduction of the Mohr–Coulomb strength parameters up to instability.

Circular economy process that aims to maintain the value of products, materials and resources as long as possible by returning them into the product cycle at the end of their use, while minimizing the generation of waste.

Clogging (of clay) the phenomenon that occurs when the cutterhead openings are completely plugged by the soil (see stickiness).

Chainage the length of or the distance along the tunnel as measured from the portal.

Chimney-like failure mechanism the typical geometry of the instable soil mass at the face of shallow tunnels.

Collapse condition *see* **failure condition**.

Computational cost the execution time for each time step during a computer simulation.

Computational method the numerical method developed to analyse HTCM processes by means of a computer simulation.

Computational time the time required to run a computational analysis by a computer.

Convergence the closure of the tunnel section induced by the radial displacement of the boundary of the tunnel cavity.

Countermeasure (see mitigation measure) action defined at the design stage, which will be activated during construction according to the predefined triggering criteria, should the key parameters reach predefined thresholds.

Cohesion the component of shear strength of a rock or soil that is independent of confining stress.

Compressibility the measure of the relative volume change of soil/rock (or fluid) as a response to a mean stress (or pressure) change.

Compressive strength capacity of a soil/rock to withstand compressive loads.

Computational fluid dynamics models a branch of fluid mechanics that uses numerical analysis and data structures to analyse and solve problems that involve fluid flows.

Conceptual design *see* **feasibility studies**.

Constitutive law *see* **constitutive relationship**.

Constitutive relationship the relation between two physical quantities, such as stress and strain in linear elasticity.

Construction design setting of the project during construction, including proper technical modifications to adapt the forecast solution to the real ground conditions, detailing side aspects with workshop drawings, writing method's statements and, possibly, designing real variants, if necessary.

Contact law the law describing the mechanical interaction between two adjacent elements or blocks mainly in the discrete element method.

Continuum approach the approach implemented to simulate HTCM processes characterized by a continuous displacement field, frequently adopted to model soils and highly fractured rock masses.

Continuous medium a medium whose behaviour can be analysed using a continuum approach.

Control variables the variables in a numerical analysis whose evolution in time and distribution in space is known.

Consolidation the hydro-mechanical process in which a soil under compression reduces its volume by expulsion of water.

Consolidation work (consolidation technique)   general definition for ground reinforcing and improving techniques.

Conditioned soil   the soil excavated by a shielded TBM operating in the earth pressure balance mode mixed with conditioning agents (usually water, foams and polymers) that fill the excavation chamber.

Conventional tunnelling method   a tunnel construction method where the excavation phase, including mucking, and the support phase are carried out in a cyclic way. The excavation phase can be carried out by explosives or punctual excavation machines.

Copy cutter   A cutter installed on the cutterhead that extends mechanically beyond the diameter of the cutterhead to create an overcut when required.

Core barrel   a hollow cylinder attached to a specially designed bit used to obtain a continuous section of the rocks penetrated in drilling.

Core (of the tunnel)   cylindrical portion of the rock/soil ahead of the tunnel face to be excavated.

Corrosion   the process converting a metal to a more stable form such as its oxide, hydroxide or sulphide state; this leads to deterioration in the material.

Coupled approach   the approach simultaneously accounting for two or more HTCM processes, as in the hydro-mechanical analyses of geotechnical applications.

Cracks   discontinuities in the concrete final lining.

Crisis management   the process by which an organization deals with a disruptive and unexpected event that threatens to harm the organization or its stakeholders. The study of crisis management originated with large-scale industrial and environmental disasters in the 1980s. It is considered to be the most important process in public relations.

Creep   the tendency of soil/rock to deform under constant stresses.

Crystalline silica   the form of silica (or silicon dioxide – SiO_2) that may be found in more than one form (polymorphism), depending on the orientation and position of the tetrahedra (i.e. the three-dimensional basic unit of all forms of crystalline silica; the natural crystalline forms of silica are quartz, cristobalite, tridymite, keatite, coesite, stishovite and melanophlogite).

Cutterhead (of a TBM)   the rotating head or wheel at the front of a TBM that cuts or excavates the tunnel face.

Cutters (cutter tools)   the tools installed on the cutterhead to excavate the ground when pushed forward and rotated. They can be pick tools, scarpers or disc cutters (also called roller cutters) or a combination of them.

Cylindrical cavity problem   the mechanical problem of the expansion/contraction of an infinitely long cylinder, to which the tunnel can be assimilated, in a continuum medium.

Débat public   in France, it is a phase of the development procedure of large organizational or infrastructure projects, which allows citizens to inform themselves and to express their point of view on the iterations and consequences of the projects.

Damage mechanics   the constitutive approach developed to describe the initiation and propagation of damage in a continuum medium.

Deformability modulus (or stiffness modulus)   the ratio of the increment of stress to the corresponding increment of strain.

Density   the mass (of soil/rock or fluid) per unit volume.

Derived solution   *see* **integrated solution**.

Detailed design   stage of design in which each aspect of the tunnel project is fully elaborated by complete description through modelling, checks, drawings and specifications.

Dilatancy law   *see* **plastic potential**.

Dilatancy   the volume change of a soil when subjected to shear deformation.

Dip   the angle formed by a rock mass discontinuity with the horizontal plane.

Dip direction   the horizontal trace of the line of dip, measured clockwise from the north.

Discontinuity   a structural weakness plane which separates intact rock blocks in a rock mass.

Discontinuum approach   the approach implemented to simulate hydro-mechanical process characterized by a discontinuous displacement field, frequently adopted to model the interaction of rock blocks in fractured rock mass.

Discontinuous medium   a medium whose behaviour can be analysed using a discontinuum approach.

Discrete element method   the numerical approach schematizing a discontinuous medium by means of rigid or deformable blocks interacting through deformable interface.

Domain reduction method   the finite element methodology for modelling earthquake ground motion in highly heterogeneous localized regions with large contrasts in wavelengths.

Drained conditions   the loading conditions where excess pore water pressures generated during construction are dissipated (either immediately, in high-permeability ground, or in the long term, in low-permeability ground).

Drainage (layout of drains)   the system of high-permeability paths obtained through perforation and installation of drains around a tunnel.

Drillability   the attitude of a rock to be drilled. It can be determined by laboratory tests.

Drill and blast (D&B)   the excavation technique based on the use of controlled explosions sequences to break the rock in a designed scheme and on the subsequent removal of the muck from the face area.

Ductility   the degree to which a material can sustain plastic deformation.

Dust   air dispersed particulates; can be classified as inspirable or respirable.

Draining box model   a numerical model with impervious bottom and lateral surfaces used to analyse the transient flow occurring after the tunnel excavation in an aquifer.

Duration   *see* **computational time**.

Ecotoxicity of a conditioned soil   the capacity of conditioned excavated soil to produce toxic effects on the living organisms (plants and aquatic and non-aquatic organisms) that contact it.

Effective stress   the stress level that is responsible for the mechanical behaviour of soil, calculated as the difference between the total stress and the pore water pressure.

Elasticity (linear elasticity)   the material behaviour that assumes the stress as a single-valued function of the strain and without any permanent effect after the unloading has occurred. It can be represented by a linear relationship between the components of stress and strain.

Element activation   the numerical technique adopted to simulate the progressive tunnel construction with new support and reinforcement elements or TBM advancement.

Element deactivation   the numerical technique adopted to simulate the progressive tunnel excavation.

Embedded element   finite element technique consisting in hosting one or more elements into other elements without shared nodes, but with a constraint in one or more degrees of freedom.

Environmental assessment   the procedure that ensures that the environmental implications of a project are taken into account before its authorization or approval. This procedure is undertaken in order to provide a high level of protection of the environment and in order to contribute to the integration of environmental considerations into the preparation of projects and plans to reduce the environmental impact.

Environmental impact   the effect (positive or negative) that a tunnelling project has on the environment.

Erector   the device installed on shielded TBMs that install the precast segments of the segment lining and place them in the correct position.

Excavation chamber (pressure chamber, working chamber and plenum)   the zone of a shielded TBM in contact with the tunnel face. This space is usually filled with pressurized conditioned soils or fluids.

Explosive atmosphere   the mixture with air, under atmospheric conditions, of flammable substances in the form of gases, vapours, mists or powders in which, after ignition, combustion propagates throughout the unburned mixture.

Extensometer   the instrument used for measuring the change in the length of an object.

Extrusometer   an instrument, installed in a borehole drilled horizontally into the nucleus/core at the tunnel front, before the excavation, for measuring longitudinal displacements beyond the face as the excavation proceeds.

Exploratory adits   a small-sized tunnel bored to obtain geological and geomechanical information about the soil/rock formations useful for the design of final tunnel.

Excavation damage zone (EDZ)   the zone around the tunnel characterized by fractures and damage due to stress redistribution typically occurring in brittle media.

Explicit integration scheme   a time discretization algorithm in which the subsequent state is a function of the current state only.

Extrados   the external surface of the lining.

Element at risk (asset)   the population, properties, economic activities, etc., at risk in a given area or job site.

Face (tunnel face or excavation face)   the vertical surface that separates the excavated area and the ground to be excavated.

Face pressure (face support pressure or operating pressure)   the pressure applied to the tunnel face to get stability and prevent water income.

Face extrusion   the movement of the excavation face, parallel to the tunnel axis due to the stress release induced by the excavation.

Face reinforcement   a layout of elements with high tensile strength (bolts, fibre-glass bars or elements) installed on the face with the aim of stabilizing it and managing the displacement.

Factor of safety   the ratio between the strength of a system and the actual applied load, often expressed as the shear strength-to-shear stress ratio in geotechnical engineering.

Failure   the limiting stress condition that cannot be exceeded.

Failure condition   the state at failure; the equation describing the failure surface, also called strength criterion.

Failure mechanism   the development of a collapse body having specific geometric and kinematic characteristics.

Failure surface   the surface in the stress space describing the failure of a material, also called strength surface/envelope.

Far field   the zones far from the tunnel boundary.

Fault   a discontinuity characterized by a significant displacement (slip) of the two sides of the fault plane.

Fault rocks   the materials originated in correspondence of friction zones (gouge, mylonites, cataclasites and breccia).

Feasibility studies (feasibility design)   the planning/conception stage in which the suitability of tunnelling and construction times/costs are parametrically assessed in relation to given constraints and estimated risks, through the comparison of different project solutions. Finally, the best possible solution is chosen to provide adequate financial resources.

Final design   intermediate stage of design, applied in some case and nations before tendering, in which the definition of the project is deepened to higher detail, also through a better knowledge of boundary conditions, geology, hydrogeology, geotechnics and geomechanics.

Filling   the material that lays in between the two faces of a rock discontinuity.

Finite difference method   the numerical method for converting the system of differential equations of a HTCM process into a system of linear equations by approximating derivatives with finite differences.

Finite element method   the numerical method for solving the system of differential equations of a HTCM process by the space discretization of the domain into fine elements.

Firedamp   the name given to a number of flammable gases, especially coalbed methane. The gas can arrive in the excavation through joints in the rock mass and, usually, accumulate in the upper areas of the underground excavations.

Flow rule   the equation describing the evolution of the plastic strain increments during a plastic process.

Folds   the plastic deformation of the earth's crust due to compression stress.

Foliation   the layering within metamorphic rocks. It occurs when a rock is being shortened along one axis during recrystallization. The minerals are elongated and rotated such that their long axes are perpendicular to the orientation of shortening.

Foam   the main conditioning element in soil conditioning for EPB-TBM tunnelling. It is made up of a foaming solution (foaming agent + water) and air turbulently mixed.

Foaming agent   the commercial product used for the foam generation. It consists mainly of water, surfactants and other chemical substances.

Fractures or joints   the result of a brittle deformation of the earth's crust. Fractures are caused by stress exceeding the rock strength and are characterized by a specific orientation with respect to the main strength.

Fracture mechanics   the numerical approach developed to describe the initiation and propagation of cracks in materials.

Free span   the maximum length of an excavation that is stable without support for a self-supporting time.

Free-field condition   the assumption of disregarding the presence of the structure in a preliminary analysis of a soil–structure interaction problem.

Free-field boundary condition   the condition applied to the lateral surfaces of the domain boundary to simulate the occurrence of an infinite domain in earthquake analyses.

Frequency domain   the representation of a signal with respect to frequency rather than time, often used in the context of earthquake analyses.

Freezing   a technique that stabilizes the soil by freezing the interstitial water.

Friction   the component of shear strength of a rock or a soil that is proportional to the normal stress acting on the failure plane.

Functional volumes   the volumes physically occupied by persons/equipment/vehicles, including the area they occupy with their movements and routes.

Full-face mechanized tunnelling method   a tunnel construction method where the excavation is carried out by a full-face tunnelling machine (see TBM). Generally, the excavation is carried out by tools carried by a circular cutterhead. In shielded machines, the final support is directly installed by the full-face machine itself (usually through an erector) under the protection of the shield.

Geomechanical survey (in rock masses)   the activity aimed to obtain quantitative information about the properties of the rock mass.

Geological model   the model that characterizes the rock mass from a geological perspective, including rock type and minerals, hydrogeology, geomorphology, geological structures and their spatial distribution, and the geological history of the rock.

Geophysical prospection   field investigation techniques consisting in the use of testing methods to measure some physical properties of the rocks/soils (i.e. seismic, thermal, electric and magnetic) and interpreting the results in terms of geological or geomechanical features.

Geotechnical characterization   the activities carried out to develop reliable geotechnical design parameters and mechanical properties and to identify the related hazards.

Geotechnical investigation   the set of investigations performed to obtain information on the mechanical properties of soil/rock.

Geotechnical model   the idealization of underground space used in design to define the underground geometric layout and the mechanical properties of the geotechnical units that exhibit similar geotechnical characteristics and natural state of stress.

Geotechnical unit   the portion of ground that can be considered as homogenous, in which the mechanical properties of the soil and rock mass are the same.

Geothermal gradient   the increase with depth of temperature in the earth's crust.

Grain size distribution   the distribution of relative proportions of different grain sizes in a soil.

Ground reaction curve (convergence–confinement curve)   the relationship between applied stresses and radial displacements at the tunnel wall.

Gripper   the system able to provide the constraint against the surrounding rock mass that is needed for applying the thrust force to advance a TBM. It consists of gripper jacks and gripper shoes.

Ground   a general expression to describe a soil or a rock to be excavated by a tunnel.

Ground conditions   the set of geological, hydrological, geotechnical and environmental conditions of the ground on the site of a construction project.

Hazard   the intrinsic property or ability of a factor to cause a damage to life, health, property, economic loss or the environment. Any hazard can have various amounts of intensity/magnitude/severity associated with different probabilities/likelihoods of occurrence and with different impacts/consequences in terms of safety, time and cost.

Hazard identification   the pinpointing of condition, material, system, process and plant characteristics that can produce undesirable consequences.

Hardening rule   the description on the changes of the yield surface with plastic deformation.

Hardness   the rock resistance to indentation.

Head (crown)   the upper portion of the tunnel boundary.

Hydraulic conductivity   the property of soil/rock mass that describes the ease with which a fluid (usually water) can move through pore spaces or fractures.

Heterogeneity   a geometrical feature in the HTCM problem that is not uniform with respect to the distribution of a specific characteristic.

Homogenized material   the equivalent continuum medium with properties derived from those of its heterogeneous components, through a homogenization technique.

Hydro-mechanical problem   a boundary value problem accounting for hydraulic and mechanical processes in the analysed domain.

HTCM (hydro-thermo-chemo-mechanical) boundary value problem   a boundary value problem accounting for hydraulic, thermal, chemical and mechanical processes in the analysed domain.

Hysteretic constitutive model   the constitutive law describing the behaviour of non-linear materials characterized by dependency on stress–strain history.

Intrados   internal part of the tunnel lining.

Igneous rocks   the rock produced by the cooling and the solidification of a magma.

Implicit integration scheme   a time discretization algorithm in which the subsequent state is a function of both the current and the subsequent states.

Inertial effect   the influence of the mass of the system in dynamic calculations.

Initial conditions   the constraints imposed to the HTCM system of differential equations at the initial time of the analysis.

Integrated solution   the closed-form expression derived from parametric numerical or stability analyses.

Interface   the zero-thickness elements between continuum subdomains in the finite element or finite difference method or among blocks or particles in the discrete element or particle finite element method.

Instantaneous face advance   the simplified approach assuming the simultaneous tunnel excavation along its entire length.

Industrial hygiene   the science devoted to the anticipation, recognition, evaluation, prevention and control of those environmental factors or stresses arising in or from the workplace which may cause sickness, impaired health and well-being, or significant discomfort among workers (or among citizens of the community).

Inspection   the visual and/or instrumental examination and assessment of the current state of an object (usually a structure) through observations and possibly measurements.

Invert   the lower portion of the tunnel boundary. It is normally an inverted arch.

Intergranular void (interparticle pore)   the voids among soil particles that are filled by one or more fluids (air, water or other fluids or gasses).

Isotropic   the characteristic of a material property to be invariant with respect to direction.

Jet grouting   injection in the soil with a cementitious mix at high pressure. A cement base grout is injected in the soil along a drilled hole at high energy with the formation of a column of treated soil with higher mechanical performances than the original soil.

Jet grouting arch (jet grouting canopy)   a pre-reinforced structure made ahead of the excavation face with an arch of soil treated with jet grouting.

Joint   *see* **discontinuity**.

Key performance indicators   a set of measurable quantities defining the system performances (for instance convergence, induced settlements and stresses in linings).

Karst   cavities and underground voids induced by a dissolution of soluble rocks such as limestone, dolomite and gypsum. It is characterized by underground systems of water circulation with voids, sinkholes and caves.

Kinematic approach of limit analysis   the approach based on the upper bound plasticity theorem considering possible failure mechanisms in the analysed domain.

Kinematic method   *see* **kinematic approach of limit analysis**.

Kinematic variables   the variables associated with the motion of a medium, such as strains and displacements in geotechnical applications.

Landfill (for muck)   a site for the disposal of the muck.

Large displacement approach   a numerical approach capable of dealing with large displacements and updated geometrical configurations as in the discrete element method or in the particle finite element method.

Large strain approach   the numerical approach considering the updated deformed configurations due to finite strains.

Limit analysis   the approach based on the upper and lower bound plasticity theorems.

Limit equilibrium method (LEM)   the simplified approach aimed at assessing the distance from failure of a system along different failure mechanisms through the definition of the factor of safety.

Lining or support (of first phase or temporary)   a not permanent system of structures designed to guarantee the immediate stability of the excavation and the safety of the workers.

Lining (final or second phase)   a structure designed to guarantee the permanent stability of the tunnel during its lifetime. It may be cast in place (reinforced or not) or assembled with precast segments (in this last case it is called **segmental lining**) or also obtained with shotcrete.

Long-term response   the response of a system at the end of the time-dependent HTCM processes.

Lower explosive level (LEL)   the lowest concentration (percentage) of a gas or a vapour in air capable of producing a flash of fire in the presence of an ignition source such as arc, flame or heat.

Lumping technique   the numerical technique reducing the mass matrix into a diagonal matrix in the finite element method.

Maintenance   the function of keeping items or equipment in, or restoring them to, serviceable condition. It includes servicing, test, inspection, adjustment/alignment, removal, replacement, reinstallation, troubleshooting, calibration, condition determination, repair, modification, overhaul, rebuilding and reclamation. Maintenance includes both corrective and preventive activities.

Manchette pipe   a pipe with several levels of holes and not-return valves used to perform soil grouting.

Management of the muck   the excavation, transport, storage, treatment, recovery (eventually) and disposal of muck, including the supervision of such operations and the aftercare of disposal sites.

Material removal   *see* **element deactivation**.

Material point method   the extension of the finite element method in which the state variable evolution is described at points moving inside a fixed mesh.

Maximization algorithm   the optimization algorithm exploited to find maxima of a function, typically adopted to search for the most probable failure condition of a geotechnical system.

Mechanical parameter   a parameter that describes the stress–strain response of a material.

Media relations   working with media with the purpose of informing the public of an organization's mission, policies and practices in a positive, consistent and credible manner. Typically, this means coordinating directly with the people responsible for producing the news and features in the mass media.

Mesh   the spatial domain discretization formed by finite elements in the finite element method.

Mesh dependence   the dependency of the numerical solution from the mesh characteristics in a finite element analysis.

Mesh distortion   the use of distorted, irregular, finite elements; the effect of large strains on mesh regularity in a finite element analysis.

Metamorphic rock   the rock type induced by the transformation of an existing rock subjected to heat and pressure (metamorphism), causing physical and/or chemical transformations (without melting).

Minimization algorithm   the optimization algorithm exploited to find minima of a function, typically adopted to search for the most probable failure condition of a geotechnical system.

Mitigation measures   the measures to be incorporated into the tunnel design in order to avoid or significantly reduce the occurrence and/or the impact of a hazardous event on the elements at risk.

Monitoring project   the complex of documents that identify the physical quantities relevant to the understanding of the overall behaviour of the system, the appropriate measuring instruments (accuracy, precision and resolution), the choice of their number, location and acquisition frequency and, finally, the list of the activities required to correctly register, process and share the recorded information, with the main aim of controlling the performance of the system under construction and under live conditions.

Muck (spoil)   the excavated ground after the excavation process has detached it from its original position underground.

Mucking   the operation of removal from the tunnel face of the excavated ground and transporting it outside the tunnel.

Multi-phase continuum   a continuum medium formed by different solid and fluid components, typically solid particles, water and air in geotechnical applications.

Multi-surface plasticity model   the strength envelope defined by more than one surface; the constitutive relationship characterized by the presence of more than one nested plastic surface that is progressively activated during the loading process.

Natural state of stress (in situ states of stress, overburden stress or overburden load)   the original stress condition inside the ground before the tunnel excavation.

Near field   the zone around the tunnel boundary.

Normally consolidated soil   a soil is said to be normally consolidated if the effective overburden pressure that it is currently experiencing is the maximum it has ever experienced in its history.

Non-linear differential equations   *see* **partial derivative differential equations**.

Numerical analysis   the computation based on algorithms using the numerical approximation to solve the system of partial derivative differential equations to analyse HTCM processes.

Numerical error   the difference between the exact and the numerical solutions.

Numerical instability   the loss of convergence in a numerical analysis.

Numerical solution   the solution obtained by a numerical analysis.

Numerical result   *see* **numerical solution**.

Observational method   the design procedure aimed at optimizing the management of the unavoidable geological and geotechnical uncertainties affecting both tunnel design and construction, according to the definition in the design phase of specific threshold value for key parameters and the control of these data with monitoring results and the consequent modification/optimization of the design solution according to the different design solutions developed in the design phase.

Occupational exposure limit (maximum exposure value)   the limit of time-weighted average of the concentration of a chemical agent in the air within the breathing zone of a worker in relation to a specific reference period or **threshold limit value**.

Occupational safety   the multidisciplinary scientific field concerned with the safety and welfare of people at occupation.

Occupational hygiene   the science dedicated to the anticipation, recognition, evaluation, communication and control of environmental stressors in, or arising from, the work place that may result in injury, illness or impairment, or affect the well-being of workers and members of the community. These stressors are normally divided into the categories of biological, chemical, physical, ergonomic and psychosocial.

OS&H occupational Safety and Health.

Opening (aperture) of a joint (or discontinuity) the perpendicular distance between the faces of the discontinuity.

Open mode shielded TBM operating modality, where there is no pressure inside the chamber.

Orientation of a joint (or discontinuity) the spatial position of a plane representing a rock discontinuity, univocally defined using the values of the dip and the dip direction.

Over-excavation (overbreak excavation) the area of the excavated section that exceeds the theoretical cross section of the tunnel (unintentional).

Overcut the area of the excavated section that exceeds the theoretical cross section of the tunnel that is intentionally obtained.

Overburden the thickness of the ground between the crown and the surface.

Overconsolidated soil a soil is said to be overconsolidated if the present in situ stress is lower than the effective stress it has experienced in the past.

Overconsolidation ratio the ratio between the maximum vertical effective stress that a soil has experienced in the past and the current vertical effective stress.

Overthrust a reverse fault in which the rocks on the upper surface of a fault plane have moved over the rocks on the lower surface.

Parametric study the repetition of analyses performed with different sets of material properties, initial values of state variables or initial conditions, or boundary conditions.

Partial derivative differential equations the system of equations describing a HTCM process.

Particle size distribution curve (grain size distribution curve) a function that defines the distribution of soil particles size in a sample as a percentage by weight.

Particle finite element method the numerical method combining the standard finite element method with remeshing procedures to overcome mesh distortion due to large strains.

Peak strength the shear strength identified at a peak point of the stress–strain curve that is followed by a drop of the maximum deviatoric stress that can be resisted.

Penalty function the algorithm used to convert constrained problems into unconstrained one by introducing an artificial penalty for violating the constraint.

Penetration (of a TBM) the excavated length for revolution of the cutting head.

Permeability *see* **hydraulic conductivity**.

Personal protection equipment a device (other than guard) which intend to reduce the risk, either alone or in conjunction with a guard (**guard**: a part of the machinery used specifically to provide protection by means of a physical barrier).

Persistence the areal extent of a rock discontinuity interface inside the rock mass.

Permeation grouting injection of grout mixes in the soil.

Pick a tooth-shaped cutting device.

Pilot tunnel (pilot drift)   the small tunnel excavated over the entire length or over part of a larger tunnel, to explore ground conditions and/or to assist in final excavation (*see also* **exploratory adit**).

Plastic hardening   the phenomenon occurring during loading a soil/rock when, upon plastic yielding, the stress needs to be continually increased in order to drive the plastic deformation.

Plastic potential   the scalar function that gives the plastic strains when differentiated with respect to the stresses; it is often represented as a surface in the stress space.

Plasticity   the attitude of a material to deform undergoing non-reversible changes in shape in response to applied stresses.

Plasticity (of clay) (see Atterberg limits)   the capacity of clayey soil for being moulded or altered   as opposite to consistency, it refers to states of soil that are dependent on the liquid content.

Plastic potential   the surface in the stress space governing the plastic flow of a material.

Plate element   a structural element suitable to model plane plate structures supporting bending forces.

Poisson ratio   the ratio of the transverse strain to the axial strain of a soil/rock element when an axial stress is applied at its ends.

Polymer   a chemical product used in the soil conditioning. Nature and type of polymers are different in function of their goal in the conditioning process. They can be used in order to increase the foam stability, to minimize wear of tools, to lower the cutterhead torque during the excavation or to reduce the clogging phenomenon in clayey soils.

Pollutant   the substance or energy introduced into the environment that has undesired effects, or adversely affects the usefulness of a resource.

Pollutant emission   any solid, liquid or gaseous substance introduced into the atmosphere which may cause air pollution and any direct or indirect discharge of volatile organic compounds into the environment (primary pollutant source   the source generating the pollutant; secondary pollution source   sources causing the emission from segregated or deposited pollution areas).

Pollutant concentration   a measure of the amount of a polluting substance in a given amount of water, soil, air, food or other media.

Pore pressure   the stress transmitted through the interstitial fluid of a soil/rock or rock mass.

Porosity   the ratio of the volume of voids in a soil/rock to the total volume of the material, including the voids.

Portal   tunnel entrance.

Precision (of an instrument)   the degree to which repeated measurements under unchanged conditions show the same result.

Pre-confinement (intervention)   pre-reinforcement technique (auxiliary method) used ahead of the tunnel face to improve the stability in the tunnel advancement span and to manage the stress release and displacements.

Preliminary design   stage of design that bridges the gaps between conceptual and detailed designs, providing definition of the project with preliminary calculations and layouts. The output of the planning/feasibility study phase and the input of the engineering phase indicate the possible mitigation measures, reducing risks to an acceptable level; in this design stage, a preliminary assessment of construction time and costs is also developed.

Pre-support (intervention)   pre-reinforcement technique (auxiliary method) used to improve the stability in the tunnel advancement span (for example the steel pipe umbrella).

Prevention (in OS&H)   the steps or measures taken or planned at all stages of work in the undertaking to prevent or reduce occupational risks.

Prevention through design (approach)   addressing occupational safety and health needs in the design and redesign processes to prevent or minimize the work-related hazards and risks associated with the construction, manufacture, use and maintenance, and disposal of facilities, materials and equipment.

Progressive failure   the progressive attainment of failure conditions by different portions of a failure surface, typically associated with a strain-softening response of the medium.

Protection   mitigation of the consequences associated with residual OS&H risks by the application of specially conceived countermeasures. General protection based on guards or personal protection.

Quasi-static problem   a problem characterized by a very low loading rate to make inertial forces negligible.

Radon   a chemical element; it is a radioactive, colourless, odourless, tasteless noble gas present in some igneous rocks (e.g. lavas, tuff and pozzolanas), granites, marbles, marls and flysches containing uraniferous minerals or radium (**radioactive minerals**).

Rate dependency   the material behaviour affected by the rate of loading.

Rayleigh damping   a type of viscous damping linearly proportional to the mass and the stiffness.

Refurbishment   the process of improvement of a structure aimed at reducing existing risks and prolonging the operational life. It usually involves more expensive and time-consuming works than periodic maintenance activities.

Relining   the construction of a further lining inside the existing one. It may be preceded by the removal of few centimetres/decimetres of the existing lining.

Remeshing   the change in shape, size and the number of mesh elements.

Representative elementary volume   the smallest volume of a porous medium for which the average macroscopic constitutive representation is a statistically accurate description of the system behaviour.

Residual strength   the stress that a yielded soil or rock can still carry without failing at very large shear strain, usually following a peak condition; in rock, it is equivalent to *post-peak strength*.

Residual risk   the risk associated with an action, event or a natural phenomenon remaining after the implementation of the mitigation measures and/or every technical, organization and protection measure.

Residual risk assessment   the assessment of whether the risk reduction objectives (zeroed or minimized risk targets) have been achieved.

Resolution (of an instrument)   the smallest change in a value that an instrument can detect/measure/display; typically, this is a feature of the instrument coupled with a reading unit.

Reuse of muck   any operation by which excavated soil is used again for civil applications.

Ring   the elementary unit of a segment lining obtained by the assembly of a set of precast segments.

Risk   the product between the likelihood of a hazard with a certain intensity and the induced damage (value of the loss caused by this event multiplied the vulnerability of the object/person potentially affected by the hazard).

Risk allocation   the phase of risk analysis in which residual risks are assigned to various parties (clients/owners, contractors, insurances, etc.) and the contract clauses are defined accordingly.

Risk analysis   process of risk evaluation and allocation.

Risk assessment   the process by which risks are identified, assessed and ranked through comparison with predefined targets.

Risk evaluation   the definition of acceptance criteria to proceed to risk mitigation, allocation or removal.

Risk mitigation   the strategies to be implemented to reduce the probability of occurrence of an event and/or its impact on the elements/people at risk.

Risk management   the systematic application of management policies, procedures and practices to the tasks of analysing, assessing and controlling risks in order to protect employees, the general public, the environment and company assets.

Rock (intact rock or matrix)   a natural aggregate of minerals and grains connected by strong and permanent bonds (unfractured element of rock).

Rock mass   a ground made of intact rock portions and the discontinuities.

Rock bolt   a structural element made of a material with high tensile strength (typically steel or fibreglass) used to bond together unstable rock portions to the rock mass. It is usually installed following a pattern to grant the stability to the tunnel.

Rock burst   a phenomenon that manifests in a sudden and violent burst of the rock mass around an underground excavation and that can lead to a seismic event.

Rock foliation   a planar arrangement of structural or textural features in rock masses.

Roughness (of a discontinuity)   the distribution of asperities along the surface of a discontinuity.

Soil sampler   a device for obtaining samples of soil from boreholes for testing purposes.

Scaling   the removal of unstable rock elements from the crown and sidewalls.

Screw conveyor   the Archimedean screw used in the earth pressure balance shielded TBM to extract the pressurized conditioned soil from the plenum.

Sedimentary rock   a rock formed by the accumulation or deposition of mineral or organic particles at the earth's surface, followed by a diagenesis process.

Sediment (soil)   a natural material broken by the processes of weathering, chemical weathering and erosion, and subsequently transported by the action of wind, water or ice, and/or by the force of gravity.

Segmental lining (or segment lining)   permanent lining made of precast segments assembled by the shielded TBM.

Sidewall   the lateral portions of the tunnel boundary.

Slurry   the viscous suspension of minerals such as bentonite or clay and/or polymers in water.

Shear band the narrow zone of intense shearing strain developing during severe plastic deformation of materials.

Shell element a structural element suitable to model curved shell structures supporting membrane and bending forces.

Shear strength the magnitude of the shear stress that a soil/rock element can sustain, as a result of friction and interlocking of particles, and possibly cementation or bonding at particle contacts.

Shield the support and protection of a TBM obtained by a steel cylinder.

Shielded TBM A full-face machine able to perform the excavation under the protection of a shield below which the segment lining is installed.

Shotcrete the method of applying a concrete to vertical or overhead surfaces. The mix of water, cement, aggregates and, eventually, fibres is projected through a nozzle that mixes the mortar with an accelerant to make it harden on the surface.

Silent boundary condition the condition applied to erase the energy of reflected waves on the boundary surfaces of the domain in earthquake analyses.

Small displacement approach a numerical approach disregarding large displacements and considering a fixed geometrical configuration.

Small strain approach a numerical approach conserving only infinitesimal strains.

Smoothed particle hydrodynamics the mesh-free particle numerical method particularly suitable for the analysis of multi-phase flows and large strain problems.

Spacing (of discontinuities) the mean distance among discontinuities belonging to the same family, measured perpendicularly to the discontinuities.

Spalling (or slabbing) the development of visible traction fractures and detachment of rock slabs near the boundary of an underground excavation.

Squeezing the time-dependent large deformations of an underground excavation associated with creep phenomena.

Stakeholder a person such as an employee, customer or citizen who is involved with an organization, society, etc., and therefore has responsibilities towards it and an interest in its success.

Stationary strength (or critical strength) the shear strength identified by the condition when the soil undergoes shear at a constant volume.

Stiffness the property of material that affects its response to stress change in terms of strain.

Standstill phase in single shield tunnelling, it is the phase necessary to complete the installation of the final lining that separates two successive excavation phases.

Steel arch (steel ribs) a curved steel beam installed on the boundary of tunnel. It may be horseshoe shaped with two foundations or be a close ring following entirely the shape of the boundary. It is installed to react to the inward movement of the soil around the void.

Steel pipe umbrella (forepoling) a pre-support technique made of set of steel pipes or steel bars installed on the upper part of the tunnel and ahead of the excavation face.

Stratification parallel layers of rock or soil that lie one upon another, originated by natural processes (changes in texture or composition during deposition or from pauses in deposition).

Strain the deformation of a continuum in response to the application of a stress.

Strength the ability of a material to withstand an applied load without failure.

Secant stiffness modulus   the slope of a line drawn from the origin of the stress–strain diagram and intersecting the curve at the point of interest.

Shear modulus   the ratio of the shear stress to the corresponding shear strain.

Strike   the line of intersection of the plane with a horizontal plane, used for the geometrical definition of the position of a discontinuity.

Swelling   the volume expansion of a rock mass or a soil due to the absorption of water, which can cause time-dependent displacement around tunnel.

Soil conditioning   the process that, through the addition of water and chemical products to the soil, changes its mechanical and hydraulic properties in order to obtain a material with high flowability, high compressibility and low permeability to be used in an Earth Pressure Balance shielded TBM.

Soil structure   the arrangement of the solid parts of the soil and of the pore space located between them.

Solid skeleton   the physical structure of soil, referring principally to the layout the soil particles are arranged.

Soil–structure interaction problem   the problem related to the interaction of soil/rock materials and structures.

Solid element   the standard continuum element in the finite element method, providing the numerical solution in terms of stresses and strains.

Solution convergence   *see* **numerical convergence**.

Spatial discretization   the numerical approach discretizing the entire domain into a number of points to replace the system of partial derivative differential equations with matrix equations.

Spherical cavity problem   the mechanical problem of the expansion/contraction of a spherical cavity, to which the tunnel face can be assimilated, in a continuum medium.

Stability analysis   the analytical or numerical procedure adopted to assess the distance of a system from failure conditions.

State variable   the variable used to describe the state of a material in a numerical analysis.

Static approach of limit analysis   the approach based on the lower bound plasticity theorem considering admissible stress field in the analysed domain.

Static method   *see* **static approach of limit analysis**.

Static problem   a problem disregarding the influence of inertial forces.

Stepwise face advance   the simulation of tunnel excavation by the progressive element deactivation over finite tunnel length.

Stickiness (sticky behaviour)   the ability of a clayey soil to adhere to a metallic surface.

Strong formulation   the exact approach imposing the satisfaction of the partial derivative differential equation system in any point of the domain.

Structural element   a special element in the finite element method with a collapsed dimension in the direction of the structural thickness, providing the numerical solution in terms of internal forces and displacements.

Structural characteristics (of rock mass)   the geological structures of a rock mass such as the bedding planes, faults, joints and folds.

Subsidence (settlement)   vertical movements induced by tunnelling.

Surfactant a chemical substance that can reduce the surface tension of a liquid and to produce the foam.

Tangent stiffness modulus the slope of the tangent line to the stress–strain diagram at the point of interest.

Tensile strength the capacity of a rock or a soil to withstand a tensile stress.

Tectonics the discipline that studies the structure of the earth crust and the deformations of the rocks through time.

Three-dimensional model a model accounting for the three-dimensional geometrical layout and all the stress and strain components.

Tied degrees of freedom boundary condition the condition imposing the same displacements to nodes located at the same elevation on opposite surfaces of the domain boundary.

Time discretization the numerical integration of the set of ordinary differential equations arising from the spatial discretization of transient or quasi-static problems in a computation analysis.

Time domain the representation of a signal with respect to time.

Time step the time increment used in a time discretization algorithm, to be carefully selected in the case of explicit integration scheme to avoid numerical instability.

Total stress the stress imposed to a unit volume of rock and soil considered as a continuum, resulting from its self-weight and the forces applied at the boundary.

Treatment (of muck) recovery or disposal operations, including preparation prior to recovery or disposal.

Treatment ratio the dosage in litres of a commercial product for cubic metres of muck to be used for soil conditioning in EPB-TBM tunnelling.

Truss element a structural element suitable to model slender linear structures that support axial loading only.

Tunnelling the process of safe excavation and permanent stabilization of an underground cavity with specific cross sections, connecting two points along a predefined alignment, providing functionality and use/operational requirements of the system during its whole design life and minimizing construction time and costs.

Two-dimensional model (plane strain model) a bi-dimensional model characterized by the hypothesis of null strain components in the direction perpendicular to the analysed section, suitable for the analysis of long structures such as tunnels.

Yield function the function describing the yield surface.

Yield surface the boundary of the elastic region behaviour in the stress space.

Yielding the departure from elastic response that occurs as loading proceeds beyond the past maximum load; onset of plastic deformation in a stress–strain curve.

Young's modulus the ratio of the increment of axial stress to the corresponding increment of axial strain in longitudinal compression/extension of a material; also, one of the elastic parameters.

Ubiquitous joint the condition accounting for the possible presence of a joint with a specified orientation in any point of the medium, irrespective of its absolute position; the constitutive formulation accounting in a continuum medium for specified low shear strength directions, corresponding to those of the joint sets in a rock mass.

Uncertainty variation range around a measured value; situation in which something is not known, or something that is not known or certain.

Uncoupled approach   the approach accounting for two or more HTCM processes in a sequential order, for example by first solving the hydraulic problem and then the mechanical one.

Undrained loading conditions   the loading conditions in low-permeability ground, characterized by no volume changes, where excess pore water pressures generated during construction are prevented to dissipate in the short term.

Undrained shear strength (undrained cohesion)   the maximum shear stress which the soil can withstand when it is sheared undrained at constant volume and constant water content.

Uniaxial compressive strength (or unconfined compressive strength)   the maximum axial compressive stress that a cylindrical sample of rock or soil can withstand when the confining stress is set to zero.

Upper explosive level (UEL)   the highest concentration (percentage) of a gas or a vapour in air capable of producing a flash of fire in the presence of an ignition source such as arc, flame or heat.

Ventilation system   a system that provides air circulation/fresh air to an enclosed space, including control and monitoring subsystems (auxiliary ventilation: a ventilation subsystem).

Ventilation duct   the pipeline used to ventilate the tunnel; eliminating polluted air can be **flexible (blowing ventilation)** or **rigid (intake ventilation)**.

Ventilation flow rate   the quantity of air moving into a duct.

Viscosity   the measure of fluid resistance to shear deformation at a given rate.

Viscous damper   a velocity-dependent rheological element.

Volume loss   the over-excavation expressed as a percentage of the theoretical cross section area of the tunnel.

Vulnerability   the expected degree of loss of something of value (i.e. life, property and environment) that can be compromised by a give action, activity or natural phenomenon.

Wall strength   the compressive strength of the discontinuity face.

Waterproofing layer   the waterproof membrane typically installed between the first-phase lining and the final lining.

Weak formulation   the approximated approach imposing the satisfaction of the partial derivative differential equation system in an integral form over the entire domain.

Wire mesh   a structural mesh made of orthogonally welded rebars. It is usually installed in adherence to the crown and the sidewalls of a tunnel to avoid the detachment of small-sized debris of rock and embedded in the shotcrete or the concrete.

Workability of the soil (pulpy behaviour)   the capacity of conditioned soil to be able to be used in an EPB-TBM.

Water table (piezometric surface or underground water level)   the upper surface of the zone of saturation (where the pores and/or the ground fractures or rock discontinuities are saturated with water).

Wear (of cutters)   the abrasion and deterioration of the tools due to the action of the soil after their use.

Work safety   *see* **occupational safety**.

Index